# Sustainable Agriculture: Technology and Management

# Sustainable Agriculture: Technology and Management

Edited by **Thelma Bosso**

R Callisto
Reference

New York

Published by Callisto Reference,
106 Park Avenue, Suite 200,
New York, NY 10016, USA
www.callistoreference.com

**Sustainable Agriculture: Technology and Management**
Edited by Thelma Bosso

© 2015 Callisto Reference

International Standard Book Number: 978-1-63239-577-1 (Hardback)

# Contents

# Preface

Sustainable agriculture is basically defined as the practice of agriculture through environment friendly techniques. In this book, topics pertaining to resource management for sustainable agricultural development have been presented. The book discusses the usage of water and waste management for viable agricultural development including factors like irrigation management to prevent soil and ground water salinization, production of solid fuel from oil palm waste, sustainable ecomaterials and biorefinery from agroindustrial waste, nonpoint pollution from agriculture and livestock activities on surface water. It further discusses about sustainable management of dryland resources especially carbon sequestration under changing climate scenario. The book also focuses on effective nutrient management for sustainable crop productivity in various agro-climatic conditions, soil quality and productivity enhancement under rainfed conditions. Impact of conservation tillage, agricultural tillage and agricultural traffic on soil properties has also been elucidated in the book.

After months of intensive research and writing, this book is the end result of all who devoted their time and efforts in the initiation and progress of this book. It will surely be a source of reference in enhancing the required knowledge of the new developments in the area. During the course of developing this book, certain measures such as accuracy, authenticity and research focused analytical studies were given preference in order to produce a comprehensive book in the area of study.

This book would not have been possible without the efforts of the authors and the publisher. I extend my sincere thanks to them. Secondly, I express my gratitude to my family and well-wishers. And most importantly, I thank my students for constantly expressing their willingness and curiosity in enhancing their knowledge in the field, which encourages me to take up further research projects for the advancement of the area.

Editor

# Water and Waste Management for Sustainable Agriculture

# Sustainable Materials and Biorefinery Chemicals from Agriwastes

M.A. Martin-Luengo, M. Yates, M. Ramos, F. Plou, J. L. Salgado, A. Civantos, J.L. Lacomba, G. Reilly, C. Vervaet, E. Sáez Rojo, A.M. Martínez Serrano, M. Diaz, L. Vega Argomaniz, L. Medina Trujillo, S. Nogales and R. Lozano Pirrongelli

Additional information is available at the end of the chapter

## 1. Introduction

Countries with economies based on agriculture generate vast amounts of low or null value wastes which may even represent an environmental hazard. In our group, agricultural industrial wastes have been converted into value added liquid substances and materials with several aims: decreasing pollution, giving added value to wastes and working in a sustainable manner in which the wastes of an industry can be used as the raw materials of the same or others, as the "cradle to cradle" philosophy states [1].

Sub-products from the agricultural food industry are being employed as renewable low cost raw materials in the preparation of Ecomaterials, designed for use in a number of industrial processes of great interest. Given their origin, these materials may compete with conventional ones since with this process a sustainable cycle is closed, in which the residues of one industry are used as raw materials in the same or other industries [2].

With regards to the composition of the residues produced from agriculture, the pH of soil is of great importance, since plants can only absorb the minerals that are dissolved in water and pH is mandatory for the physical, chemical and biological properties of soil and the main cause of many agronomic questions related to nutrient assimilation [3-5]. Variations of pH modify the solubility of most elements necessary for the development of crops and also influence the microbian activity of soil, which will affect the transformation of elements that are liberated to the soil and can be assimilated to form crops or not [3]. For example at pH lower than 6 or higher than 8 bacterian activities are lowered, the oxidation of nitrogen to nitrate is reduced and the amount of nitrogen available for plant food is decreased. However Al, Fe and manganese are more soluble at low pHs, reaching even toxic concentrations. Potassium and sulphur are easily adsorbed at pH higher than 6, calcium and magnesium between 7 and 8.5

and iron at pH lower than 6. For alkaline pH in soil, the availability of $H_2PO_4^-$ can be reduced through precipitation of phosphorous containing salts with cations such as calcium $Ca^{2+}$ or magnesium $Mg^{2+}$. However when soils have acid pH other compounds with $HPO_4^{2-}$ and iron ($Fe^{2+}$), aluminium ($Al^{3+}$) and manganese ($Mn^{2+}$) can form, with increased solubility.

The main factors that influence soil pH are the mineral composition and how it meteorizes, the decomposition of organic matter, how nutrients are partitioned among the solution and aggregates and of course the pluviometry of the zone and atmospheric contamination. Lower pHs are found in places with high pluviometry, with high organic matter decomposition, young soils developed on acid substrates, and places with high atmospheric contamination (acid rain).

Depending on the species, crops can benefit from calcareous soils with high calcium carbonate content such as alfalfa, but other plants prefer soils with acid pH such as potatoes, coffee or tobacco. It is clear that different seasons will produce plants with a varying composition depending on the atmospheric conditions and therefore the materials derived from them need to be characterised and analysed to determine their possible uses.

Variation in content of components in fruits and vegetables depends upon both genetics and environment, including growing conditions, harvest and storage, processing and preparation. For example in broccoli, low soil water content during plant growth and post harvest cold storage were conditions that, combined, gave higher amounts of l-ascorbic acid [6,7]. Higher polyphenol contents are found in organic tomato juices compared to non-organic ones due to a higher phosphorus uptake and limited nitrogen availability in the first case [8]. Therefore, thorough characterisation of the residue composition is a key step before determination of the possible uses of a given residue.

Among the applications being developed in our group, Bio-refinery processes (preparation of sustainable *p*-cymene and hydrogen, avoiding the use of petroleum derivatives and synthesis of pharmaceutical and fine chemical intermediates), design of structured materials capable of effluent decontamination and preparation of biomaterials to act as scaffolds for cell growth towards development of prostheses and implants, will be considered here.

Given its multidisciplinary approach, this work is being carried out through the collaboration among national (Institute of Materials Science of Madrid (ICMM, CSIC), Institute of Catalysis (ICP, CSIC), Centre of Molecular Biology Severo Ochoa (UAM-CSIC), Polytechnic University of Madrid (UPM), University at distance (UNED), University Complutense of Madrid (UPM) and international (University of Sheffield and University of Ghent) research groups, in addition to various industries interested in the transformation of their residues and or sub-products into "value added materials", with whom various research projects have been and are being sponsored by the MICINN and CDTI.

## 2. Production of sustainable *p*-cymene and hydrogen

Environmental problems pose a great challenge, particularly in countries such as Spain, where water use, residues desertic and contaminated soils have become a matter of the utmost concern. Efforts from academia, industry and government are mainly based on technological changes that improve chemical processes to avoid negative environmental consequences. New renewable and sustainable chemicals are now being obtained from agricultural wastes, applying "biorefinery technologies", which reduce the need for non-renewable fossil fuel resources to help solve many environmental problems. Spain is the third producer of citrus fruits in the world [9]. Limonene is a six membered ring terpene, present in agricultural wastes derived from citrus peels, with a purity of more than 95% in orange peel oil. Sustainable p-cymene and hydrogen were prepared form limonene, comparing commercial and agricultural waste derived catalysts and conventional and non-conventional activation paths. These studies show interesting results through the development of solids that activate certain kinds of reactions shown in Figure 1.

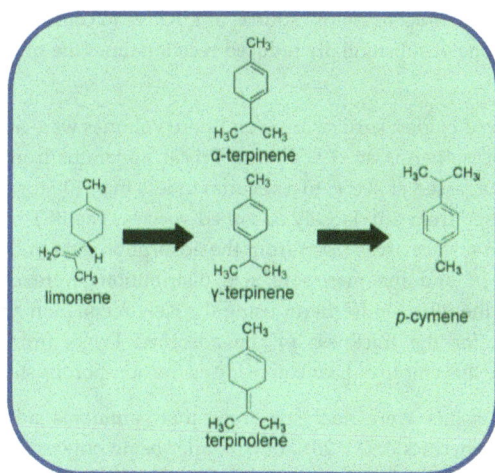

**Figure 1.** Conversion of limonene to p-cymene and reaction intermediates

This work is based on designing a clean process to transform limonene, into p-cymene, in a rapid, simple and economic way. Limonene, a cyclic terpene with empiric formula $C_{10}H_{16}$, is extracted with a purity of *ca.* 95% from citrus fruit processing. Although initially used in pharmaceutical industries, food is the main means for human exposure, although due to its low toxicity [10] it does not constitute a risk to human health at the actual exposure levels. This compound has various uses, as a biodegradable solvent for resins (replacing organic solvents such as mineral oils, methyl-ethyl ketone, toluene, glycolic ethers, CFCs, or as an additive in pigments, inks and adhesives. However, new uses are being sought to give added value to this subproduct, thus increasing the income of these industries, but also with an obvious benefit to society. From the chemical structure of limonene, shown in Figure 1, it may be appreciated that having a six membered ring this substance has the potential to obtain compounds that are usually produced employing petroleum derivatives, with the added bonus of it being a less toxic renewable intermediate. In this study the production of p-cymene, an important intermediate in industrial fine chemicals syntheses, for fragrances, flavourings, herbicides, pharaceuticals, *p*-cresol production, syntheses of not nitrated musk's (i.e. tonalide), etc. [11] will be described.

The preparation of *p*-cymene is usually carried out by Friedel-Crafts alkylation of toxic benzene or toluene (from petroleum), with $AlCl_3$ as catalyst and the respective halides or propanol [12]. In these processes mixtures of *ortho* and *para* isomers, are found and therefore further separation processes are required. Furthermore, the use of benzene or toluene and $AlCl_3$ are restricted by environmental legislation in industrialised countries. In this study a series of commercial silica–alumina mixed oxides, supplied by Sasol, with silica contents ranging from 1 to 40 wt.% designated as SIRAL 1 to SIRAL 40, accordingly, were employed. The textural and acid characteristics of these materials were determined and related with their catalytic activities in the transformation of limonene to *p*-cymene using microwave irradiation as a rapid and efficient energy source, alternative to conventional heating, where apart from the advantage of substantially reduced reaction times the product selectivity was also enhanced [10].

The texture of the mixed oxides: surface areas and pore volumes were analysed by nitrogen adsorption/desorption isotherms at $-196°C$, in a Tristar apparatus from Micromeritics, on samples previously outgassed at 300°C to a vacuum of less than $10^{-4}$ torr to ensure that they were clean, dry and free from any loosely adsorbed species. The BET method was used to determine the specific surface areas ($S_{BET}$) from the adsorption data in the relative pressure range of 0.05–0.30 $p/p°$ and the mesopore size distributions were calculated from the desorption branch of the nitrogen isotherm using the Kelvin equation and the BJH method with the parameters for the thickness of the adsorbed layers from the Harkins–Jura equation, chosen since this employed a metal oxide as the non-porous standard [13].

The acidities of the solids were analysed from their ammonia adsorption capacities, determined on a Micromeritics ASAP 2010 device, after being outgassed overnight at 300°C to a vacuum of $>10^{-4}$ torr, then the ammonia adsorption isotherms at 30°C were determined up to a pressure of about 350 torr, obtaining the total chemisorption plus physisorption capacities. Subsequently the sample was outgassing at 30°C and a second adsorption isotherm at 30°C determined to measure the physisorption capacity of the sample. The chemisorption capacity was calculated from the differences in the ammonia uptakes between the first and second isotherms. The total acidities were subsequently calculated assuming that each molecule of ammonia reacted with one acid site [14].

Pyridine adsorption coupled with infrared analysis was used to qualitatively determine the Lewis ($Al^{3+}$) and Brønsted (Al-OH) acid sites of the materials, these being capable of catalysing the conversion of limonene to *p*-cymene. The catalysts were pressed as self-supporting discs, outgassed under high vacuum, contacted with pyridine vapour and outgassed to obtain the infrared spectra corresponding to Lewis and Brønsted acid sites with adsorption peaks at 1455 and 1545 $cm^{-1}$, respectively [15]. The microwave reactions were carried out with a programmable focalised monomodal microwave apparatus, Synthewave 402 from Prolabo, under both dry media and reflux conditions. Dry media microwave induced reactions can be considered a clean technology, since no solvents are used. In those experiments 50 μl of limonene were physically mixed with 200 mg of solid, placed in a glass reactor and irradiated at 300 W for 5, 10 or 20 min. In reflux conditions 500 mg of solid was mixed with 5 ml of limonene and heated with a reflux column to 165°C for 10 or 20 min, avoiding overheating of the reaction mixture. The final temperature was chosen to be slightly lower than the boiling points of the reactant and products (limonene 175°C, p-cymene 177°C) in order to control the reaction. At the end of the experiments the reaction mixtures were cooled, extracted and dissolved in ethanol and analysed by gas chromatography coupled to a mass spectrometer (GC–MS). Following the extraction of the reaction products with ethanol, the catalytic activities of the samples were redetermined to ascertain if the materials had suffered any change in their activities or selectivities.

It has been previously proposed that the available surface area coupled with the accessibility to the active acid sites play important roles in controlling the catalytic process in this reaction [16]. The specific surface areas, pore volumes and average mesopore diameters of the solids are summarised in Table 1. The nitrogen adsorption/desorption isotherm for SIRAL 1 was of type IV with a well-defined plateau at high relative pressures with a type H1 hysteresis loop, characteristic of a mesoporous solid with a narrow well defined mesopore size distribution [13]. On increasing the silica content the specific surface area, mesopore volume and average mesopore diameters increased progressively as evidenced by a widening of the hysteresis loop between the adsorption and desorption branches and with the highest silica contents the plateau at high relative pressure became less well defined as the pore sizes shifted into the narrow macropore range. From the isotherms obtained for SIRAL 1 and SIRAL 40 presented in Figure 2 the change in the hysteresis loop due to the widening of the pores may be appreciated while the upward displacement of the curve with the increased silica content was indicative of the greater specific surface area. The other samples with intermediate silica contents, not shown here for clarity, gave rise to adsorption/desorption curves that lay between these two extremes.

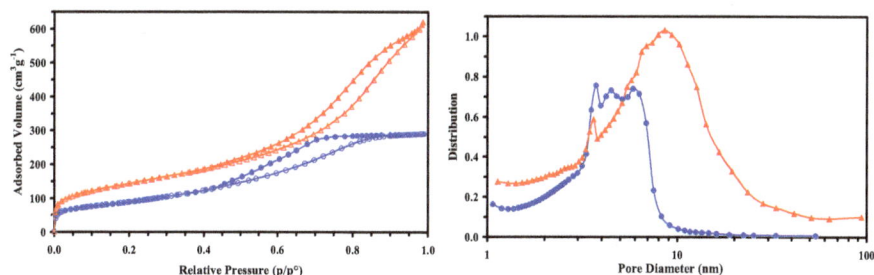

**Figure 2.** Textural characteristics of mixed oxides SIRAL 1 ○● and SIRAL 40 △▲

Since the reaction is catalysed by the acid sites, the higher activity as the number and accessibility of the surface acid sites rose with increased silica content was to be expected. Greater activities were found as the total acidity was increased (measured by ammonia adsorption, shown in Figure 3).

| Sample | Surface area ($m^2g^{-1}$) | Pore volume ($cm^3g^{-1}$) | Average mesopore diameters (nm) | |
|--------|-----------|-----------|------|------|
| SIRAL 1 | 321 | 0.45 | 4.5 | 5.8 |
| SIRAL 5 | 364 | 0.57 | 4.8 | 5.9 |
| SIRAL 10 | 422 | 0.59 | 4.8 | 6.8 |
| SIRAL 20 | 432 | 0.68 | 5.4 | 9.4 |
| SIRAL 40 | 506 | 0.95 | 5.8 | 12.7 |

**Table 1.** Textural properties of the mixed oxides

The catalytic activity results obtained with microwave irradiation of limonene are presented in Table 2.

| | | | Selectivity | | | |
|--------|-----------|------------|-----------|-----------|-----------|-----------|
| Sample | Time (min) | Conversion (%) | α-terpinene (%) | Γ-terpinene (%) | α –terpinolene (%) | p-cymene (%) |
| | 5 | 15 | 85 | 0 | 15 | 0 |
| **SIRAL 1** | 10 | 56 | 52 | 14 | 0 | 34 |
| | 20 | 100 | 0 | 0 | 0 | 100 |
| | 5 | 42 | 60 | 12 | 21 | 7 |
| **SIRAL 5** | 10 | 63 | 27 | 25 | 0 | 48 |
| | 20 | 100 | 0 | 0 | 0 | 100 |
| | 5 | 65 | 48 | 11 | 31 | 10 |
| **SIRAL 10** | 10 | 69 | 15 | 0 | 14 | 71 |
| | 20 | 100 | 0 | 0 | 0 | 100 |
| | 5 | 88 | 31 | 6 | 15 | 48 |
| **SIRAL 20** | 10 | 100 | 0 | 0 | 0 | 100 |
| | 20 | 100 | 0 | 0 | 0 | 100 |
| | 5 | 93 | 0 | 0 | 3 | 87 |
| **SIRAL 40** | 10 | 100 | 0 | 0 | 0 | 100 |
| | 20 | 100 | 0 | 0 | 0 | 100 |

**Table 2.** Limonene conversions and selectivities under dry media conditions

During the reactions (SIRAL 1 shown as example in Figure 4), only α-terpinene, γ-terpinene, γ-terpinolene (from isomerisation) and p-cymene (from dehydrogenation) were found as products, with no other compounds such as menthanes or menthenes, produced by disproportionation or polymerisation, detected. In Table 2 it should be noted that the selectivity to p-cymene increased with both longer irradiation times and increased silica contents. For the two samples with the highest silica contents reaction times of just 10 min gave selectivities to p-cymene of greater than 90%. Since these samples had the most acid centres, calculated on both per gram or per $m^2g^{-1}$ basis, it would appear that the selectivity to p-cymene was due to the rapid aromatisation of the intermediates produced by isomerisation over the acid centres. Although conversion of limonene to p-cymene over Lewis acid sites with conventional heating is known, the much longer reaction times necessary to increase the overall conversion (3 h) leads to reduced selectivities due to the formation of undesirable mentanes, etc. [17].

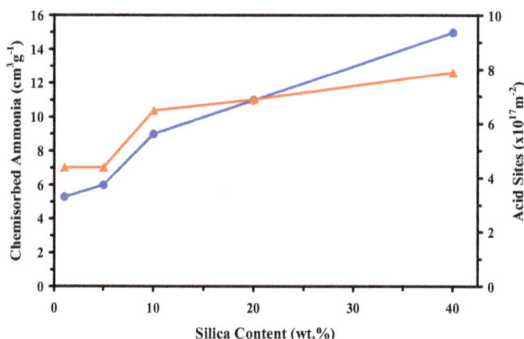

**Figure 3.** Ammonia chemisorption $cm^3g^{-1}$ • and number of acid sites $m^{-2}$ ▲ vs. silica content of the mixed oxides.

In order to carry out the reaction in conditions similar to those used industrially the two samples with the highest activities in dry media were chosen for further study in reflux conditions, with the ratio between the reactants and the solid: 0.00025 and 0.01 $cm^3g^{-1}$, respectively. From Table 3 it may be observed that the intermediate isomerisation product α-terpinene was only found for SIRAL 20 and at a very low level (4%) and 100% conversion of limonene to p-cymene was achieved with Siral 40 after 10 min. The speed of the reaction when heated by microwave irradiation is possibly the reason for the high selectivities, since the short reaction time necessary to attain these high conversions and selectivity to p-cymene avoid the formation of undesirable by-products such as mentanes (products of disproportionation) or polymers, that are found with the longer reaction times employed with conventional heating [18]. Thus, microwave irradiation favoured the production of p-cymene from limonene, avoided the use of highly toxic benzene, toluene and aluminium trichloride and allowed high conversions and excellent selectivities towards the desired product due to the accelerated reaction rates. As the reaction is governed by the number and accessibility of the acid sites greater activities were achieved with increased silica contents, which led to higher specific surface areas, pore volumes and average pore sizes in addition to an increased number of acid sites.

## 2.1. Modified clays

Sepiolite modified with sodium hydroxide and impregnated with either iron or manganese salts were also used as catalytic supports for conversion of limonene to p-cymene. The use of an inexpensive natural clay for the catalyst preparations reduces costs and the need for commercial synthesised solids is avoided. The sepiolite used was from Tolsa SA (hydrated magnesium silicate of the philosilicate type 2:1 with a layer of magnesium between two layers of silica tetrahedra. The octahedral sheet is composed mainly of $Mg^{2+}$, mainly composed of $SiO_2$, 62%, MgO 25%, $Al_2O_3$ 1.2%, $Fe_2O_3$ 0.5%). This natural clay of high abundance in Madrid (Spain), has a specific surface area ($S_{BET}$) close to 300 $m^2g^{-1}$, of which 150 $m^2g^{-1}$ is external (pores with diameters >2 nm ø) and the rest is due to the micropores of the material (< 2nm ). It has a high density of –SiOH groups originated at the edges due to breakage of Si-O-Si bonds at *ca.* 0.5 nm intervals, having a density of *ca.* 2.2 groups/10 $nm^2$[19].

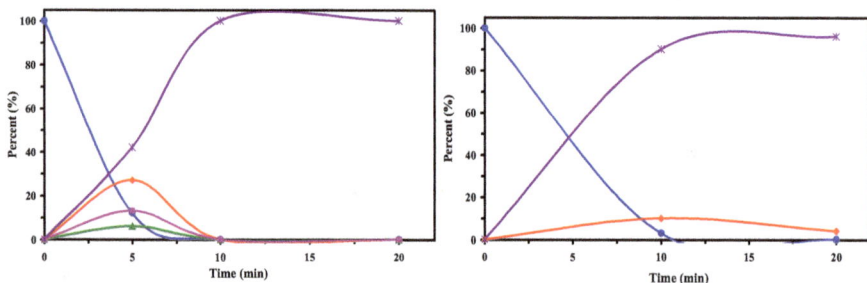

**Figure 4.** Reactant and product distributions vs. time for (a) Siral 1 in dry media and (b) Siral 20 under reflux: limonene (●), α-terpinene (♦), γ-terpine (▲), α-terpinolene (■) and *p*-cymene (✶)

The activity of the parent sepiolite was modified by adding 5.75 wt.% of iron or manganese, in the form of their nitrate or acetate, respectively, since their oxides efficiently absorb microwave radiation and have low toxicity compared to other metals commonly used as catalysts (*i.e.* Pd, Cd, Cr etc.), or price (Au or Ag). A further sample was pretreated with sodium hydroxide to introduce a similar amount of sodium.

The best procedure to decompose the impregnated compounds to their corresponding supported oxides was determined with TG-DTA analyses in air flow of the precursors in a Stanton model STA 781 thermogravimetric analyser up to 1000ºC. The thermograms obtained show 8% loss of adsorbed water in the interval 20-120ºC, 5% at 120-250ºC loss of sepiolite (zeolitic) water, and 8% loss for decomposition of the anions at *ca.* 400ºC for iron nitrate and at ca. 300ºC for the manganese acetate. In accordance with these findings all the precursor solids were calcined at 400ºC for 4 h in a 50 $cm^3min^{-1}$ air flow [20].

The nitrogen adsorption isotherms for the parent sepiolite heat treated at 400°C gave a mixed type I/II form, due to the presence of both micropores (0-2 nm) and mesopores (2-50 nm) that extend into the macropore region (> 50 nm). For the SepFe (Figure 5), SepMn and SepNa there was a loss in the specific surface area due to the collapse of the microporous structure of sepiolite by folding due to the thermal treatment. For all samples the hysteresis loops were of type H3, typical for solids with slit-shaped pores, commonly found with clay materials.

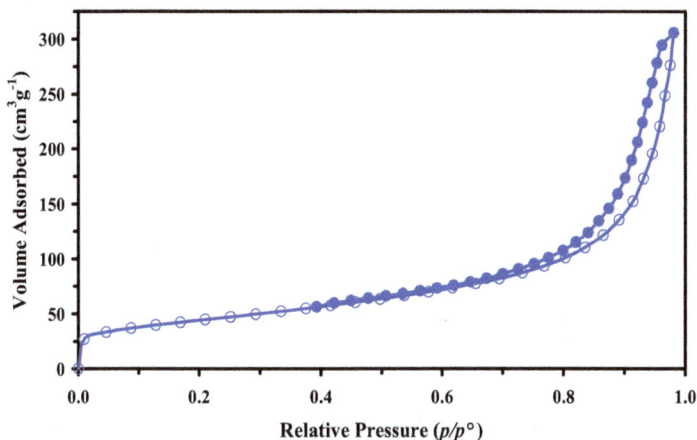

**Figure 5.** Nitrogen isotherm for SepFe

Sample SepNa had the lowest pore volume and surface area, probably due to blocking of pores with sodium species. The textural data from the corresponding isotherms are presented in Table 3.

| Solid | $S_{BET}$ | $S_{EXT}$ | $V_{mic}$ | $V_{mes}$ | $V_t$ |
|---|---|---|---|---|---|
| Sepiolite | 298 | 149 | 0.038 | 0.417 | 0.455 |
| SepNa | 100 | 93 | 0.002 | 0.362 | 0.364 |
| SepFe | 155 | 149 | 0.002 | 0.416 | 0.418 |
| SepMn | 153 | 138 | 0.008 | 0.431 | 0.439 |

$S_{BET}$ = Specific surface area (m$^2$g$^{-1}$) $S_{EXT}$ = External surface area (m$^2$g$^{-1}$) Vmic = Micropore volume (0-2nm pore diameter, cm$^3$g$^{-1}$), V$_{mes}$ = Mesopore volume (2-50nm pore diameter, cm$^3$g$^{-1}$), Vt = Total pore volume (cm$^3$g$^{-1}$).

**Table 3.** Textural characteristics of the solids

A JEOL model FXII electron microscope operating at 200kV was used to study the structure of the materials. The basic sites of the solids were quantified by the adsorption-desorption of carbon dioxide in a Coulter Omnisorp 100 apparatus on solids previously outgassed to clean the surface at different temperatures to quantify the amount and strength of basic sites. The results obtained by this procedure are included in Figure 5 [21].Transmission electron microscopy showed the oxide particles present in the structures of SepFe and SepMn with sizes of 4-5 nm but with no other observable alteration to the fibrilar sepiolite structure and a more homogeneous distribution of the oxide particles for SepFe compared to SepMn, which reduce the exposed oxide area during reaction compared to SepFe. Homogeneous particle sizes have been claimed to be a positive effect for the reactivity in these kinds of reactions.

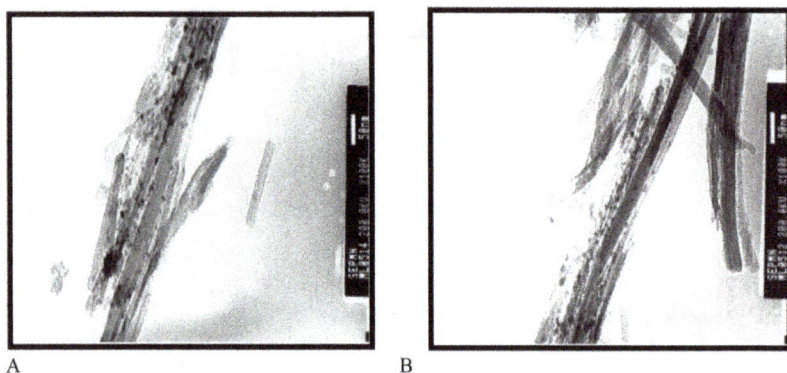

**Figure 6.** Transmission electron micrographs of (a) SepMn and (b)SepFe.

The amount and strength of acid sites determined by TPD of pyridine (Figure 7) showed the measured acidities followed the order: SepFe > SepMn > Sep. As expected no pyridine adsorption was found for the sample pretreated with sodium hydroxide (SepNa). The increase in the amount and strength of acid sites by addition of iron or manganese oxides to these supports was mainly related to the presence of $Fe^{x+}$ and $Mn^{x+}$ ions (Lewis acidity) and to surface OH groups (Brönsted acidity) [22].

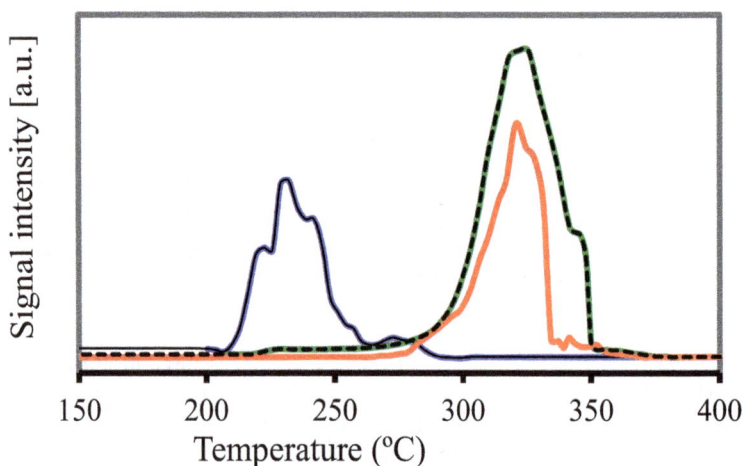

**Figure 7.** Pyridine determination of acid sites on Sepiolite, SepMn and SepFe.

From the adsorption-desorption of carbon dioxide shown in Figure 8 the basicity of Sepiolite (due to the existence of basic sites mainly $Mg^{2+}$ ions and oxygen on the surface) is low, while temperatures in excess of 400ºC were needed to desorb carbon dioxide from the other three materials. The solid with greater amount of stronger basic sites was SepNa. Samples SepFe and SepMn were more basic than Sepiolite and had greater total amounts of basic sites of lower strength.

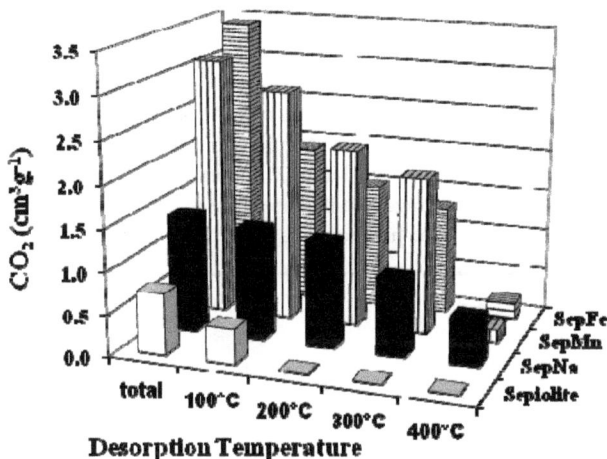

**Figure 8.** Basicity of the catalysts by $CO_2$ adsorption-desorption

For the reactions under dry media (200 mg of catalyst mixed with 0.057 ml of limonene) and solvent free conditions, after irradiation the system was cooled and eluted with 2 ml ethanol and filtered to extract the reactants and products and subsequently analysed as described in section 1.1. In the solvent free reactions, a temperature of 165ºC, was chosen to transform limonene, based on the results obtained under dry media conditions. In this case 5 ml of limonene and 500 mg of catalyst were used and a reflux attachment included in the microwave oven. At the end of the experiment 0.057 ml of liquid were dissolved in 2 ml of ethanol to follow the analysis procedure as described above. The results for the dry media reactions are shown in Table 4.

| Sample | Reaction Time (Min) | Conversion (%) | Product Distribution (%) | | | |
|---|---|---|---|---|---|---|
| | | | α-terpinene | γ-terpinene | α-terpinolene | p-cymene |
| | 5 | 35 | 25 | 8 | 31 | 36 |
| Sepiolite | 10 | 73 | 26 | 0 | 0 | 74 |
| | 20 | 100 | 0 | 0 | 0 | 100 |
| | 5 | 75 | 0 | 0 | 0 | 100 |
| SepFe | 10 | 100 | 0 | 0 | 0 | 100 |
| | 20 | 100 | 0 | 0 | 0 | 100 |
| | 5 | 89 | 0 | 0 | 0 | 100 |

| Sample | Reaction Time (Min) | Conversion (%) | Product Distribution (%) | | | |
|--------|---------------------|----------------|α-terpinene|γ-terpinene|α-terpinolene|p-cymene|
| SepMn | 10 | 100 | 0 | 0 | 0 | 100 |
| | 20 | 100 | 0 | 0 | 0 | 100 |

**Table 4.** Conversions and selectivities under dry media conditions

SepNa did not show activity, with high amounts of basic sites and practically non-existent acid sites, even though it reached the highest temperature during microwave heating. With sepiolite, selectivity to p-cymene rose, reaching 100% after 20 min and for SepFe and SepMn after only 10 min due to iron or manganese that are microwave adsorbing centres compared to the parent sepiolite, but not to the final temperature attained (Figure 9).

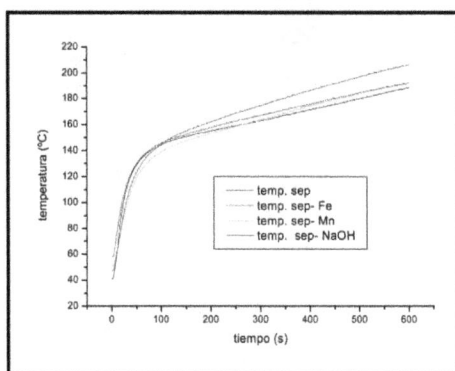

**Figure 9.** Temperatures reached under dry media conditions

The solids that presented activity under dry media conditions (Sepiolite, SepFe and SepMn) were tested in the liquid phase reactions that allow higher liquid to solid ratios (Table 5) and the temperatures reached are included in Figure 10.

| Sample | Reaction Time (Min) | Conversion (%) | Product Distribution (%) | | | |
|--------|---------------------|----------------|α-terpinene|γ-terpinene|α-terpinolene|p-cymene|
| Sepiolite | 10 | 100 | 0 | 0 | 0 | 100 |
| | 20 | 100 | 0 | 0 | 0 | 100 |
| SepFe | 10 | 85 | 32 | 19 | 35 | 14 |
| | 20 | 100 | 9 | 3 | 0 | 88 |
| SepMn | 10 | 10 | 34 | 8 | 58 | 0 |
| | 20 | 75 | 33 | 17 | 33 | 17 |

**Table 5.** Conversions and selectivities in liquid phase

As in dry media, the selectivity to p-cymene increased with longer reaction times and higher temperatures, in agreement with the fact that the formation of this product is an endothermic reaction. Under these reaction conditions Sepiolite gave the best results for both activity and selectivity, reaching 100% conversion and selectivity after only 10 minutes. The differences in the temperature profiles for these three materials during the reaction, were not so great as to explain such differences in their catalytic activities and selectivities.

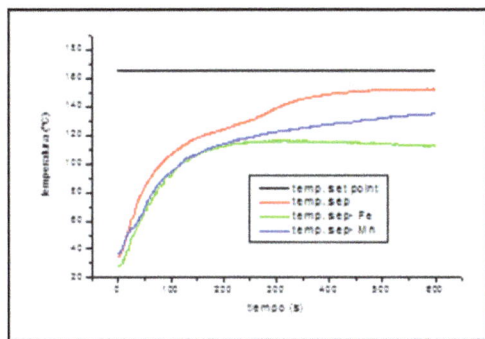

**Figure 10.** Temperatures reached under solvent free conditions.

The strong basic sites present in SepNa, SepMn and SepFe may be capable of adsorbing limonene to such an extent that it was not readily available for reaction under the reaction conditions, this being the reason why Sepiolite is the most active solid for the dehydrogenation of limonene to p-cymene. Previous work by the authors showed that the acid sites of silica–alumina mixed oxides could dehydrogenate limonene in conditions similar to those used here [10]. Thus, the acid sites present on the Sepiolite surface, although weaker than those found with iron or manganese containing solids, were of the right strength to convert limonene to p-cymene. Further experiments showed that 300 mg of Sepiolite were enough to convert 5 ml of limonene to p-cymene in 10 min with a 100% conversion and selectivity. Similar conversions and selectivities to those found in this study have recently been reported for the transformation of limonene, although dangerously high hydrogen pressures, greater than atmospheric and the use of toxic and expensive palladium catalysts were required [23].

Conventional heating of limonene using acid solids as catalysts in liquid-solid conditions leads to lower activities and selectivities than those found in this work. We believe the short reaction times required using dielectric heating were responsible for the higher conversions and selectivities, due to the increased activity of these paramagnetic absorbing centres, thus avoiding other products such as dimers and polymers, that are formed during the longer reaction times necessary with conventional heating [17].

## 3. Sustainable fine chemicals from agri wastes

A vast amount of work has been carried out regarding the preparation of fine chemicals, with oxidation being one of the main paths followed. Renewable value added chemicals were prepared in this work using solid and liquid agricultural industrial wastes from rice and citrus production, as renewable raw materials, avoiding the use of substances toxic to the environment and achieving a reduction in energy expenditure. The whole process is consistent with a sustainable development. The present investigation has demonstrated how the transformation of a low value subproduct (limonene) into high value materials (carvona, carveol and limonene epoxides) can be achieved with similar conversions and selectivities to those found in the literature for catalysts that have higher toxicity and are less environmentally friendly.

After a thorough review of literature data, careful design of oxidation conditions of limonene allows the production of carvone and carveol through allylic oxidation and limonene oxides through double bond epoxidation (Figure 11). These products have interesting pharmaceutical properties and also are chemical intermediates with prices from 5 to 10 times higher than limonene [9].

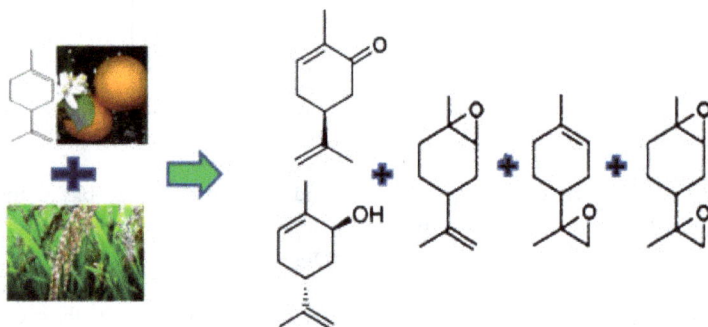

**Figure 11.** Renewable carvone, carveol and epoxides prepared from limonene. Source of photographs: http://www.finecooking.com/assets/uploads/posts/5552/ING-oranges_sql.jpg http://1.bp.blogspot.com/_siEMqPzFQAY/TJO2Pt4tX3I/AAAAAAAABW0/gWJ_-7IP9co/s1600/arroz9.jpg

A clean process for this transformation was developed here by using iron oxide supported on silica (RHS) from rice husk (RH) to catalyse the reaction. RH is produced in huge amounts annually and its ash contains *ca.* 94% silica, with the added bonus of zero $CO_2$ energy being produced during the calcination of RH to produce RHS.

The organic reactions carried out in this work were activated with dielectric heating that allows higher yields under mild reaction conditions, avoiding thermal decomposition of sensitive products or reagents, and therefore affecting selectivity [10]. Tert-butyl hydroperoxide (tBHP) was used as oxidant due to its easy handling characteristics and high thermal stability. The tert-butanol and its by-products on oxidation, can be distilled and recycled, a particularly important step when industrial amounts are required. Silica from rice husk (RHS) was used as support and catalyst. It was prepared by calcining rice husk in air at 500°C for 4 h. For comparison reasons, silica Aerosil Evonik was used since it had similar textural properties to RHS (surface area of 91 $m^2g^{-1}$). The supported catalysts were prepared by dry impregnation of RHS with iron nitrate ($Fe(NO_3)_3 \cdot 9H_2O$) (> 99% purity, Sigma-Aldrich) solutions, the impregnated precursors were air dried overnight at 100ºC, prior to TG-DTA and FTIR studies of the calcination procedures necessary to produce FeOx/RHS catalysts (400 or 600ºC, 4 h in air). Since treating silica with low pH iron nitrate solutions can produce structural changes, fresh RHS was treated with nitric acid solutions of the impregnating pH (1.4) and then calcined at 400 or 600ºC, for comparison purposes

The basicity of the solids was measured by decomposition of acetic acid, in a gas analysis system with quadrupole mass spectrometry, model M3 QMS200 Thermostar coupled to a thermogravimetric/differential thermal analysis equipment, Stanton STA model 781. Increasing the temperature of the spiked solids under nitrogen flow at a heating rate of 5ºCmin$^{-1}$ up to 700ºC, recording the evolution of mass 44, ascribed to $CO_2$. The amount and temperature of evolution of the $CO_2$ signal gave an indication of the strength and amount of basic sites. The $CO_2$ signal was calibrated from the decomposition of a known amount of calcium oxalate.

The catalytic oxidations of limonene were studied in a programmable focussed microwave oven Syn402 from Prolabo, described above in section 1.1. Preliminary experiments were carried out in dry media conditions on the Fe-containing catalysts, to determine the best conditions to use in further liquid/solid experiments, choosing the conversion of limonene and selectivity to carvone as parameters for comparison of the effects caused by the various conditions studied. Studied parameters were the limonene: tBHP volume ratio, temperature and reaction time. Based on these preliminary experiments, the conditions chosen for the liquid phase work were 300W microwave power, with a reflux attachment, reaction temperature (150ºC) and reaction time up to 120 min, using 4 ml limonene, 14 ml of TBHP solution and 150 mg of catalyst. Conventional heating results were carried out for comparison with similar amounts of reactants and catalyst, starting the reaction time from the moment the reaction temperature was reached and the catalyst added. The effects of reaction temperature, amount of catalyst and oxidant on both the reactivity and selectivity were studied using the most promising catalyst. The reactants and products were analysed in a GC-MS system, as described above.

The chemical composition of the RHS used in this work was 6% S, 2% K, 1% Cl, P and Ca, 0.1% Mn and Fe and 0.04% Zn. The iron containing precursors were calcined at 400 or 600ºC for 4 h, according to TG-DTA and FTIR data. The iron containing catalysts (FeOx/RHS) had 4.8% or 8.9% Fe. XRD patterns indicated the amorphous nature of RHS and the FeOx/RHS show crystalline iron oxide particles, more common on increasing the amount of iron and calcination temperature, due to sintering, corresponding to maghemite ($\gamma$-$Fe_2O_3$: 36º, 44º, 54º, 58º and 63º) and hematite ($\alpha$-$Fe_2O_3$: 24º, 33º, 36º, 50º and 62.5º).

Textural analyses gave rise to the results included in Table 6, showing nitrogen isotherms of type IIb with hysteresis loops type H3, typical of samples with slit-shaped mesopores that extended into the macropore region, caused by the spaces between the plates of material and with no microporosity according to the t-plot analyses [13]. It may be observed that higher calcination temperatures caused a fall in the surface area and a shift in the hysteresis loop to higher relative pressure/wider pores, due to sintering of the samples, in agreement with XRD results.

| Solid calc. T °C | $S_{BET}/m^2g^{-1}$ | $V_{mes}/mLg^{-1}$ |
|---|---|---|
| RHS | 85 | 0.172 |
| RHS HNO₃ 400 | 93 | 0.203 |
| RHS HNO₃ 600 | 93 | 0.210 |
| 4.8%Fe/RHS 400 | 118 | 0.200 |
| 4.8%Fe/RHS 600 | 92 | 0.185 |
| 8.9%Fe/RHS 400 | 122 | 0.181 |
| 8.9%Fe/RHS 600 | 87 | 0.163 |

**Table 6.** Textural analyses of catalysts

Most silicas are almost transparent under the electron beam of the transmission microscope and therefore easy to distinguish from oxide particles deposited on their surfaces [24]. However, the RHS particles prepared in this work showed an unexpected dense structure under study by TEM and SEM (Figure 12), formed by lamellar entities, in agreement with the shapes of the nitrogen isotherms, explained above, that made it difficult to distinguish the iron oxide particles.

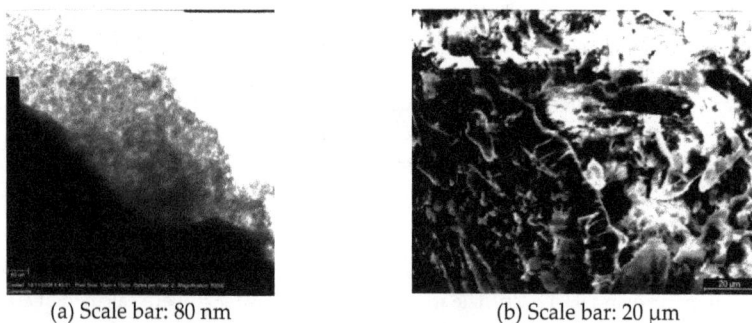

(a) Scale bar: 80 nm                    (b) Scale bar: 20 μm

**Figure 12.** Electron micrographs for 8.9% Fe/RHS 400 a. TEM        b. SEM

The adsorption of acetic acid used to characterise the catalysts basic sites produces mainly carbonate, bidentate or bridged acetate species, that decompose to $CO_2$, giving rise to bands due to the interaction with basic centres of higher basicity with increasing temperature, i.e. T< 150°C for the desorption of physically adsorbed $CO_2$, those at T = 150-250°C corresponding to molecules of $CO_2$ evolved from basic sites of low to medium strength, whilst those at T = 300-400°C correspond to the interaction with basic sites of medium to high strength (Figure 13)[25]

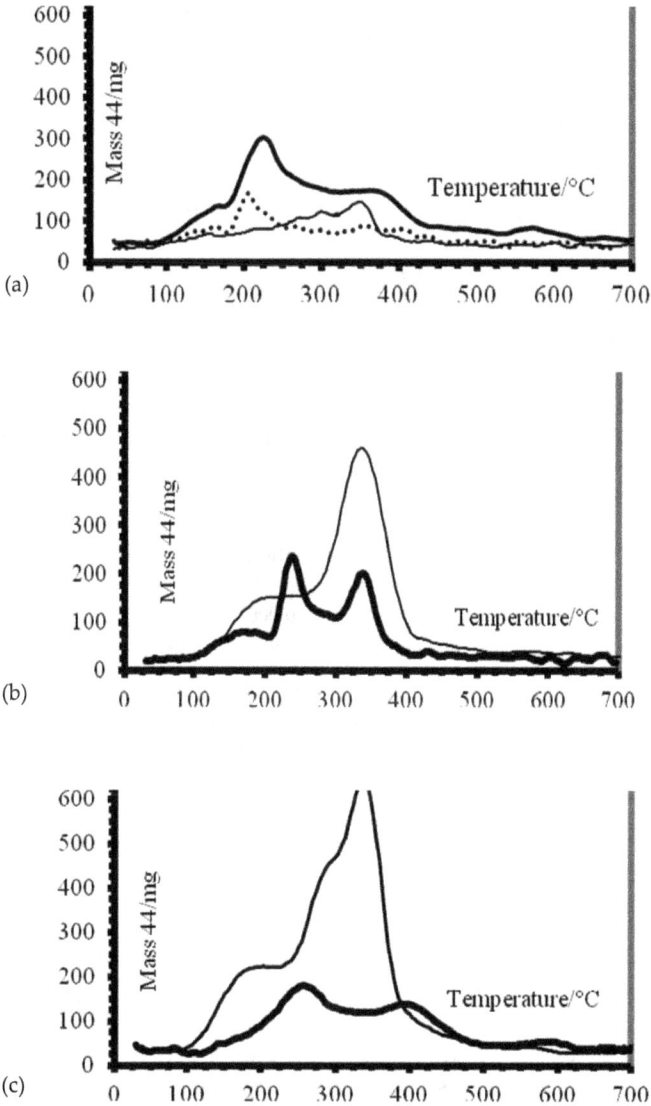

**Figure 13.** Mass 44/mg evolved from acetic acid decomposition on:
4a. Original and modified RHS (dotted line: RHS, thick line RHS HNO₃ 400, thin line RHS HNO₃ 600)
4b. 8.9%Fe/RHS (thick line calcined at 600°C, thin line calcined at 400 °C)
4c. 4.8%Fe/RHS (thick line calcined at 600°C thin line calcined at 400 °C)

From these data it was concluded that the materials employed had the following order of basicity:

Silica Evonik <<< RHS HNO₃ 600 < RHS < RHS HNO₃ 400

The traces obtained with silica Evonik were within the noise level of the technique, indicating that the acetic acid was only physisorbed, due to its low basicity. These results can be explained taking into account that the basicity, related to both the strength and number of basic sites of RHS was mainly due to the presence of basic oxides. On treating RHS with nitric acid, particles are disaggregated leading to an increase of basic sites available on the surface, but at higher calcination temperatures sintering of the particles occurs decreasing the amount of surface basic centres. The Fe-containing catalysts present higher basicities than the RHS due to the presence of iron oxide, but those calcined at 600ºC have lower basicity than their analogues calcined at 400ºC, in agreement with the sintering occurring in the particles of iron oxide, XRD, textural and published data [26].

Epoxidation and allylic oxidation reaction pathways of limonene are competitive, with valuable epoxides, carvone and carveol being produced. The selective oxidation of alkenes in the presence of peroxides and basic site containing materials is advantageous, since epoxides undergo breakage on acid sites, which decreases the selectivity of the processes. A thorough analysis of literature on limonene oxidative transformations data provide an insight into the conditions needed to convert limonene into value added and renewable limonene oxides (mono or diepoxides) from epoxidation through electrophilic attack at the double bonds, and allylic oxidation when hydrogen abstraction is the dominant reaction towards carvone and carveol, both interesting molecules, since they retain the olefinic functionality, which allows further useful transformations [27, 28].

The catalytic activity was first measured under dry media conditions, reacting 150 mg of catalysts and 48 µL of limonene:tBHP solution, with different limonene:tBHP volume ratios and temperatures, in order to determine optimal conditions for further liquid phase reactions, that allow treating higher amounts of limonene. Experiments with no control of solid temperature and times over 20 min produced evaporation of reactants, giving rise to irreproducible results, which were avoided by maintaining the temperature of the solid constant throughout the reaction, leaving the microwave power to vary accordingly and limiting the reaction times to 20 min. Under these conditions, the experiments were carried out in duplicate and the results obtained with the Fe-containing catalysts under dry media conditions at 120ºC and 20 min irradiation time, with limonene:tBHP solution= 1:1 (volume ratios) are included in Table 7, showing that under equal conditions, the best catalyst for conversion and selectivity to carvone was 8.9%Fe/RHS 400. The catalysts with higher iron content and lower calcination temperatures present higher activities although the selectivities were similar for catalysts with similar iron contents.

With other reaction variables maintained constant, Table 8 shows that by increasing the amount of oxidant the conversion of limonene increases, but that at high amounts the selectivity to carvone starts to decrease and the more oxidised product carvacrol was found. From these results an optimum limonene:tBHP solution=1:3.5 (volume ratio) was chosen.

| Catalyst | Conversion (%) | $S_{carvone}$ (%) |
|---|---|---|
| No catalyst | 0 | |
| 4.8%Fe/RHS 400 | 35 | 35 |
| 4.8%Fe/RHS 600 | 25 | 36 |
| 8.9%Fe /RHS 400 | 48 | 45 |
| 8.9%Fe/RHS 600 | 39 | 48 |

S = selectivity

**Table 7.** Conversions and selectivities of Fe-containing catalysts under dry media conditions (limonene:tBHP ratio=1:1, 120ºC, 20 min, free microwave power, 0.15 mg catalyst)

| Limonene:tBHP solution | Conversion (%) | $S_{carvone}$ (%) |
|---|---|---|
| 1:1 | 48 | 45 |
| 1:2 | 55 | 37 |
| 1:4 | 63 | 23 |
| 1:3.5 | 60 | 36 |

**Table 8.** Conversions and selectivities of 8.9%Fe/RHS 400 catalyst at 120ºC under dry media conditions with different Limonene:tBHP ratios (20 min reaction, free microwave power, 0.15mgs catalyst)

With regards to the reaction temperature, from Table 9, a value of 150ºC was favoured, since higher temperatures produced over oxidation of carvone, decreasing the selectivity to this compound and at lower temperatures lower conversions were attained.

| Temperature (ºC) | Conversion (%) | $S_{carvone}$ (%) |
|---|---|---|
| 140 | 45 | 38 |
| 150 | 60 | 36 |
| 160 | 70 | 21 |

**Table 9.** Conversions and selectivities of 8.9%Fe 400 catalyst at different temperatures under dry media conditions (limonene:tBHP solution ratio=1:3.5, 20 min reaction time, free microwave power, 0.15mg catalyst)

Based on these results, the parameters chosen for the experiments carried out under liquid conditions were a reaction temperature of 150ºC, reaction times up to 2 h (with reflux attachment), limonene:tBHP solution=1:3.5 (volume ratio), 18 ml total volume, 0.15 g of catalyst. The experimental results obtained under these conditions are presented in Table 10. The main reaction products were the epoxides, carvone and carveol in different amounts. The epoxide found was mainly the endo, with ratios endo/exo+diepoxide close to 2/1+1.

| Catalyst/$T_{calc}$ (°C) | Reaction time min | Conversion % | $S_{epoxides}$ % | $S_{carvone}$ % | $S_{carveol}$ % |
|---|---|---|---|---|---|
| **Blank 14mL decane** | 30 | 5 | 91 | 9 | 0 |
| | 60 | 5 | 86 | 14 | 0 |
| | 90 | 6 | 85 | 15 | 0 |
| | 120 | 6 | 84 | 16 | 0 |
| **Silica Evonik** | 120 | 6 | 81 | 19 | 0 |
| | 30 | 12 | 62 | 24 | 14 |
| **RHS** | 60 | 16 | 50 | 27 | 23 |
| | 90 | 22 | 47 | 29 | 24 |
| | 120 | 22 | 45 | 31 | 24 |
| | 30 | 16 | 74 | 15 | 11 |
| **RHS HNO₃ 400** | 60 | 23 | 59 | 19 | 22 |
| | 90 | 33 | 55 | 24 | 21 |
| | 120 | 35 | 49 | 30 | 21 |
| | 30 | 16 | 69 | 31 | 0 |
| **RHS HNO₃ 600** | 60 | 18 | 55 | 31 | 14 |
| | 90 | 24 | 50 | 34 | 16 |
| | 120 | 25 | 43 | 35 | 22 |
| | 30 | 17 | 76 | 14 | 10 |
| **4.8%Fe/RHS 400** | 60 | 23 | 72 | 16 | 12 |
| | 90 | 35 | 67 | 18 | 15 |
| | 120 | 37 | 55 | 26 | 19 |
| | 30 | 16 | 77 | 13 | 10 |
| **4.8%Fe/RHS 600** | 60 | 20 | 69 | 18 | 13 |
| | 90 | 29 | 61 | 22 | 17 |
| | 120 | 30 | 54 | 27 | 19 |
| | 30 | 16 | 88 | 12 | 0 |
| **8.9%Fe/RHS 400** | 60 | 25 | 68 | 19 | 13 |
| | 90 | 41 | 64 | 21 | 15 |
| | 120 | 43 | 58 | 23 | 19 |
| | 30 | 19 | 73 | 16 | 11 |
| **8.9%Fe/RHS 600** | 60 | 21 | 68 | 18 | 14 |
| | 90 | 31 | 63 | 20 | 17 |
| | 120 | 33 | 54 | 24 | 22 |
| **8.9%Fe/RHS 400ᶜ** | 30 | 5 | 75 | 20 | 5 |
| | 60 | 9 | 70 | 21 | 9 |
| | 90 | 20 | 65 | 22 | 10 |
| | 480* | 22 | 55 | 20 | 10 |

ᶜ Conventional heating (* Other products found, mainly carvacrol)

**Table 10.** Conversions and selectivities under liquid conditions. (Limonene:tBHP ratio=1:3.5, 0.15g catalyst, 150°C, reflux.

Repeat experiments with fresh catalysts showed a 1-2% difference in conversions and 2-3% in selectivities in both the dry media and liquid phase conditions. It can be seen that in the absence of catalyst, a conversion of ca. 5% after 30 min of reaction was reached and maintained throughout the experiment, with a selectivity of 91% to limonene oxides and 9% carvone. RHS gave a 12% conversion after 30 min, reaching 22% at 90 min with no further increase by the final reaction time of 2 h. The selectivity to epoxides, after 30 min was higher than 60%, decreasing to *ca* 45% by the end of the reaction, with increasing amounts of carvone and carveol. This reactivity was as expected, due to the intrinsic basicity of the oxides contained in this silica (see Figure 14).

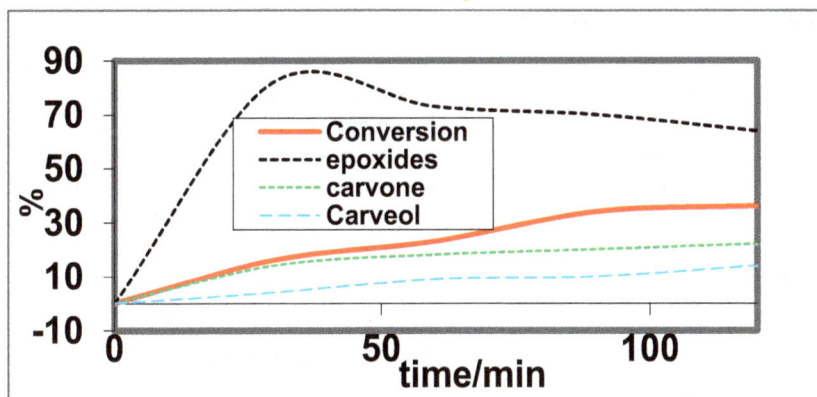

**Figure 14.** Conversions and selectivities of 8.9%Fe/RHS with orange peel oil, under the conditions shown in Table 10.

The reactivity of RHS $HNO_3$ 400 was higher than that of RHS, reaching *ca* 34% after 90 min, in agreement with the higher basicity, surface area and pore volume of the former. The reactivity of RHS $HNO_3$ 400 was higher than that of RHS $HNO_3$ 600, for the same reasons. The activity of commercial silica Aerosil Evonik, chosen for its similar textural properties (91 $m^2g^{-1}$), to RHS but with no basic oxides content was practically that observed with no catalyst, showing the importance of basic sites on the reaction. With regards to the catalytic activity of the Fe-containing catalysts, higher conversions were reached with those catalysts prepared with greater amounts of iron and lower calcination temperatures, as seen in the dry media results, in agreement with their higher amounts of basic sites (Figure 13) and better accessibility, as shown in the study of their textural properties (Table 7).

The low reaction times needed in this work due to optimisation of the catalysts and dielectric activation avoids the formation of undesired products, such as polymers sometimes found in the oxidation of limonene at longer reaction times [24]. These workers find one of the technological advantages to be that, since the final mixtures contain high concentrations of valuable oxygenated products of interesting organoleptic properties, they can be used directly, for example in fragrance compositions where separation is unnecessary.

As in Menini's work [29] at ca. 40% conversion, the reactions become stagnant. Addition of fresh catalyst to the mixture of the substrate and the products does not promote further conversion of the substrate, which seems to indicate that the products accumulated are probably acting as radical scavengers. No iron leaching occurred under the reaction conditions and the catalysts could be recovered by thorough washing with decane and reused at least three times with no appreciable loss of activity. Under the conditions used in Table 10, the behaviour of catalyst 8.9%Fe 400 was determined:

1.  adding a small amount (70 µg) of radical scavenger hydroquinone
2.  with conventional heating
3.  with orange oil, instead of limonene

Addition of hydroquinone greatly decreased the activity, down to that found with no catalyst, indicating the radical nature of the reaction. Under conventional heating after 2 h refluxing, only 22% conversion was reached with 55% selectivity to the epoxides, 20% to carvone, 10% to carveol and 15% to carvacrol, the latter compound due to the extended oxidation under the conditions used. When orange oil was used (Figure 14), the compounds present other than limonene (myrcene, alpha pinene, beta pinene, linalool and decanal) were recovered unreacted and thus the results were promising, especially bearing in mind its lower price, compared to commercial limonene.

Comparing the results obtained here with some from the literature, it can be seen that optimised mixed oxides of iron, cobalt and manganese manage to convert 0.5 ml of limonene after 7 h with an oxygen atmosphere, whereas in this work 1.6 ml limonene were converted with tBHP in 1.5 h. As in our work, these authors obtained as products limonene epoxides, carvone and carveol, with selectivity towards limonene oxide, although the presence of cobalt in the composition of those catalysts had a negative influence on their environmental impact [30,31]. Similar conversions were found with $V_2O_5/TiO_2$ and tBHP. Again, vanadim pentoxide is more toxic than the iron oxide used in our work. Furthermore, limonene glycol and polymers decrease the selectivities to the desired compounds in that work [32]. Laborious and time consuming preparation and optimisation of synthetic hidrotalcites with immobilised palladium and copper in their compositions achieve similar conversions and selectivities to those found here, however reaction times up to 6 h were required in that case and palladium and copper, although quite reactive, are not very environmentally friendly [33].

The main advantage of our process is that the renewable raw materials used for the preparation of the catalysts, with adequate design, can give rise to catalysts that can compete with synthetic ones, prepared usually in more expensive and time consuming ways and therefore less environmentally friendly [34].

## 4. The use of agricultural residues to improve the textural characteristics of structured solids to decontaminate effluents

Atmospheric contamination from industrial processes that contain VOCs which lead to the formation of photochemical smog is of great important with respect to air quality. The most usual remedy for this type of pollution are the so called "end of pipe" methods in which catalysts, adsorption beds or a combination of the two are employed. When large volumes of gas are to be treated, the conformation of the catalysts into open channel monoliths or extrudates is a necessity to avoid or reduce the pressure drop across the catalytic bed. However, if incorporated catalysts are used due to their robust nature and abrasion resistance a reduced activity compared to powder catalysts of similar composition is generally encountered due to diffusion limitations of the gas to be treated into the conformed catalyst. This limitation can be reduced by controlling the texture, surface area and porosity, of the incorporated catalyst.

Toluene was chosen as a typical aromatic VOC to be eliminated by its total oxidation from a spiked air stream. The subproducts from the rice industry chosen to modify the physical characteristics of the conformed catalysts were rice husk ash (RHA), from burning the husks to produce electricity, with the added advantage of drastically reducing the volume of this residue, and rice bran (Bran) that is separated from the rice grain during the whitening process. The Bran is a Pore Generating Agent (PGA) that during the firing of the extruded green body at 500°C to form the conformed ceramic is burnt out with the subsequent increase in the overall porosity of the structure. Whilst the RHA was used to both modify the porosity of the structure and hopefully increase the mechanical strength of the ceramic due to its high silica content [35].

Different compositions and textural properties can be achieved by careful design of the structured solids prepared using rice production wastes (rice husk ash and rice bran). Optimisation of these processes lowers the temperature and therefore the economic expenditure for decontamination of effluents spiked with toluene, chosen as standard for comparison purposes and for being a classical example of a VOC. The textural developement of the solids is a key point to change their behaviour, where the use of rice wastes are a convenient, cheap and ecologic way to improve the activities of the final materials.

The clay used as the agglomerating agent and as the final support for the catalyst was α-sepiolite (Sep), due to its exceptional rheological properties that allow extrusion of the paste produced by mixing with water [36] and the formation of a stable ceramic body with an acceptable strength when heat treated at temperatures above 330°C. Iron nitrate was used as the precursor to the catalyst due to its low cost, toxicity and environmental impact [37]. Four incorporated catalysts [38] were prepared in which the iron to sepiolite ratio was maintained constant but different amounts of Bran and RHA were also incorporated into the original paste before mixing, extrusion and firing at 500°C. Heat treatment at 500°C was chosen since at this temperature the sepiolite forms a ceramic structure that binds all of the other components in the extruded paste. Also this temperature is sufficient for the decomposition of the iron salt to iron oxide which is complete by 200°C, and the thermal decomposition of the Bran. The catalytic activity studies were carried out by passing an air flow spiked so as to contain 100 ppm of toluene through the catalyst bed with a Gas Hourly Space Velocity (GHSV) similar to that expected for the decontamination of an industrial effluent stream. The temperature of the catalyst bed was raised from ambient to 500°C in a number of steps, maintaining each for a period of 20 min to ensure that the result was in equilibrium and monitoring the destruction of the organic by the use of a flame ionisation detector. The gas balance was confirmed by the amount of $CO_2$ produced, monitored by infra red spectroscopy.

The original compositions of the green bodies and the textural, mechanical and catalytic activities of these materials after heat treatment at 500°C to produce the conformed ceramic catalyst as a solid extrudate with a diameter of 3 mm are collated in Table 11.

| Sample | Composition Sep/Fe/Bran/RHA Parts by Weight | Crushing Strength MPa | Surface Area $m^2g^{-1}$ | Pore Volume $cm^3g^{-1}$ | Catalytic Activity $T_{50}$ °C |
|--------|--------------|---------|---------|--------|----------|
| 1 | 17/1/0/0 | 18.3 | 143 | 0.576 | 382 |
| 2 | 17/1/2/0 | 13.8 | 140 | 0.734 | 350 |
| 3 | 17/1/0/2 | 19.0 | 125 | 0.656 | 377 |
| 4 | 17/1/2/2 | 15.8 | 124 | 0.797 | 343 |

**Table 11.** Compositional and Textural Characteristics of the catalysts heat treated at 500°C.

From the results presented in Table 11 it may be observed that the mechanical strength of the materials was reduced by the incorporation of Bran into the original paste before extrusion and its subsequent removal by heat treatment at 500°C comparing the result obtained with samples 1 and 2 and that of samples 3 and 4. This was to be expected since with the removal of the PGA on firing at 500°C there is a significant increase in the total pore volume. The mechanical strength of brittle ceramics is related to both the total pore volume and the pore size distribution [39] where an increase in the pore volume or the size of the pores is detrimental to the strength development. From the porosimetry curves presented in Figure 15 it may be seen how the thermal decomposition of the Bran led to both an increase in the total pore volume and a shift to wider pores. Thus, sample 1 had a monomodal pore size distribution curve with practically the whole pore volume located in pores of less than 200 nm. When Bran was incorporated and subsequently burnt out of the conformed material there was a 27% increase in the total pore volume located in the pores between 100 nm and 1 μm.

With the incorporation of RHA the most obvious change in the porosity of sample 3 compared to that of sample 1 was the development of a bimodal curve with the pores close to 900 nm being due to the particle size of the RHA. The slight reduction in the porosity in pores up to about 100 nm was due to the reduction in the amount of Sep per gramme of material. The effects of the incorporation of both RHA and Bran may be seen by comparing the curves obtained with sample 3 with that of sample 4. The curves were similar up to about 100 nm then their was a gradual increase in the pore volume that reached about 21% and a shift in the diameter of the wider pores to approximately 1.5 μm.

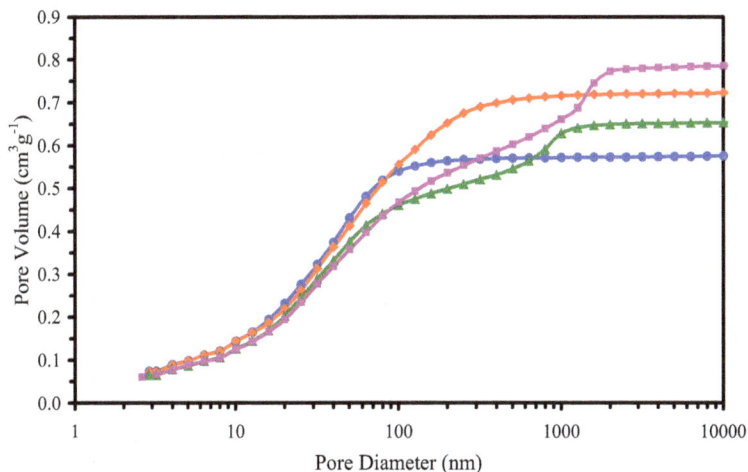

**Figure 15.** Cumulative pore volume curves for samples heat treated at 500°C: 1 ●, 2 ♦, 3 ▲ and 4 ■.

The catalytic oxidation activity of the four samples is presented in Figure 16 where it may be seen that the least active was that based on sample 1 which had the lowest total pore volume and narrowest pores. With the introduction of Bran into the original composition and its subsequent removal by heat treatment at 500°C which led to a significant increase in both the pore volume and the connectivity the temperature at which 50% of the toluene was decomposed ($T_{50}$) was reduced by 32°C. For sample 3, although the pore volume and width of the largest pores was increased, which should have lead to less diffusion limitation, the $T_{50}$ was only reduced by 5°C. The highest catalytic activity was found with sample 4 which had the

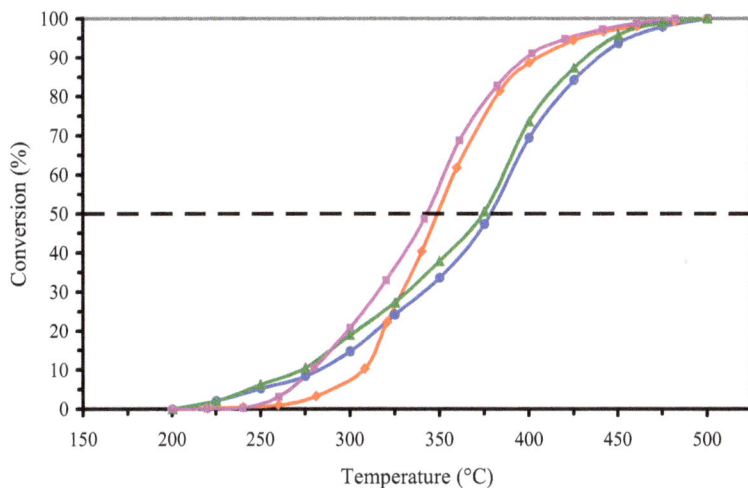

**Figure 16.** Catalytic activity curves for samples heat treated at 500°C: 1 ●, 2 ♦, 3 ▲ and 4 ■.

advantage of the highest pore volume and the widest pore diameters due to the presence of both Bran and RHA in the original composition. The $T_{50}$ was reduced by 39°C and the material also had a respectable mechanical strength due to the incorporation of the RHA.

## 5. Preparation of renewable biocompatible scaffolds for bone replacement therapy

Under this heading we present how agricultural wastes have been analysed, modified and changed in order to have properties similar to bone growth scaffolds. SEM micrographs (Figure 17) allow the observation of the porous nature of the inorganic skeleton that remains after the treatment of the waste, and its similarity to bone.

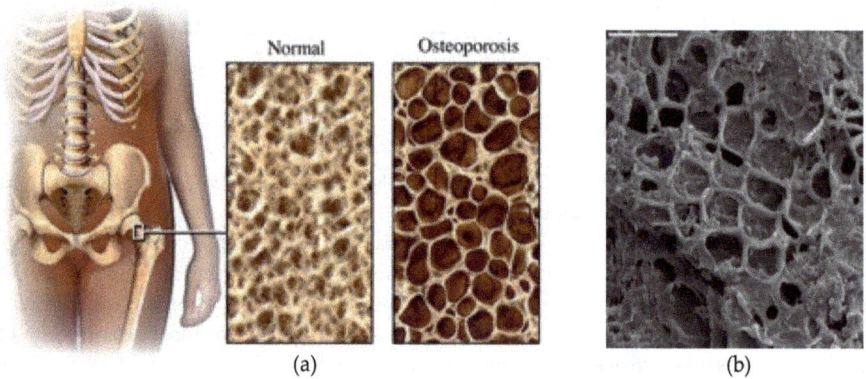

Source a: http://blogdefarmacia.com/tipos-de-osteoporosis/          b: Bioecomaterial (from our own production)

**Figure 17.** Scanning electron micrograph of treated beer bagasse (scale bar 20 µm) and similarity with bone structure

XRD analysis of these materials show tricalcium phosphate and calcium silicate of structure and composition similar to the synthetic materials presently used in bone and tooth replacement (Figure 18). These materials have proven to be biocompatible and capable of promoting bone growth as confirmed by similar growth rates in vitro of osteoblasts to those of hydroxyapatite used as standard.

It is estimated that approximately a million bone grafts are performed each year to treat bone defects resulting from trauma and diseases in the United States. Various strategies have been used to solve this problem. Autografts are used to treat these defects, but available bone can be inadequate and difficult to shape and obtain. Allografts and xenografts must be processed to eliminate the risk of transmission of live viruses [40]. These difficulties have been the impetus for research into a variety of bone grafting materials.

Bone tissue engineering techniques have become an expanding research area in regenerative medicine. Bone and tooth replacement require materials that act as scaffolds or artificial extra-cellular matrix, directing tissue formation and allowing the transport of biological nutrients to restore the structure and function of damaged tissues. These materials, which require tailored porosity, surface chemistry, and mechanical strength, are typically produced from animal bone, organic oil-derived polymers, inorganic materials, or complex mixtures of all the above [41]. Figure 19 includes the mercury intrusion porosimetry results for a beer bagasse derived bioecomaterial.

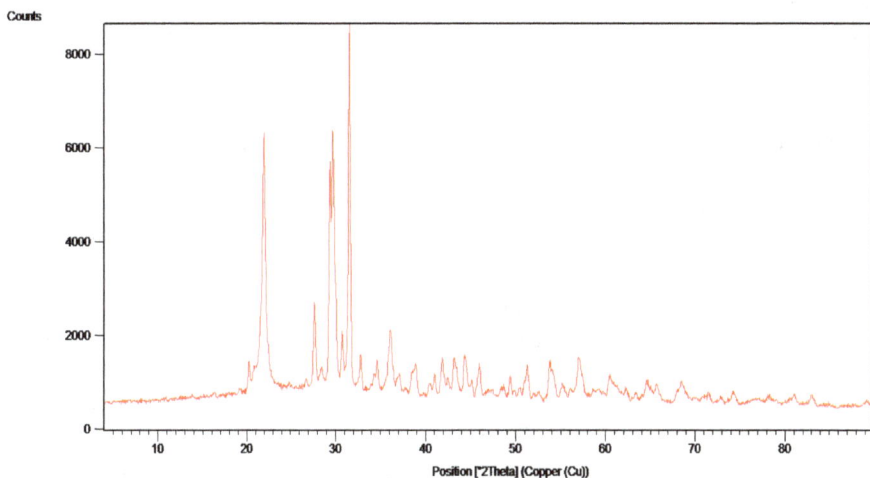

**Figure 18.** X-ray diffraction of beer bagasse based Bioecomaterial

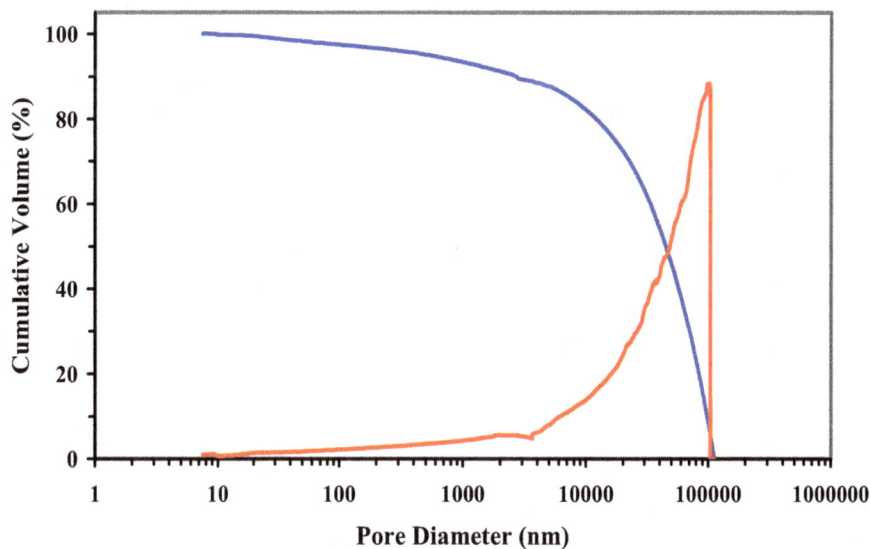

**Figure 19.** Cumulative Pore Volume (blue line) and pore size distribution (red line) of beer bagasse based bioecomaterial.

The SEM image of the treated beer bagasse (Figure 17b) displayed pores of about 10μm. However, from the mercury intrusion porosimetry curve presented in Figure 19 it may be appreciated that a continuously increasing curve was observed from the largest pores down to those at the physical limit of this technique (4 nm). Thus, although the cumulative intruded volume in the largest pores was due to filling of the intraparticulate space, while at diameters below 10 μm the porosity must be due to the interparticulate porosity. However, what is most important from this figure is that the porosity between the particles and within them formes a continuous network that favours cell growth when used as a scaffold for bone regeneration.

Materials used as bone substitutes need to embrace several important requirements: i) biocompatibility; ii) osteoinduction and osteopromotion, iii) porosity, iv) stability under stress, v) resorbability and degradability, vi) plasticity, vii) sterility and viii) stable and long-term integration of implants [42]. All these requirements serve as a basis for effective long-term tolerance in bone replacement therapies [43].

The first approach in the selection of a suitable scaffold for bone regeneration is the evaluation of its in vitro cytotoxicity using appropriate cell lines in culture, such as osteoblasts. Thereby, we can determine the biocompatibility of each scaffold under study. A good biocompatibility could be expected from materials whose physicochemical properties promote cell adhesion and differentiation into mature osteoblasts.

In the present study we aimed to develop scaffolds for bone regeneration obtained from agricultural wastes from several crops. These materials have proved to be good replacement candidates for use as biomaterials for the growth of osteoblasts and could be used in bone replacement therapies.

In this study, beer bagasse, a low-cost waste material from beer production, was employed as a renewable raw material that was tailored for use as biocompatible scaffolds for osteoblast growth, given its structure and composition [44-46]. To the best of our knowledge this is the first time that agricultural waste has been modified to produce solids that can act as scaffolds for tissue regeneration.

To study cell adhesion to the materials, their biocompatibility and efficiency for bone growth, MC3T3-E1, an osteoblast-like cell line, was chosen because they are well characterised for modelling endogenous osteoblasts [47]. A commercial synthetic scaffold, hydroxyapatite (HA) nanopowder (B200 nm) from Sigma Aldrich, was used for comparison purposes.

The osteoblastic cells were seeded onto material-coated plates in DMEM-10%FBS and allowed to adhere for 2 h at 37ºC in a 5% $CO_2$ atmosphere. Plates were then washed with phosphate-buffered based saline solution (PBS) to remove non-adherent cells. Adherent live cells were quantified using the live/dead viability assay kit (Molecular Probes [40]). The number of live adhered cells was evaluated 2h after seeding by fluorescence microscopy, counting at least 10 representative fields per well. The cell adhesion process was studied after 2h of incubation at 37ºC. The adhesion of MC3T3E1 cells to the various materials evaluated is shown in Figure 20. As can be seen, the number of osteoblast cells which adhered to the materials derived from beer bagasse was comparable to that of the commercial material used as control (HA). The cell adhesion rate was even higher in the case of sample BB47. However, the adhesion of MC3T3E1 cells to the materials was slightly lower than that on plastic, especially for cells seeded on HA and BB48. Assuming 100% as the rate of cell adhesion for the commercial material (HA), those for BB47, BB48 and BB410 were found to be 141.5%+/- 6.1, 92.5%+/-7.4 and 118.1%+/-5.8, respectively. Thus, confirming similar adhesion rates for the studied materials, demonstrating that they possess an appropriate porosity to allow cell adhesion.

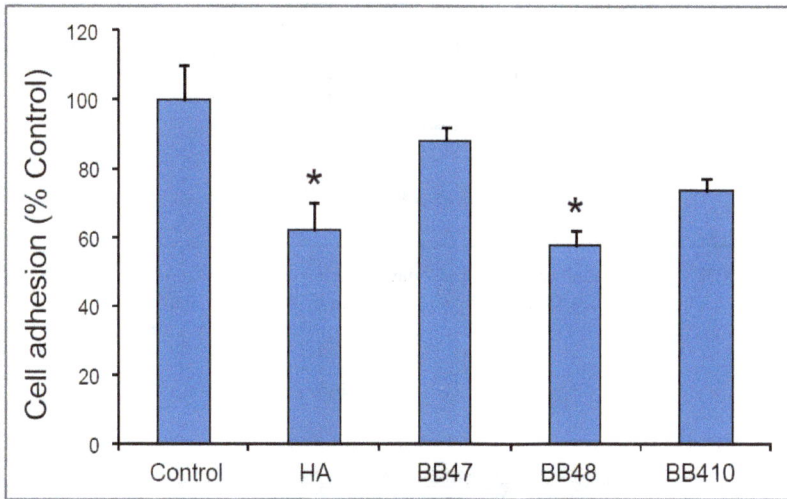

**Figure 20.** Adhesion of MC3T3E1 cells. Cells were plated on a plastic control dish (Control) and on plates coated with the control material (HA) and with materials derived from beer bagasse (BB47, BB48, BB410). Cells were incubated for 2h at 37°C. Then, adherent live cells were quantified as described in the text. Results are expressed as a percentage of the control (cells seeded on plastic plates). Data represent mean+/-S.E.M. of three independent experiments. (p<0.05, ANOVA, post-hoc Tukey HSD test, * vs. Control).

The biocompatibility of the materials derived from beer bagasse, was assessed with the same MC3T3-E1 cells after different periods of exposition, comparing the results to those obtained with the commercial material HA and to regular tissue culture polystyrene plates, used as controls. The viability of MC3T3-E1 cells growing on plastic plates, BB47, BB48, BB410 and the commercial material HA was determined at one, three and seven days after seeding by the live/dead viability assay [48], to distinguish dead cells (red) from live ones (green), as observed by inverted fluorescence microscopy (Figures 21 and 22).

Calculating the percentage of live cells over total cells (live cells plus dead cells), a high cell viability rate was observed for all scaffold materials produced from beer bagasse, similar to that of the commercial product HA and the control cell culture plastic plates (Figure 21).

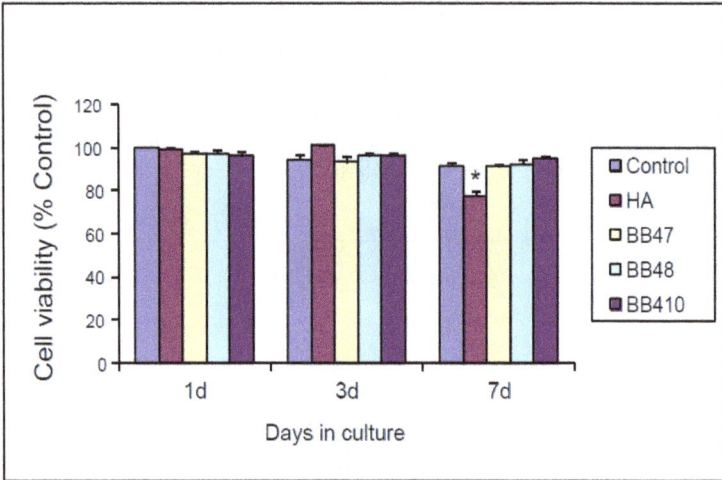

**Figure 21.** MC3T3-E1 cell viability after 1, 3 and 7 days growing on plastic plates or on plates coated with HA, BB47, BB48 and BB410. Cell viability was evaluated by live/dead viability assay kit. Data represent mean+/-S.E.M. of three independent experiments. ($p<0.05$, ANOVA, post-hoc TukeyHSD test, * vs. Control).

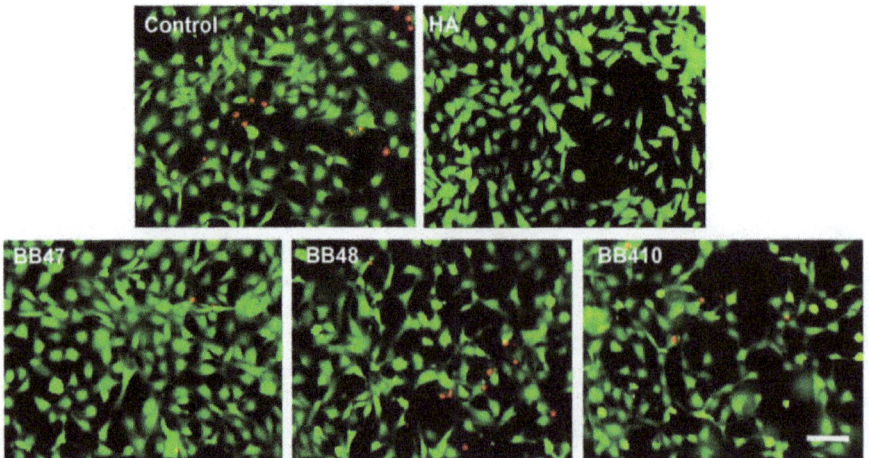

**Figure 22.** Fluorescence microscopy images showing MC3T3-E1 cells growing on different materials 3 days after seeding and stained with Live (green)-Dead (red) viability assay kit. Scale bar: 100µm.

None of the materials analysed resulted in a reduction of cell viability at any time points, more than 90% of the cells growing on these materials remained viable over the whole evaluation period. Only cells growing on HA for 7 days showed a significant reduction in the rate of cell viability.

Since bone formation is a lengthy process, the most relevant aspect is that even after seven days of culture, all materials displayed similar biocompatibilities rates to those observed on the commercial sample and the plastic plates.

In summary, murine-derived osteoblastic cells (MC3T3-E1) actively adhere to the beer bagasse-derived materials, with no significant changes in cell viability of cells growing on these materials compared to those observed with either the commercial material or the cells seeded directly onto plastic plates. Such characteristics are desired in dental and orthopaedic prostheses [49].

We conclude that our renewable raw material (RRM) are scaffolds that have the right characteristics to support adhesion and survival of the osteoblast-derived cell line MC3T3E1, strongly suggesting that they could be employed in oral and/or bone tissue regeneration.

## 6. Conclusions

1.  Agro-industrial wastes are a renewable source of many solid and liquid substances and furthermore, their use leads to a reduction in environmental hazards. These wastes are an inexhaustible and sustainable source of materials and substances in a biorefinery based economy that is constantly gaining interest in industrial sectors, especially because the actual petroleum based will come to an end.
2.  Some agricultural wastes, can be used as catalysts for production of fine chemicals. For example, silica prepared from rice husk, with a lamellar structure, basicity due to oxides from its natural origin and unusual high density, can act both as support and catalyst for the oxidation of limonene. Also, sustainable chemicals can be produced from agricultural wastes.
3.  The use of agricultural wastes allows the production of extrudates that can be employed to optimise cleaning of contaminated effluents. The catalytic activity of conformed catalysts may be significantly enhanced by the use of low cost sub-products and residues from the rice production industry. Rice bran is a useful PGA that improves the connectivity of the porous matrix in these incorporated catalysts while the RHA both improves the strength development of the ceramic body while also improving the texture and thus reducing the diffusion limitations of the gas to be treated into the porous structure of the catalyst.
4.  Some waste derived solids, after the appropriate pretreatment, can be used for bone growth. The biocompatibility of some modified agro-industrial wastes has shown to be interesting in their use in fields like release of medicines, supports for enzymes and food supplements. Recent studies "in vivo" are showing promising results for these materials as economical sustainable scaffolds to be used in tissue engineering.

## Author details

M.A. Martin-Luengo, E. Sáez Rojo, A.M. Martínez Serrano,
M. Diaz, L. Vega Argomaniz, L. Medina Trujillo and S. Nogales
*Department of New Architectures, Institute of Materials Science of Madrid, CSIC, Spain*

M. Yates, F. Plou, E. Sáez Rojo, M. Diaz,
L. Vega Argomaniz, S. Nogales and R. Lozano Pirrongelli
*Institute of Catalysis and Petroleochemistry, CSIC, Spain*

M. Ramos and A.M. Martínez Serrano
*Centre for Biomedical Technology, Polytechnical University of Madrid, Spain*

J.L. Salgado
*AIZCE Technical Committee, Interprofesional Asociation of Juices and Citric Concentrates*

J.L. Lacomba and and A. Civantos
*Institute of Biofunctional Studies, Complutense University of Madrid, Madrid, Spain*

G. Reilly
*Department of Materials Science and Engineering, Kroto Research Institute, Broad Lane, University of Sheffield, Sheffield, United Kingdom*

C. Vervaet
*Laboratory of Pharmaceutical Technology, Gent, Belgium*

## 7. References

[1]   Kem-Laurin K, (2012) Approaches to A Sustainable User Experience User Experience in the Age of Sustainability, Pages 31-68.

[2]   Mahesh K, Utkarsh Ravindra M, Adinpunya M (2012) Rapid separation of carotenes and evaluation of their in vitro antioxidant properties from ripened fruit waste of Areca catechu – A plantation crop of agro-industrial importance. Ind. Crops and Products, 40: 204-209.

[3]   Brady NC, Weil, RR (1999) The Nature and Properties of Soils (12th ed), Prentice Hall, Upper Saddle River, New Jersey.

[4]   Notarnicola B, Kiyotada H, Curran MA, Huisingh D (2012) Progress in working towards a more sustainable agri-food industry. J. of Cleaner Production, 28: 1-8.

[5]   Karlen DL, Ditzler CA, Andrews SS (2003) Soil quality: why and how? Geoderma, 114: 145-156.

[6]   Cogo SLP, Chaves FC, Schirmer MA, Zambiazi RC, Nora L, Silva JA, Rombaldi CV (2011) Low soil water content during growth contributes to preservation of green colour and bioactive compounds of cold-stored broccoli (Brassica oleraceae L.) florets. Postharvest Biology and Technology, 60:158-163.

[7] Pérez-Balibrea S, Moreno DA, García-Viguera C (2011) Improving the phytochemical composition of broccoli sprouts by elicitation. Food Chemistry, 129: 35-44.

[8] Vallverdú-Queralt A, Medina-Remón A, Casals-Ribes I, Lamuela-Raventos RM (2012) Is there any difference between the phenolic content of organic and conventional tomato juices? Food Chemistry, 130: 222-227.

[9] Salgado JL, AIZCE Technical Committee (Interprofessional association of Citric Juices and concentrates of Spain).

[10] Martin-Luengo MA, Yates M, Martínez Domingo MJ, Casal B, Iglesias M, Esteban M, Ruiz-Hitzky E (2008) Synthesis of p-cymene from limonene, a renewable feedstock. Appl. Catal. B: Env. 81: 218-224.

[11] Aschmann SM, Arey J, Atkinson R (2011) Formation of p-cymeme from OH + γ-terpinene: H-atom abstraction from the cyclohexadiene ring structure. Atmospheric Env. 45: 4408-4411.

[12] Yadav GD, Purandare SA (2007, Vapor phase alkylation of toluene with 2-propanol to cymenes with a novel mesoporous solid acid UDCaT-4, Microp. and Mesop. Materials, 103: 363-372.

[13] Rouquerol F, Rouquerol J, Sing K.S.W (1999) Adsorption by Powders and Porous Solids: Principles, Methodology and Applications.

[14] Lavalley JC (1996) Infrared spectrometric studies of the surface basicity of metal oxides and zeolites using adsorbed probe molecules, Catal. Today 27: 377-401.

[15] Leyva C, Ancheyta J, Travert A, Maugé F, Mariey L, Ramírez J, Rana MS (2012) Activity and surface properties of NiMo/SiO$_2$–Al$_2$O$_3$ catalysts for hydroprocessing of heavy oils. Appl. Catal. A: Gen. 425–426: 1-12.

[16] Fernandes C, Catrinescu C, Castilho P, Russo PA, Carrott MR, Breen C, (2007) Catalytic conversion of limonene over acid activated Serra de Dentro (SD) bentonite. Appl. Catal. A: Gen. 318: 108-120.

[17] Catrinescu C, Fernandes C, Castilho P, Breen C (2006) Influence of exchange cations on the catalytic conversion of limonene over Serra de Dentro (SD) and SAz-1 clays: Correlations between acidity and catalytic activity/selectivity. Appl. Catal. A: Gen 311: 172-184.

[18] Schneider RCS, Baldissarelli VZ, Martinelli M, von Holleben MLA, E.B. Carama EB (2003) Determination of the disproportionation products of limonene used for the catalytic hydrogenation of castor oil. J. Chromatogr. A 985: 313-319.

[19] Martin–Luengo MA, Pajares JA, González TejucaL (1985) Particle size determination of palladium supported on sepiolite and aluminum phosphate. J. of Colloid and Interface Sci. 107: 540-546.

[20] Oh S-T, Kim M-S, Choa Y-H, Kim KH, Lee S-K (2012) Preparation of Fe–50 wt.% Co nanopowders by calcination and hydrogen reduction of nitrate powders. Microelectronic Eng. 89: 97-99.

[21] Xu W, Liu X, Ren J, Liu H, Ma Y, Wang Y, Lu G (2011) Synthesis of nanosized mesoporous Co–Al spinel and its application as solid base catalyst. Microp. and Mesop. Materials 142: 251-257.

[22] Gervasini C, Messi P, Carniti A, Ponti N, Ravasio F, Zaccheria F (2009) Insight into the properties of Fe oxide present in high concentrations on mesoporous silica. J. of Catal. 262: 224-234.

[23] Zhao C, Gan W, Fan X, Cai Z, Dyson PJ, Kou Y. (2008) Aqueous-phase biphasic dehydroaromatization of bio-derived limonene into *p*-cymene by soluble Pd nanocluster catalysts. J.Catal. 254: 244-250.

[24] Lee OC, Oh Y-G. and S-G (2009) Synthesis and characterization of hollow silica microspheres functionalized with magnetic particles using W/O emulsion method. Coll. and Surf. A: Physicochem. and Eng. Aspects 337: 208-212.

[25] Baraton M, X. Chen X, Gonsalves KE (1997) Ftir study of a nanostructured aluminum nitride powder surface: Determination of the acidic/basic sites by CO, $CO_2$ and acetic acid adsorptions. Nanostr. Mater. 8: 435-445.

[26] Ross PF, Busca G, Lorenzelli V, Lion M, Lavalley C (1988) Characterization of the surface basicity of oxides by means of microcalorimetry and fourier transform infrared spectroscopy of adsorbed hexafluoroisopropanol. , J. Catal. 109: 378-386.

[27] Oliveira P, Machado A, Ramos AM, Fonseca I, Fernandes FMB, Botelho do Rego AM, VitalJ (2009) MCM-41 anchored manganese salen complexes as catalysts for limonene oxidation Microp. and Mesop. Materials 120: 432-440

[28] Martin-Luengo MA, Yates M, Diaz M, Saez Rojo E, Gonzalez Gil L (2011) Renewable fine chemicals from rice and citric subproducts: Ecomaterials. Appl. Catal. B: Env. 106: 488-493.

[29] Menini L, Pereira MC, Parreira LA, Fabris JD, Gusevskaya EV (2008) Cobalt- and manganese-substituted ferrites as efficient single-site heterogeneous catalysts for aerobic oxidation of monoterpenic alkenes under solvent-free. J. Catal. 254: 355-364.

[30] Menini L, Pereira MC, Parreira LA, Fabris JD, Gusevskaya EV (2008) Cobalt- and manganese-substituted ferrites as efficient single-site heterogeneous catalysts for aerobic oxidation of monoterpenic alkenes under solvent-free conditions. J. Catal. 254:355-364.

[31] Robles-Dutenhefner PA, da Silva MJ, Sales LS, Sousa EMB, Gusevskaya EV (2004) Solvent-free liquid-phase autoxidation of monoterpenes catalyzed by sol–gel $Co/SiO_2$. J. Mol. Catal. A: Chem. 217: 139-144.

[32] Oliveira P, Rojas-Cervantes ML, Ramos AM, Fonseca IM, Botelho do Rego AM, Vital J (2006) Limonene oxidation over $V_2O_5/TiO_2$ catalysts. Catal. Today 118: 307-314.

[33] Bussi J, López A, Peña F, Timbal P, Paz D, Lorenzo D, Dellacasa E (2003) Liquid phase oxidation of limonene catalyzed by palladium supported on hydrotalcites. Appl. Catal. A: Gen. 253: 177-189.

[34] Guidotti M, Ravasio N, Psaro R, Ferraris G, Moretti G (2003) Epoxidation on titanium-containing silicates: do structural features really affect the catalytic performance? J. Catal. 214: 242-250.

[35] Jimmy Nelson Appaturi, Farook Adam, Zakia Khanam (2012) A comparative study of the regioselective ring opening of styrene oxide with aniline over several types of mesoporous silica materials. Microporous and Mesoporous Materials 156:16-21.

[36] Murray HH (1991) Overview — clay mineral applications. Applied Clay Science 5: 379–395.

[37] Kim SC, Shim WG (2008).Influence of physicochemical treatments on iron-based spent catalyst for catalytic oxidation of toluene. Journal of Hazardous Materials 154: 310-316.

[38] Cybulsky A and Moulin JA (1998) Structurated Catalysts and Reactors. Ed. Marcell and Dekker, Inc New York.

[39] Yates M (2006) Application of mercury porosimetry to predict the porosity and strength of ceramic catalyst supports. Particle & Particle Systems Characterization 23: 94-100.

[40] Simon, C.G.; Guthrie, W.F.; Wang, F.W. (2004) Cell Seeding into Calcium Phosphate Cement. J. Biomed. Mater. Res. Part A 68A: 628-639.

[41] Cao, H.; Kuboyama, N. A (2010) Biodegradable Porous Composite Scaffold of PGA/b-TCP for Bone Tissue Engineering. Bone 46: 386-395.

[42] Kolk A, Handschel J, Drescher W, Rothamel D, Kloss F, Blessmann M, Heiland M, Wolff KD, Smeets R (2012) Current trends and future perspectives of bone substitute materials - From space holders to innovative biomaterials. J Craniomaxillofac Surg. (in press, corrective proof, available online 30th Jan)

[43] Horch HH, Sader R, Pautke C, Neff A, Deppe H, Kolk A (2006) Synthetic, pure-phase beta-tricalcium phosphate ceramic granules (Cerasorb) for bone regeneration in the reconstructive surgery of the jaws. Int. J. Oral Maxillofac Surg, 35: 708–713.

[44] Harish Prashanth, K.V.; Tharanathan, R.N (2007) Chitin/Chitosan: Modifications and Their Unlimited Application Potential-an Overview. Trends Food Sci. Technol. 18: 117-126.

[45] Al Seadi, T.; Holm-Nielsen B (2004) III. 2 Agricultural Wastes. Waste Management Series 4:207-215.

[46] Martin Luengo M.A.; Yates M.; Casal B. (2010) Preparation of Biocompatible Materials from Beer Production and Their Uses. Spanish patent PCT/ES2009/070475.

[47] Chen, F.; Ouyang, H.; Feng, X.; Gao, Z.; Yang, Y.; Zou, X.; Liu, T.; Zhao, G.; Mao, T. Anchoring Dental Implant in Tissue-Engineered Bone Using Composite Scaffold: A PreliminaryStudy in Nude Mouse Model. J. Oral Maxillofac. Surg. 2005, 63, 586-595.

[48] Chou YF; Huang W; Dunn JC, Miller TA; Wu BM (2005) The Effect of Biomimetic Apatite Structure on Osteoblast Viability, Proliferation, and Gene Expression. Biomaterials 26: 285_295.

[49] Bertazzo, S.; Zambuzzi, W.F.; da Silva, H.A.; Ferreira, C.V.; Bertran, C.A. Bioactivation of Alumina by Surface Modification: A Possibility for Improving the Applicability of Alumina in Bone and Oral Repair. Clin. Oral Implants Res. 2009, 20 (3), 288-293.

# Chapter 2

# Efficacy of Hydrothermal Treatment for Production of Solid Fuel from Oil Palm Wastes

Ahmad T. Yuliansyah and Tsuyoshi Hirajima

Additional information is available at the end of the chapter

## 1. Introduction

Oil palm is a perennial crop that has higher productivity than other oil-producing crops. One hectare of oil palm plantation produces approximately 3.68 ton oil/year, which is much higher than for rapeseed (0.59 ton/year), soybean (0.36 ton/year), and sunflower (0.42 ton/ha) [1]. Crude palm oil (CPO) has become a more valuable commodity owing to recent price increases on international markets.

Until 2006, Malaysia was a leader in crude palm oil (CPO) production in the world. However, their limited land for expansion of plantation slows down the production rate. In contrast, Indonesia still has a huge amount of uncultivated land enabling extensive development of palm oil industry. Within 1995-2006, Indonesian plantation area jumped around three times into 6.6 million Ha. Such rapid expansion enhanced its annual CPO production. In 2008 Indonesia produced 19.33 million ton of CPO whereas Malaysia produced 17.63 million ton. Totally, both countries contributed around 85.9 % of world's palm oil production [2].

Despite its benefit, the palm oil industry generates a huge amount of biomass waste. As the continued growth of the industry, the amount of waste produced also significantly increases. Based on location where the waste is collected from, it can be divided into two types: waste from harvesting and replanting activity in plantation fields and waste from the milling process in palm oil mill.

Oil palm frond and trunk belong to first type of biomass waste. Zakaria [3] mentions that around 10.4 ton/ha of frond is generated annually from regular pruning, while around 14.4 ton/ha of frond and 75.5 ton/ha of trunk are obtained from replanting (once in 25 years). The second type of biomass waste are the empty fruit bunch (EFB), fiber, and shell, corresponding to approximately 22%, 11% and 8%, respectively, of the amount of fresh fruit bunch processed in a mill. In case of Indonesia, a total of 43.05 million ton of frond and 13.94 million ton of trunk were approximately generated in 2005 [4]. In addition, the Indonesian Oil Palm Research Institute (IOPRI) estimates that Indonesian mills generated approximately 15.9 million ton of EFB, 9.0 million ton of fiber and 4.8 million ton of shell during 2005.

In order to reduce negative impact of the waste into the environment, as well as to obtain more added values, extensive works have been conducted. For example, frond, trunk, and EFB are simply returned to plantation fields as an organic fertilizer, either directly or by pre-composted. In contrast, fiber and shell are used for boiler fuel in palm oil mill. However, such utilization methods face a common problem.

Decomposition of waste in plantation fields generates $CO_2$ and $CH_4$ (considered as green house gas) coupled with unpleasant smells, which can last for up to 1 year. In addition, high transportation and distribution cost, water pollution by the rest oil (in EFB), and its attractiveness for beetles and snake have been a number of obstacles that hardly are solved [5]. On the other hand, the direct use of shell and fiber in boiler is constrained by their relatively high moisture content which lowers the heating value. They also can not be stored for a long period of time due to decomposition. Since the amount of waste increases rapidly recently, a more effective technology to handle the waste is highly desired.

This chapter discusses on application of hydrothermal treatment for handling of oil palm waste at laboratory scale. Hydrothermal treatment refers to a thermochemical process for decomposing carbonaceous materials such as coal and biomass with water in a high temperature and high pressure condition. The use of such method for processing biomass was firstly developed by Bobleter and his co-workers [6] and had attracted much attention for hydrolysis of lignocellulosic materials since then.

This method relies on fact that water in subcritical/supercritical condition has outstanding characteristics. At ambient temperature, water is polar and it has an infinite network of H-bonding, and does not solubilize most organics. As water is heated, the H-bonds start weakening, allowing dissociation of water into hydronium ions ($H_3O^+$) and hydroxyl ions ($OH^-$). In the subcritical region, the ion product ($K_w$) of water increases with temperature and it is about 3 orders of magnitude higher than that of ambient water and the dielectric constant ($\varepsilon$) of water drops from 80 to 20. A low dielectric constant allows subcritical water to dissolve organic compounds, while a high ion product allows subcritical water to provide an acidic medium for the hydrolysis of biomass components. Although this dielectric constant of the subcritical region is low enough for organic to be soluble, it remains high enough to allow salt dissolution. In addition, the physical properties of water, such as viscosity, density, dielectric constant, and ion product, can be tuned by changes in temperature and/or pressure in the subcritical region [7].

Compared to other thermo-chemical conversion methods such as pyrolysis and gasification, the temperature for hydrothermal treatment is much lower (200–500°C for hydrothermal, compared with 450–550 °C for pyrolysis and 900–1200°C for gasification) [8,9]. In addition, biomass conversion takes place in a wet environment, so high moisture content of feed biomass is not an issue. The role of water in the treatment is not only as a medium, but also a chemical reactant on decomposition. A contrast situation is found on pyrolysis and gasification which have a limitation on moisture content of the feed [9,10]. Therefore, such method is suitable for treating biomass with high moisture content, such as agricultural wastes which contain more than 50 wt. % of moisture in fresh condition.

Many studies using hydrothermal treatment have been conducted, but most of these used the method as a biomass pretreatment step in bio-ethanol production [11–14]. Few studies have considered benefits of the resulting solid. The focus of the present chapter is experimental study on upgrading of oil palm waste into solid fuel by hydrothermal treatment.

## 2. Experimental

### 2.1. Materials

Oil palm waste is collected from an oil palm plantation and oil palm mill in southern Sumatra, Indonesia. Prior to use, each material is air-dried and is then pulverized to form powder with a maximum particle size of 1 mm by cutting-grinding (except for oil palm shell by impact-grinding) using IKA MF apparatus. The composition of the waste material is determined using a procedure recommended by the US National Renewable Energy Laboratory that is substantially similar to ASTM E1758-01. The detailed procedure of the chemical analysis will be explained further in section 2.3. The analysis results are listed in Table 1.

| Component | Frond | Trunk | Fiber | Shell | EFB |
|---|---|---|---|---|---|
| Cellulose (wt. %, d.b) | 31.0 | 39.9 | 19.0 | 14.7 | 35.8 |
| Hemicellulose (wt. %, d.b) | 17.1 | 21.2 | 15.2 | 16.4 | 21.9 |
| Klason lignin (wt. %, d.b) | 22.9 | 22.6 | 30.5 | 53.6 | 17.9 |
| Wax (wt. %, d.b) | 2.0 | 3.1 | 9.1 | 2.3 | 4.0 |
| Ash (wt. %, d.b) | 2.8 | 1.9 | 7.0 | 2.3 | 3.0 |
| Others (by difference) | 24.2 | 11.3 | 19.2 | 10.7 | 17.4 |

d.b, dry basis

**Table 1.** Composition of raw materials

### 2.2. Apparatus and experimental procedure

Experiments are carried out in a 500-mL batch-type autoclave (Taiatsu Techno MA 22) equipped with a stirrer and an automatic temperature controller. The autoclave has a maximum temperature of 400°C and a maximum pressure of 30 MPa. It is made from hastelloy with the dimension 6.0 cm ID x 20.6 cm L. A slurry of 300 mL of water and 30 g of waste material is loaded into the autoclave. A stream of $N_2$ gas is used to purge air from the autoclave and to maintain an initial internal pressure of 2.0 MPa. With stirring at 200 rpm, the autoclave is heated by an electric furnace to the target temperature at an average rate of 6.6°C/min. The target temperature, ranging from 200 to 380°C, is automatically adjusted. The reaction temperature is measured by K-thermocouple inserted into vessel of autoclave. Once the target temperature is reached, the sample is held for a further 30 min before the autoclave is cooled to ambient conditions by air blow using an electric blower.

After cooling, the gas products are fed into a gasometer (Shinagawa DC-1) to measure the volume. The gas is sampled using a microsyringe (ITO MS-GANX00) and its composition is determined by gas chromatography with thermal conductivity detection (Shimadzu GC-4C). The remaining slurry is filtered using an ADVANTEC 5C filter and a water aspirator. The solid part is dried to a constant weight in an oven at 105°C to yield the final solid product.

## 2.3. Analysis

The solid products are characterized using several techniques. The elemental composition is measured using Yanaco CHN Corder MT-5 and MT-6 elemental analyzer. The cellulose, hemicellulose, and lignin contents are measured using a procedure recommended by the US National Renewable Energy Laboratory [15]. In brief, the wax content is determined by a soxhlet extraction of 5.0 g of sample (particle size ≤0.5 mm) with ethanol. Approximately 150 mg of dewaxed and dried sample is then treated with 1.5 mL of 72 w/w% $H_2SO_4$ at 30°C for 1 h. Further, 42 mL of $H_2O$ is added to the mixture followed by hydrolysis at 121°C for 1 h in an autoclave. The resulting mixture is washed several times using a hot water and filtered by a GP glass filter (pore size of 16 μm) in a vacuum condition. The obtained solid residue is dried at 105°C to a constant weight and its weigh is noted as klason lignin. Meanwhile, prior to being analyzed in HPLC, the filtrate is treated by an On Guard IIA column and 0.2 μm membrane filters to assure no residual solid and $H_2SO_4$ contained. The HPLC system is consisted of a KC-811column (JASCO) and refractive index (RI) detector (RI-2031, JASCO) and is operated under a 2 mM $HClO_4$ flowing at a rate of 0.7 ml/min and oven temperature of 50°C. The polysaccharide concentration is determined by measuring the corresponding sugar concentration with a correlation:

$$\text{cellulose (wt.\%)} = \text{glucose (wt.\%)} \times 0.9 \tag{1}$$

$$\text{hemicellulose (wt.\%)} = (\text{xylose} + \text{arabinose}) \text{ (wt.\%)} \times 0.88 \tag{2}$$

Additionally, proximate, total sulfur and calorific analyses are carried out according to JIS M 8812, JIS M 8819, and JIS M 8814, respectively. The equilibrium moisture content (EMC) of raw materials and the corresponding solid products is determined according to JIS M 8811. Hence, an aliquot of the sample is placed in a desiccator containing saturated salt solution. The relative humidity inside desiccator is maintained at 75 %. After equilibrium is reached, it is quickly measured using a Sartorius MA 150 moisture content analyzer. Identification of the chemical structure and functional groups is performed on a Fourier-transform infrared (FTIR) spectrometer (JASCO 670 Plus) using the KBr disk technique. Cross polarization/magic angle spinning (CP/MAS) $^{13}$C NMR spectra are measured on a solid-state spectrophotometer (JEOL CMX-300) with the following conditions: 10,000 scans; contact time, 2 ms; spinning speed, >12 kHz; pulse repetition time, 7 s. The spectrum is calibrated using hexamethyl benzene. Curve fitting analysis of the spectrum is performed using Grams/AI 32 ver. 8.0 software.

# 3. Results

## 3.1. Solid products and properties

The yield and properties of the solid products for different treatment temperatures are presented in Table 2 and 3. Due to the degradation reactions, the solid product yield decreases at elevated temperature, indicating the degradation reactions accomplish more completely. For instance, the solid yield at 200°C for all materials ranges from 58.3 to 67.7 wt.%. Meanwhile, their yields are only 31.7–37.3 wt.% at 380°C. Solid yield for oil palm shell and fiber are the highest among the oil palm wastes, particularly for ≥ 270°C treatments.

| Properties | Raw | Treatment temperature (°C) | | | | | | |
|---|---|---|---|---|---|---|---|---|
| **(a) Frond** | | 200 | 240 | 270 | 300 | 330 | 350 | 380 |
| Solid yield (wt.%, db) | | 58.3 | 52.0 | 42.5 | 38.4 | 36.7 | 35.1 | 34.4 |
| Fixed carbon (wt.%, daf) | 17.5 | 20.5 | 29.7 | 45.9 | 48.1 | 52.3 | 54.8 | 62.6 |
| Volatile matter (wt.%, daf) | 82.5 | 79.5 | 70.3 | 54.1 | 51.9 | 47.7 | 45.2 | 37.4 |
| Ash (wt.%, db) | 1.8 | 1.3 | 1.3 | 1.2 | 1.0 | 0.7 | 0.8 | 1.8 |
| Equilibrium Moisture (wt.%, ar) | 14.7 | 7.6 | 6.4 | 5.5 | 5.2 | 5.0 | 5.2 | 5.0 |
| Gross calorific value (MJ/kg, db) | 18.4 | 20.2 | 22.6 | 26.3 | 27.1 | 28.8 | 29.4 | 30.9 |
| **(b) Trunk** | | | | | | | | |
| Solid yield (wt.%, db) | | 67.7 | 56.9 | 41.7 | 38.7 | 36.9 | 35.3 | 33.7 |
| Fixed carbon (wt.%, daf) | 16.1 | 16.2 | 26.6 | 45.1 | 48.8 | 52.8 | 55.0 | 63.6 |
| Volatile matter (wt.%, daf) | 83.9 | 83.8 | 73.4 | 54.9 | 51.2 | 47.2 | 45.0 | 36.4 |
| Ash (wt.%, db) | 2.2 | 1.8 | 1.8 | 2.2 | 2.1 | 1.9 | 2.1 | 2.9 |
| Equilibrium Moisture (wt.%, ar) | 13.6 | 7.5 | 6.5 | 5.1 | 4.8 | 4.6 | 4.5 | 4.1 |
| Gross calorific value (MJ/kg, db) | 18.3 | 19.5 | 22.1 | 26.4 | 27.4 | 28.9 | 29.0 | 30.0 |
| **(c) Fiber** | | | | | | | | |
| Solid yield (wt.%, db) | | 67.5 | 61.9 | 43.3 | 42.5 | 41.0 | 37.3 | 39.6 |
| Fixed carbon (wt.%, daf) | 17.8 | 23.8 | 28.5 | 37.4 | 36.8 | 37.2 | 46.6 | 48.2 |
| Volatile matter (wt.%, daf) | 82.2 | 76.2 | 71.5 | 62.6 | 63.2 | 62.8 | 53.4 | 51.8 |
| Ash (wt.%, db) | 7.0 | 6.5 | 6.4 | 9.7 | 9.0 | 10.2 | 11.0 | 8.6 |
| Equilibrium Moisture (wt.%, ar) | 13.0 | 6.9 | 6.4 | 5.1 | 5.1 | 4.2 | 4.3 | 3.6 |
| Gross calorific value (MJ/kg, db) | 19.8 | 22.1 | 23.7 | 27.4 | 29.4 | 29.9 | 30.0 | 30.6 |
| **(d) Shell** | | | | | | | | |
| Solid yield (wt.%, db) | | 63.2 | 57.6 | 48.3 | 45.6 | 39.6 | 38.9 | 37.3 |
| Fixed carbon (wt.%, daf) | 19.9 | 25.3 | 31.7 | 44.0 | 50.2 | 56.2 | 59.1 | 65.6 |
| Volatile matter (wt.%, daf) | 80.1 | 74.7 | 68.3 | 55.7 | 49.8 | 43.8 | 40.9 | 34.4 |
| Ash (wt.%, db) | 2.3 | 1.6 | 2.0 | 1.9 | 1.6 | 1.3 | 1.1 | 2.7 |
| Equilibrium Moisture (wt.%, ar) | 9.9 | 6.7 | 5.5 | 4.6 | 4.0 | 3.6 | 3.4 | 3.1 |
| Gross calorific value (MJ/kg, db) | 20.8 | 22.8 | 24.8 | 27.4 | 28.6 | 29.2 | 30.0 | 32.1 |
| **(e) EFB** | | | | | | | | |
| Solid yield (wt.%, db) | | 64.1 | 55.2 | 40.2 | 37.0 | 36.0 | 33.5 | 31.7 |
| Fixed carbon (wt.%, daf) | 18.6 | 17.3 | 26.0 | 42.8 | 42.5 | 49.3 | 51.7 | 59.8 |
| Volatile matter (wt.%, daf) | 81.4 | 82.7 | 80.0 | 57.2 | 57.5 | 50.7 | 48.3 | 40.2 |
| Ash (wt.%, db) | 3.0 | 2.6 | 2.4 | 2.9 | 3.5 | 2.7 | 3.1 | 4.2 |
| Equilibrium Moisture (wt.%, ar) | 11.8 | 8.1 | 5.8 | 5.2 | 4.1 | 4.0 | 3.5 | 3.2 |
| Gross calorific value (MJ/kg, db) | 18.4 | 20.1 | 22.6 | 26.8 | 28.9 | 30.0 | 30.8 | 31.1 |

db, dry basis; ar, as received basis; daf, dry ash-free basis

**Table 2.** Yield, proximate analysis and gross calorific value of the raw materials and solid products.

| | Percentage component (wt.%, daf) | | | | | | | |
|---|---|---|---|---|---|---|---|---|
| (a) Frond | Raw | 200°C | 240°C | 270°C | 300°C | 330°C | 350°C | 380°C |
| C | 47.2 | 53.6 | 58.6 | 69.4 | 71.1 | 73.9 | 75.1 | 78.5 |
| H | 5.9 | 5.7 | 5.4 | 4.9 | 4.9 | 4.9 | 4.8 | 4.6 |
| N | 0.2 | 0.2 | 0.3 | 0.4 | 0.4 | 0.4 | 0.4 | 0.5 |
| O (difference) | 46.6 | 40.4 | 35.7 | 25.3 | 23.5 | 20.7 | 19.5 | 16.2 |
| S | 0.1 | 0.1 | 0.1 | 0.1 | 0.1 | 0.1 | 0.1 | 0.1 |
| (b) Trunk | | | | | | | | |
| C | 47.5 | 51.4 | 57.5 | 69.3 | 71.4 | 73.4 | 75.3 | 78.2 |
| H | 5.9 | 5.9 | 5.6 | 5.1 | 5.0 | 4.9 | 4.9 | 4.5 |
| N | 0.5 | 0.4 | 0.6 | 0.8 | 0.8 | 1.0 | 1.0 | 1.1 |
| O (difference) | 45.9 | 42.1 | 36.2 | 24.6 | 22.6 | 20.6 | 18.6 | 16.1 |
| S | 0.1 | 0.1 | 0.1 | 0.1 | 0.1 | 0.1 | 0.2 | 0.1 |
| (c) Fiber | | | | | | | | |
| C | 50.7 | 59.2 | 61.0 | 71.9 | 73.0 | 76.1 | 77.0 | 80.0 |
| H | 6.9 | 6.6 | 6.4 | 6.6 | 7.1 | 7.3 | 6.7 | 5.9 |
| N | 2.5 | 1.3 | 1.2 | 1.6 | 2.2 | 2.4 | 2.7 | 1.5 |
| O (difference) | 39.2 | 32.7 | 31.3 | 19.8 | 17.5 | 13.9 | 13.4 | 12.5 |
| S | 0.6 | 0.1 | 0.1 | 0.2 | 0.2 | 0.2 | 0.2 | 0.2 |
| (d) Shell | | | | | | | | |
| C | 50.6 | 57.1 | 62.2 | 70.8 | 72.4 | 72.2 | 74.3 | 80.9 |
| H | 5.9 | 5.7 | 5.6 | 5.4 | 5.0 | 4.7 | 4.5 | 4.8 |
| N | 0.4 | 0.4 | 0.4 | 0.6 | 0.5 | 0.6 | 0.6 | 0.7 |
| O (difference) | 43.0 | 36.8 | 31.8 | 23.1 | 22.1 | 22.5 | 20.5 | 13.5 |
| S | 0.0 | 0.0 | 0.0 | 0.1 | 0.0 | 0.0 | 0.0 | 0.1 |
| (e) EFB | | | | | | | | |
| C | 46.9 | 51.6 | 57.1 | 70.5 | 71.8 | 74.1 | 77.0 | 79.9 |
| H | 6.3 | 6.3 | 6.0 | 5.6 | 5.4 | 5.7 | 5.7 | 5.1 |
| N | 0.5 | 0.5 | 0.5 | 0.8 | 0.8 | 0.9 | 1.0 | 1.1 |
| O (difference) | 46.2 | 41.6 | 36.3 | 23.0 | 22.0 | 19.2 | 16.2 | 13.8 |
| S | 0.1 | 0.0 | 0.0 | 0.1 | 0.0 | 0.1 | 0.1 | 0.1 |

daf, dry ash-free basis.

**Table 3.** Ultimate analysis of the raw materials and solid products.

The reaction has also changed both the physical and the chemical properties of the materials. Like other biomass materials, oil palm wastes have very high volatile content (between 80.1 and 83.9 wt.%), in contrast to the low fixed carbon (between 16.1 and 19.9 wt.%). Progressive decomposition reactions occur at higher temperature, leading to an increase in fixed carbon content and a decrease in volatile content. Treatment at 380°C increases the fixed carbon content to approximately 48.2–65.6 wt.% and decreases the volatile content to 34.4–51.8 wt.%. The data suggest that carbonization and devolatilization occur during treatment. This leads to an increase in gross calorific value of the solid product (Table 2). Among the products, solids obtained from treatment of shell have the highest fixed carbon and the lowest volatile content.

The calorific value is correlated with the elemental composition of a solid. The data in Table 2 show that an increase in calorific value is correlated with an increase in carbon content and a decrease in oxygen content (Table 3). Compared to the raw material, the solid produced at 380°C has approximately ~67% higher carbon content and ~65% lower oxygen content. These results suggest that components degraded and removed from the materials are mainly oxygen-rich compounds. The data in Tables 2 and 3 suggest that remarkable changes in solid properties occur in 200–270°C range. For instance, a 270°C treatment increases the gross calorific value of fiber from 19.8 to 27.4 MJ/kg, while additional heating process into 380°C only raises the gross calorific value to 30.6 MJ/kg. In other words, a 270°C treatment contributes ~70.4% of total change equivalent of gross calorific value. Similar trend is observed for other solid properties.

Due to thermal degradation, the content of cellulose, hemicellulose, and lignin in the solid decrease. Figure 1 shows the percentage of these components in the solid products after treatment at 200–300°C. The data suggest that hemicellulose and cellulose are relatively easier to degrade than lignin. The treatment significantly degrades both hemicellulose and cellulose to produce a more ligneous solid. Lignin identified in this analysis comprises not only original lignin from feed materials, but also precipitate derived from polymerization of water soluble compounds. This is obviously indicated from an increase in lignin content in 200–270°C range for solid products of frond, trunk, and EFB.

Slightly different behavior is observed for hemicellulose decomposition among materials. The frond and shell solid produced at 200°C still have a small amount of hemicellulose, which completely vanishes on treatment at 240°C. By contrast, no hemicellulose is found for the trunk, fiber and EFB products, even for treatment 200°C. This suggests that hemicellulose decomposition starts at temperatures <200°C. On the other hand, cellulose is gradually degraded at higher temperature and <0.5 wt.% of it (on a solid product basis) is remained after treatment at 270°C. This behavior is in agreement with other earlier reports [16,17].

## 3.2. Coal bands

Under hydrothermal treatment, all materials undergo a coalification-like process, as demonstrated in the van Krevelen diagram in Figure 2. The raw materials have high atomic H/C and O/C ratios, which both gradually decreases during treatment. The slope of the trajectories suggests that the O content decreases in proportion to the H content, probably due to dehydration. It is clear that the decrease in O and H content occurs mainly in the range 200–270°C. Less significant changes are observed at higher temperature. The products after treatment at ≥300°C have almost identical compositions with sub-bituminous and bituminous coals.

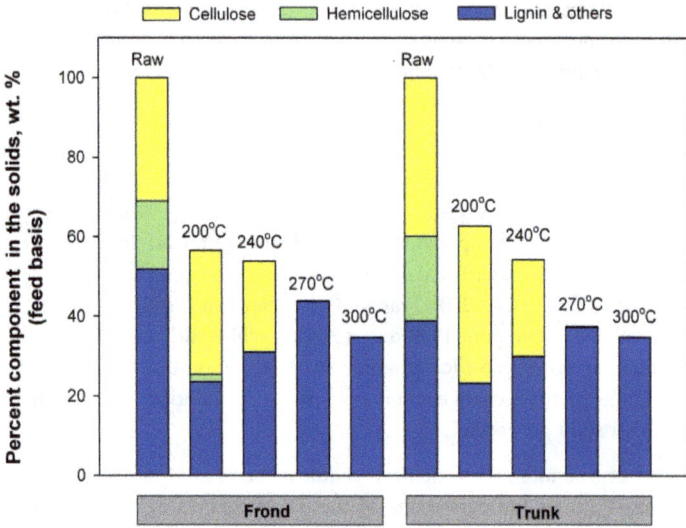

(a)

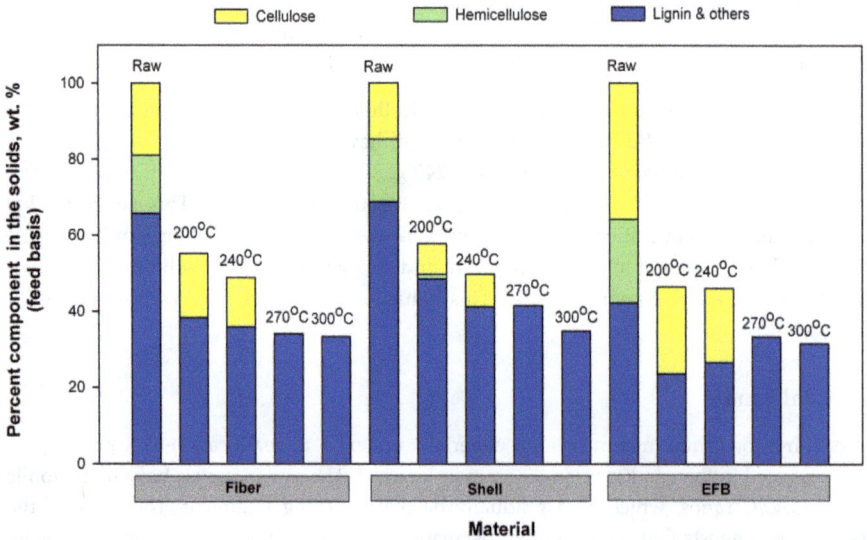

(b)

**Figure 1.** Percentage components in the products of low temperature treatment: (a) Frond and Trunk; (b) Fiber, Shell, and EFB

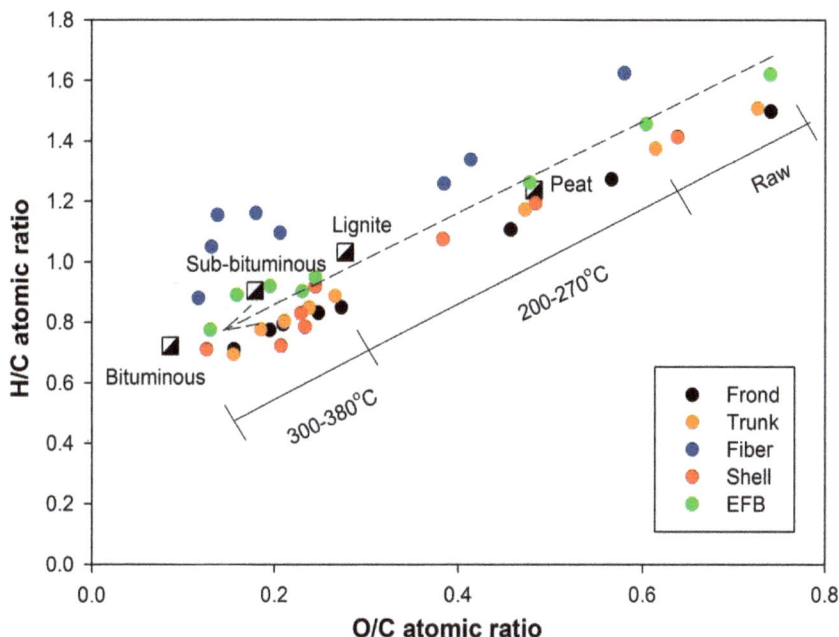

**Figure 2.** Van Krevelen diagram for materials and their corresponding products in comparison with some solid fuels

## 3.3. Fourier Transform Infrared (FT-IR) analysis

To understand changes in functional groups during hydrothermal treatment, FTIR analysis of the products is performed. Peaks are assigned based on literature data [18,19]. Figure 3–5 show spectra of the raw fiber, shell, EFB and the corresponding solid products. The intensity of the peak ~3500 cm$^{-1}$ attributed to -OH groups decreases at elevated temperature, indicating that water molecules within the solids are gradually expelled. In other words, dehydration of the feed material occurs. The peak at ~2900 cm$^{-1}$ attributed to aliphatic CH$_n$ groups also weakens, indicating that several long aliphatic chains are broken down. More distinctive peaks are observed in the region below 2000 cm$^{-1}$. The peak at ~1700–1740 cm$^{-1}$ represents carbonyl (C=O) stretching vibrations. The peak at ~1050 cm$^{-1}$ attributed to glycosidic bonds, indicating the presence of cellulose, steadily weakens and completely disappears for temperatures >270°C, indicating that cellulose is totally degraded at this temperature. The decrease in intensity for both aromatic skeletal vibrations at ~1515 cm$^{-1}$ and C-O-C aryl-alkyl ether linkages at ~1230 cm$^{-1}$ suggest lignin decomposition. Conversely, solids derived from polymerization of intermediate compounds in the liquid phase increase the aromatic content, particularly at temperatures >300°C, as indicated by the increase in intensity for the peak at 1600 cm$^{-1}$ attributed to aromatic skeletal vibrations and CO stretching.

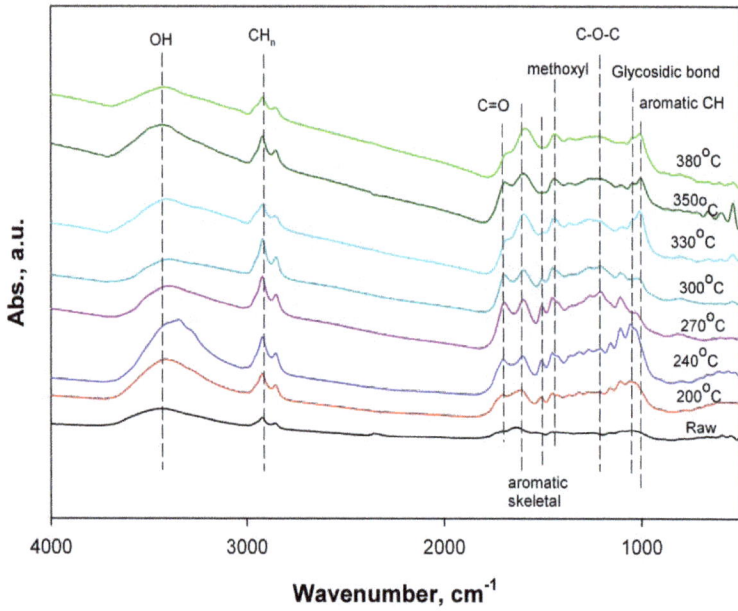

**Figure 3.** FT-IR spectra for raw fiber and solid products obtained at various temperatures

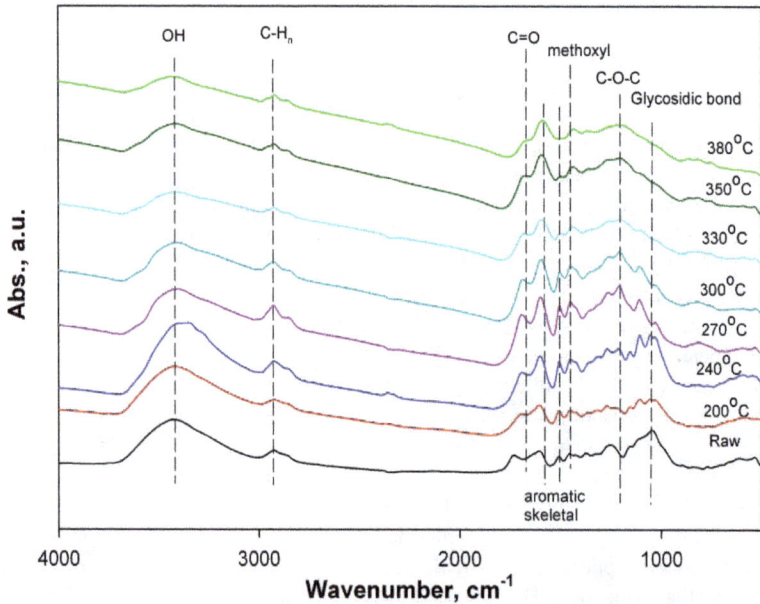

**Figure 4.** FT-IR spectra for raw shell and solid products obtained at various temperatures

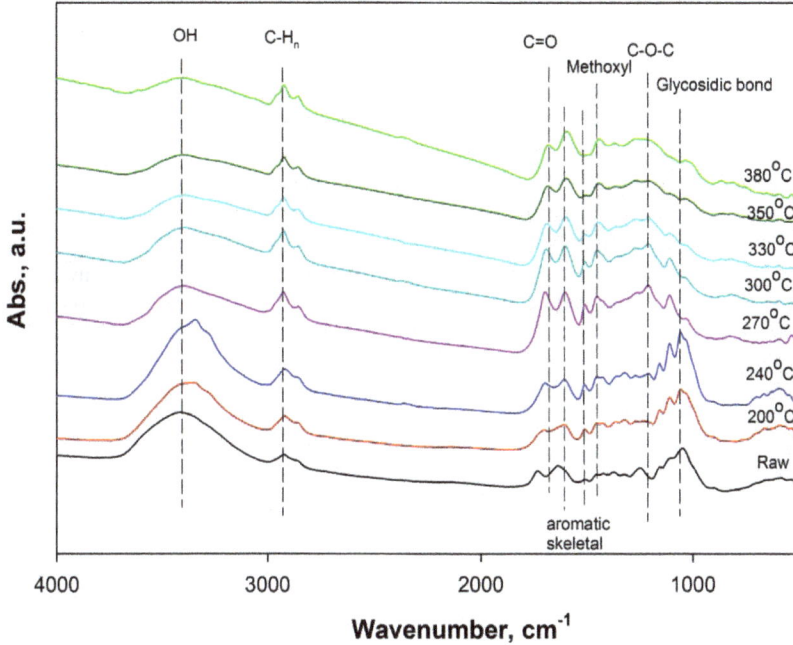

**Figure 5.** FT-IR spectra for raw EFB and solid products obtained at various temperatures

## 3.4. The $^{13}$C NMR analysis

$^{13}$C NMR measurements are conducted to complement FTIR in characterizing the molecular structure of the solid products. NMR is useful for making comparisons without the need for peak ratios. Each resonance peak can be measured relative to the total resonance intensity to give the relative amount of individual molecular groups. Typical $^{13}$C NMR spectra for raw biomass with peak assignment can be found in the literature [20–24].

In brief, resonance peaks in spectra for raw oil palm wastes are assigned to CH$_3$ in acetyl groups (21 ppm), methoxyl groups in lignin (56 ppm), C-6 carbon atoms in cellulose (62–65 ppm), C-2/C-3/C-5 atoms in cellulose (72–75 ppm), C-4 atoms in cellulose (84–89 ppm), C-1 atoms in hemicellulose (102 ppm), C-1 atoms in cellulose (105 ppm), unsubstituted olefinic or aromatic carbon atoms (110–127 ppm), quaternary olefinic or aromatic carbon atoms (127–143 ppm), olefinic or aromatic carbon atoms with OH or OR substituents (143–167 ppm), esters and carboxylic acids (169–195 ppm) including acetyl groups in hemicellulose (173 ppm), and carbonyl groups in lignin (195–225 ppm). Despite the various resonance peaks observed, for semi-quantitative analysis Wikberg and Maunu [21] and Wooten et al. [22] simply classify spectra into aliphatic (0–59 ppm), carbohydrate (59–110 ppm), aromatic (110–160 ppm), carboxyl (160–188 ppm) and carbonyl (188–225 ppm) regions.

Data for raw and treated fiber reveal that the peak resonance for hemicellulose and cellulose progressively decrease (Figure 6). A similar pattern is observed for other oil palm wastes (Figure 7–8). The spectra suggest that peaks corresponding to CH₃ (21 ppm) and COOH (173 ppm) in acetyl groups of hemicellulose are eliminated at 200°C. Carbon atoms in cellulose and hemicellulose (62–105 ppm) progressively decrease at 200–270°C. Thus, neither cellulose nor hemicellulose carbon atoms remain after treatment at ≥270°C. Lignin decomposition is indicated by a gradual decrease in methoxyl lignin (56 ppm) along the temperature range. Treatment at ≥270°C leads to extreme spectral changes to a more aromatic nature. Hence, aromatic carbon atoms from lignin structures (31 and 110–160 ppm) dominated the entire spectra. This is in good agreement with the component analysis, which suggests that lignin is the predominant component for treatment at ≥270°C (Figure 1).

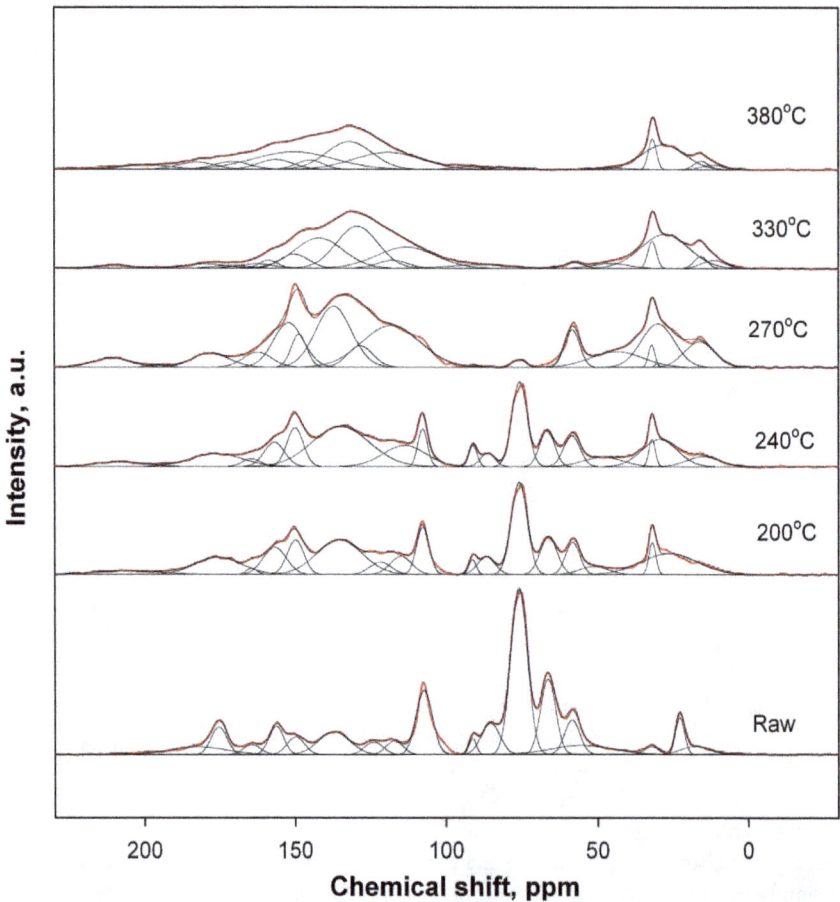

**Figure 6.** ¹³C NMR spectra for raw fiber and its treated products.

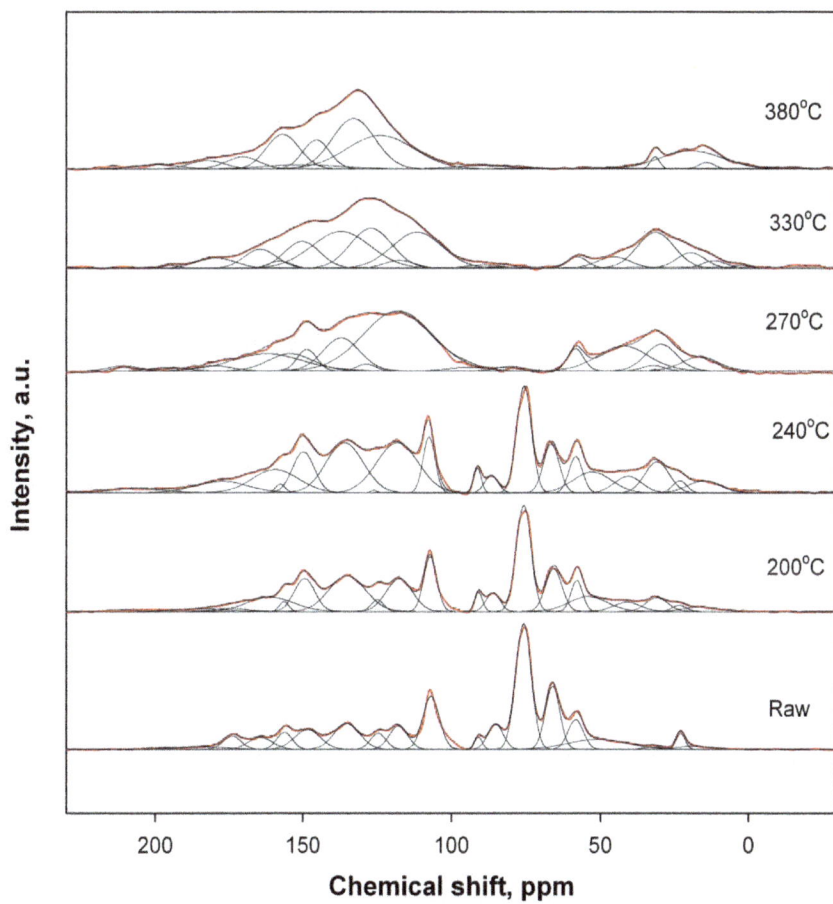

**Figure 7.** ¹³C NMR spectra for raw shell and its treated products

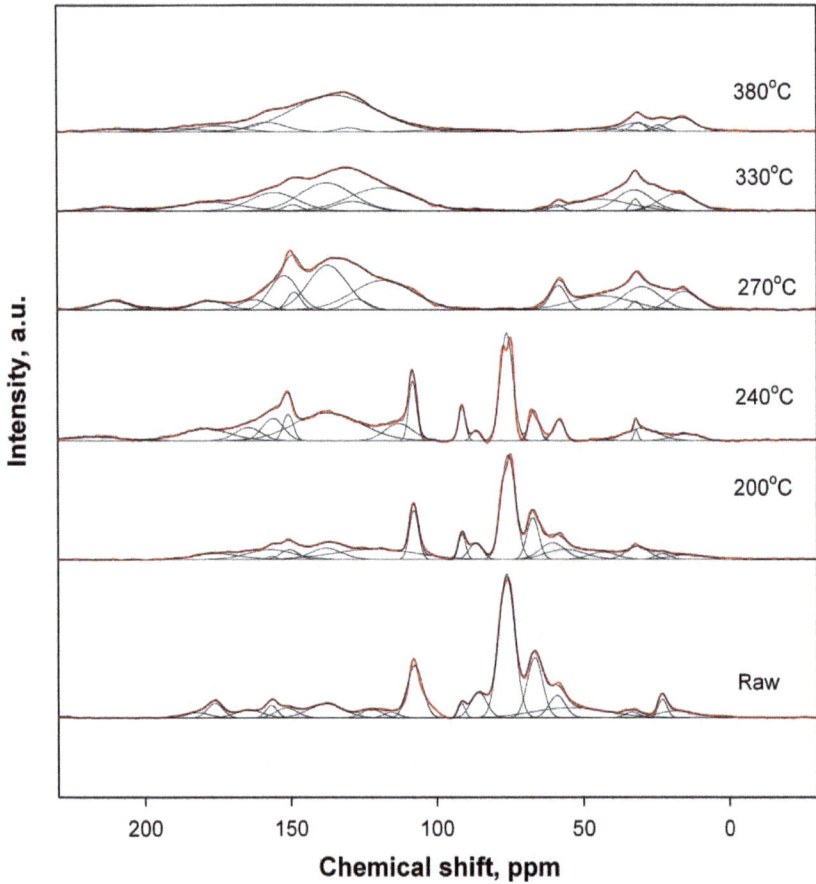

**Figure 8.** 13C NMR spectra for raw EFB and its treated products.

### 3.5. Equilibrium moisture content analysis

Hydrothermal treatment greatly reduces the EMC of materials. As shown in Table 2.2, treatments at 200°C reduces the EMC from 14.7 to 7.6 wt.% for frond, and from 13.6 to 7.5 wt.% for trunk. Similarly, the EMC of fiber, shell, and EFB decrease from 13.0 to 6.9 wt.%, from 9.9 to 6.7 wt.%, from 11.8 to 8.1 wt.%, respectively after treatment at 200°C. Further treatment at 380°C leads to EMC as low as 5.0 wt.% (frond), 4.1 wt.% (trunk), 3.6 wt.% (fiber), 3.1 wt.% (shell), and 3.2 wt.% (EFB), respectively. However, the decrease in EMC mainly occurrs in the range 200–270°C, with only small changes observed at higher temperatures. These results correspond to the changes in solid components shown in Figure 1.

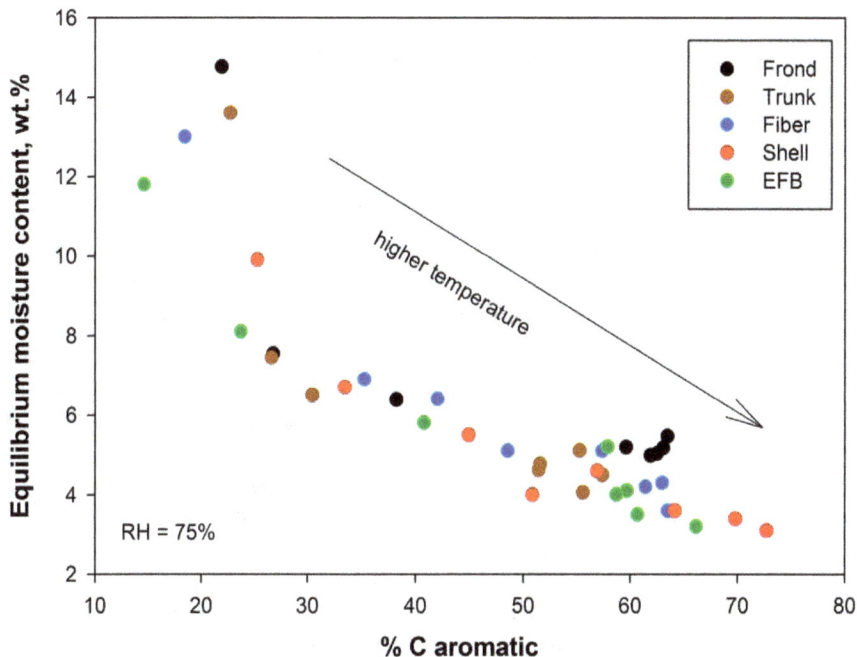

**Figure 9.** Relationship between percentage aromatic carbon and equilibrium moisture content

Based on the component characteristic on water adsorption, hemicellulose is the strongest, followed by cellulose and lignin [25]. Since hemicellulose is removed first from the solid at low temperature, it is reasonable that the EMC of the material dramatically decreases in this range. By contrast, solids with high lignin content adsorb only a small amount of moisture.

The EMC results are confirmed by NMR results demonstrating an increase in aromatic content in the solid material. The presence of aromatic compounds, which are hydrophobic, results in resistance to humidity and water adsorption from air. Therefore, a higher aromatic content is correlated with lower EMC. The relationship between the relative amount of aromatic carbon and the EMC is presented in Figure 9.

EMC and calorific value are two important properties of solid fuels. When material is burned, some of the energy released by combustion is consumed to vaporize the water contained in the material. Material with a higher EMC will require more energy for moisture evaporation. Thus, a good solid fuel should have a high calorific value and a low EMC. Our experiments demonstrate that both properties are improved by hydrothermal treatment

## 4. Conclusions

Upgrading of oil palm waste is investigated for hydrothermal treatment at 200–380°C and initial pressure of 2.0 MPa for a residence time 30 min. Approximately 30–60 wt.% of the original material is recovered after the process. The very high oxygen and volatile matter content of the original material are significantly reduced. By contrast, the fixed carbon content increases sharply due to carbonization. A van Krevelen diagram reveals that solids resulting from treatment at >330°C have a composition comparable to that of sub-bituminous and bituminous coals. FTIR analysis confirms that oxygen elimination due to dehydration in conjunction with decomposition of hemicellulose and cellulose occurr at 200–270°C. At temperatures >270°C, the structure of the solid dramatically changes and is dominated by lignin. This is indicated by an increase in aromatic compounds, as determined by $^{13}$C NMR spectroscopy.

Hydrothermal treatment progressively changes the calorific value and EMC of materials. Treatment at 380°C produces solid with a gross calorific value of 30.0–32.1 MJ/kg and EMC of 3.1–5.0 wt.%. However, approximately 65% of the total increase in calorific value and 92% of the total decrease in EMC take place within the range 200–270°C, which can be attributed to complete removal of hemicellulose and cellulose. Presence of hydrophobic aromatic compounds in these solids rejects moisture adsorption from atmosphere that potentially reduces the net energy produced. In addition, as solid fuel, which may be transported from one location to other, their higher energy density will affect on reduction of storage, as well as transportation cost. Based on these results, it is proposed that hydrothermal treatment could become an advantageous technology for producing solid fuel from biomass wastes. Among the wastes treated, solid products of oil palm shell demonstrate the best results.

## Author details

Ahmad T. Yuliansyah
*Dept. of Chemical Engineering, Faculty of Engineering, Gadjah Mada University, Yogyakarta, Indonesia*

Tsuyoshi Hirajima
*Dept. of Earth Resources Engineering, Faculty of Engineering, Kyushu University, Nishi-ku, Fukuoka, Japan*

## Acknowledgement

The authors are grateful for support of this research by a Grant-in-Aid for Scientific Research No. 21246135 and No. 24246149 from the Japan Society for the Promotion of Science (JSPS) and the Global COE program (Novel Carbon Resources Sciences, Kyushu University).

# 5. References

[1] Basiron Y (2007) Palm Oil Production through Sustainable Plantations. Eur j. lipid sci. tech. 109(4): 289-295.

[2] Oil World in Bulletin of PT Astra Agro Lestari (2009) Jakarta: March 2.

[3] Zakaria ZZ, ed. (2000) Agronomic Utilization of Wastes and Environmental Management. 2 ed. In: Basiron Y, Jalani BS, Chan KW, editors. Advances in Oil Palm Research, Vol. 2. Kuala Lumpur: Malaysian Palm Oil Board.

[4] Yuliansyah AT, Hirajima T, Rochmadi (2009) Development of the Indonesian Palm Oil Industry and Utilization of Solid Waste. Journal of MMIJ. 125: 583-589.

[5] Schuchardt F, Darnoko D, Guritno P (2002) Composting of Empty Oil Palm Fruit Bunch (EFB) with Simultaneous Evaporation of Oil Mill Wastewater (POME). in Proceeding of the International Oil Palm Conference. Bali, Indonesia.

[6] Bobleter O, Niesner R, Rohr M (1976) Hydrothermal Degradation of Cellulosic Matter to Sugars and Their Fermentative Conversion to Protein. J. appl polym sci. 20(8): 2083-2093.

[7] Kumar S, Gupta RB (2009) Biocrude Production from Switchgrass Using Subcritical Water. Energ fuel. 23: 5151-5159.

[8] Arvanitoyannis IS, Kassaveti A, Stefanatos S (2007) Current and Potential Uses of Thermally Treated Olive Oil Waste. Int. j. food sci. technol. 42: 852–867.

[9] Zhang LH, CB Xu, Champagne P (2010) Overview of Recent Advances in Thermo-Chemical Conversion of Biomass. Energ convers manage. 51: 969–982.

[10] Kruse A (2009) Hydrothermal Biomass Gasification. J supercrit fluid. 47: 391–399.

[11] Allen SG, Schulman D, Lichwa J, Antal MJ, Laser M, Lynd LR (2001) A Comparison between Hot Liquid Water and Steam Fractionation of Corn Fiber. Ind eng chem res. 40: 2934–2941.

[12] Ehara K, Saka S (2002) A Comparative Study on Chemical Conversion of Cellulose between the Batch-Type and Flow-Type Systems in Supercritical Water. Cellulose. 9: 301–311.

[13] Hamelinck CN, Van Hooijdonk G, Faaij APC (2005) Ethanol from Lignocellulosic Biomass: Techno-Economic Performance in Short-, Middle- and Long-Term. Biomass bioenerg. 28: 384–410.

[14] Laser M, Schulman D, Allen SG, Lichwa J, Antal MJ, Lynd LR (2002) A Comparison of Liquid Hot Water and Steam Pretreatments of Sugar Cane Bagasse for Bioconversion to Ethanol. Bioresource technol. 81: 33–44.

[15] Sluiter A, Hames B, Ruiz R, Scarlata C, Sluiter J, Templeton D, Crocker D (2005) Determination of Structural Carbohydrates and Lignin in Biomass. The US National Renewable Energy Laboratory.

[16] Ando H, Sakaki T, Kokusho T, Shibata M, Uemura Y, Hatate Y (2000) Decomposition Behavior of Plant Biomass in Hot-Compressed Water. Ind eng chem res. 39(10): 3688-3693.

[17] Ehara K, Saka S (2005) Decomposition Behavior of Cellulose in Supercritical Water, Subcritical Water, and Their Combined Treatments. J wood sci. 51(2): 148-153.

[18] Yang H P, Yan R, Chen H P, Lee D, Liang DT, Zheng CG (2006) Pyrolysis of Palm Oil Wastes for Enhanced Production of Hydrogen Rich Gases. Fuel process technol. 87(10): 935-942.

[19] Kobayashi N, Okada N, Hirakawa A, Sato T, Kobayashi J, Hatano S, Itaya Y, and Mori S (2009) Characteristics of Solid Residues Obtained from Hot-Compressed-Water Treatment of Woody Biomass. Ind eng chem res. 48(1): 373-379.

[20] Capanema EA, Balakshin MY, Kadla JF (2005) Quantitative Characterization of a Hardwood Milled Wood Lignin by Nuclear Magnetic Resonance Spectroscopy. J. agr food chem. 53(25): 9639-9649.

[21] Wikberg H, Maunu SL (2004) Characterisation of Thermally Modified Hard- and Softwoods by C-13 CPMAS NMR. Carbohyd polym. 58(4): 461-466.

[22] Wooten JB, Kalengamaliro NE, Axelson DE (2009) Characterization of Bright Tobaccos by Multivariate Analysis of C-13 CPMAS NMR Spectra. Phytochemistry. 70(7): 940-951.

[23] Atalla RH, VanderHart DL (1999) The Role of Solid State C-13 NMR Spectroscopy in Studies of the Nature of Native Celluloses. Solid state nucl mag.15(1): 1-19.

[24] Liitia T, Maunu SL, Sipila J, Hortling B (2002) Application of Solid-State C-13 NMR Spectroscopy and Dipolar Dephasing Technique to Determine the Extent of Condensation in Technical Lignins. Solid state nucl mag. 21(3-4): 171-186.

[25] Morohoshi N (1991) Mokushitsu Biomass no Riyou Gijyutsu. Tokyo: Buneido Shuppan (in Japanese).

# Chapter 3

# Irrigation Management in Coastal Zones to Prevent Soil and Groundwater Salinization

Nicolas Greggio, Pauline Mollema,
Marco Antonellini and Giovanni Gabbianelli

Additional information is available at the end of the chapter

## 1. Introduction

Soil salinization is one of the most widespread soil degradation processes on earth and, worldwide, one billion hectares are affected, mainly in the arid–semiarid regions of Asia, Australia and South America [1]. In Europe, soil salinity has effects on one million hectares mainly in the Mediterranean countries [1]. There are two types of salinization: primary salinization caused by natural events such as sea spray or rock weathering or seepage [2] and secondary salinization that is caused by human activities such as irrigation with salty water, groundwater overexploitation and excessive drainage [1].

Along the Adriatic coast of the Po Plain, freshwater resources are becoming increasingly scarce, because of irrigation and other intense water use, salinization and long periods of drought [3]. Custodio [4] underlines that, especially in southern Europe, the irrigation practices and the water requirements to sustain the coastal tourism industry exhort a strong pressure on water resources.

The impact of groundwater salinization in coastal areas affects both natural vegetation biodiversity and agricultural production, through soil salinization and reduction of freshwater availability for irrigation. Salinization is closely associated with the process of desertification, because salinity may have direct negative effects on crop yields by reducing the ability of plant roots to take up water [5]. The most common salinity effect is a general stunting of plant growth, but not all plants respond in the same way. Grain and corn may reduce their seed production without appreciably plant dimensions reduction. Wheat, barley, cottons seed productions, on the other hand, are decreased less than their vegetative growth [6]. The only agronomical significant criterion for evaluating soil salinization effects is the commercial crop yield decrease[6]. This encourages farmers in certain areas to cultivate salt resistant crops to adapt to high salinity of the soil and groundwater [5]In our area fresh irrigation water until now was readily available and no salt resistant crops are grown, or other adaptation measures were taken against soil and groundwater salinization of the farmland.

An increase in water and soil salinity within a coastal habitat has direct effects on the natural vegetation and agriculture [5],[7], including decrease in plant species richness, decrease in wetland dry biomass, changes in plant communities, decrease in seedling germination, decrease in surface area of the leaves, decrease in stem length, N2-fixation inhibition, increasing mortality, and indirect effects such as habitat loss for some animal species [8],[9],[10].

In the Ravenna province, more than 68 square kilometers of farmland are at a risk for soil salinization (Figure 3). These are primarily the areas near the Pialasse lagoons and near the rivers open to sea (Figure 1). This agricultural land is at sea level and drained by pumping machines. Many of the pumping machines are located 5 km far from the shoreline and they maintain the water table at 2 meters below sea level during the year, creating hydraulic gradients land inwards and promoting salt-qwater intrusion from the Adriatic Sea [3].

High rates of anthropogenic and natural subsidence and artificial drainage, among others causes, have caused groundwater salinization in the coastal unconfined and semi-confined aquifer near Ravenna, [3],[11] and a subsequent loss of fertile soils and vegetation species richness in the natural areas [7].

In the Ravenna Municipality just about 35000 hectares are agricultural land, of these 30000 are arable land and 5000 are orchards. Irrigation is widely diffuse and more than 50% of farms are equipped for irrigation. Until two decades ago the main source of irrigation water was groundwater from the phreatic and confined aquifers, but the increase of subsidence rates, forced to change the water supply from groundwater to surface water [12],[13], mostly from the Po River and some from the minor rivers flowing from the Apennines.

The phreatic aquifer consists of sandy beach and dune deposits and is unconfined close to the sea and confined by a clay layer further inland (Figures 11and 12). Two dune belts parallel to the coast covered by pine forests form the only topography above sea level. The area in between the dunes belts is used for irrigated and rain-fed farmland, as well as the land behind the older dune belt, 5 km from the coast (Figure 1).

Most of the coastal aquifer near Ravenna is very salty (above 10 g/l) [3][11][14] but, besides some small shallow freshwater lenses at the top of the aquifer, we found freshwater below an irrigated field. In some ways this result is worrying, because irrigation water should not reach the groundwater, so that leaching of nitrates is avoided and waste of resources prevented. Mollema et al. ([15]) concluded from a water budget analysis over the *Quinto* basin watershed (south of the Ravenna urban area) that the amount of water used for irrigation was abundant but only looked at total volumes for the whole basin and did not quantify the contribution of irrigation to groundwater.

The objective of this study is to quantify in detail the effect of irrigation water on the groundwater hydrology with the help of geochemical analysis, geophysical profiling and infiltration measurements. We address, in particular, the mechanism of how excess irrigation water ends up in the aquifer and how that affects the salinity of the groundwater. By doing all this, we assess whether irrigation practices can help to counteract further groundwater and soil salinization.

## 2. Study area

The study area (red rectangle in Figure 1) is part of the *Quinto* basin, the watershed that is confined between two rivers flowing from the Northern Apennines to the Adriatic Sea: the *Uniti* River in the north and the *Bevano* River in the south. The eastern border of the basin is the Adriatic Sea and the western border follows, in part, the *Ronco* River, a tributary of the *Uniti* River.

An east-west profile through the area starting at the shoreline sea includes a beach near the town of *Lido di Dante*, a recent dune belt covered by a coastal pine forest called *Ramazzotti*, planted at the beginning of the 1900's, agricultural land, an older dune belt (paleodunes) covered with a pine forest (*Classe* pine forest), a zone with many active and abandoned gravel quarries that form lakes and more agricultural land (Figures 1 and 5).

Our focus is on a plot of agricultural area (500 by 500 m) in the eastern part of the profile (Figure 2). The area is around a ditch (shown in red). The green dots are wells used to monitor the freshwater lenses, in particular P1S, P2S and P3S are fully penetrating the aquifer (-27m), the others are shallow piezometers (-6m).

The ditch was dug in 1981 as a reservoir for irrigation water that was withdrawn from the *Fiumi Uniti* River. In 2008, the supply for irrigation changed and now it comes through an underground pipe from the *Po* River. The ditch is 500 m long, 5 m wide and 2,5m deep, resulting in a total volume of 6250 m$^3$; it is completely disconnected from the drainage channels network.

In 2011, the planting schedule was as follows: in the east parcel tomato plants were grown from April to August; Alfa Alfa was contemporaneously grown in the west parcel. Because of the particularly dry 2011 weather conditions, tomato plants needed irrigation water from the early stage of growth onward.

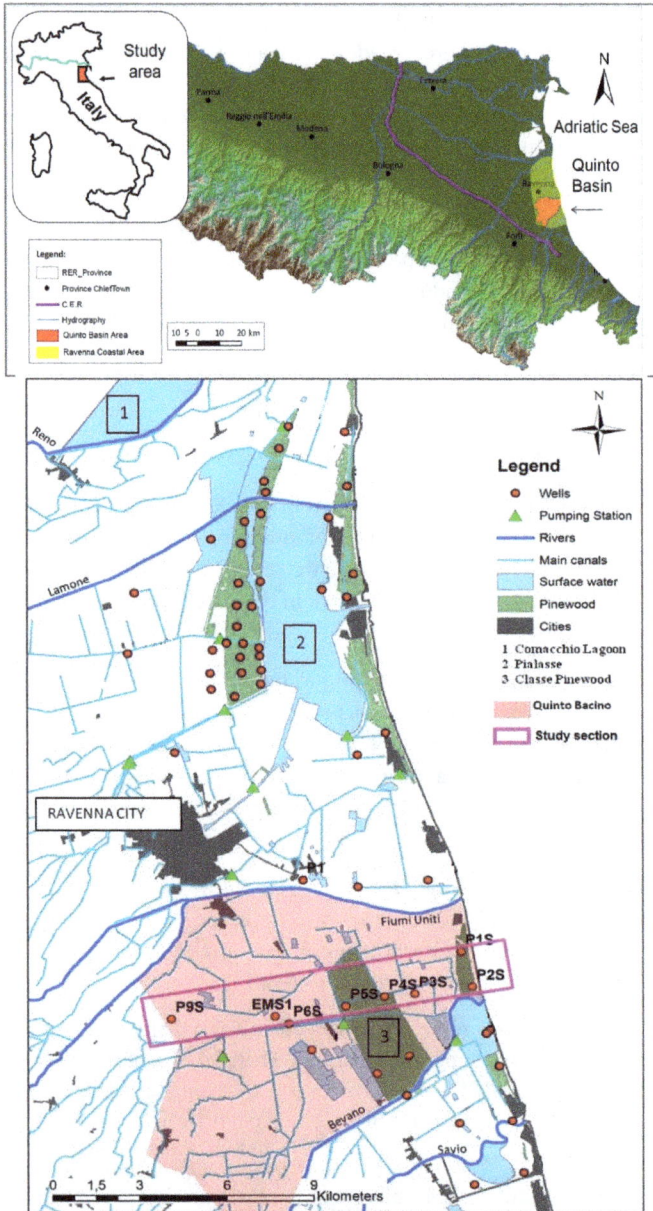

**Figure 1.** Index map and detail of the Ravenna coastal zone and location of study area (*Quinto* basin); the map reports also monitoring wells, rivers, channels, pumping stations and main environmental features.

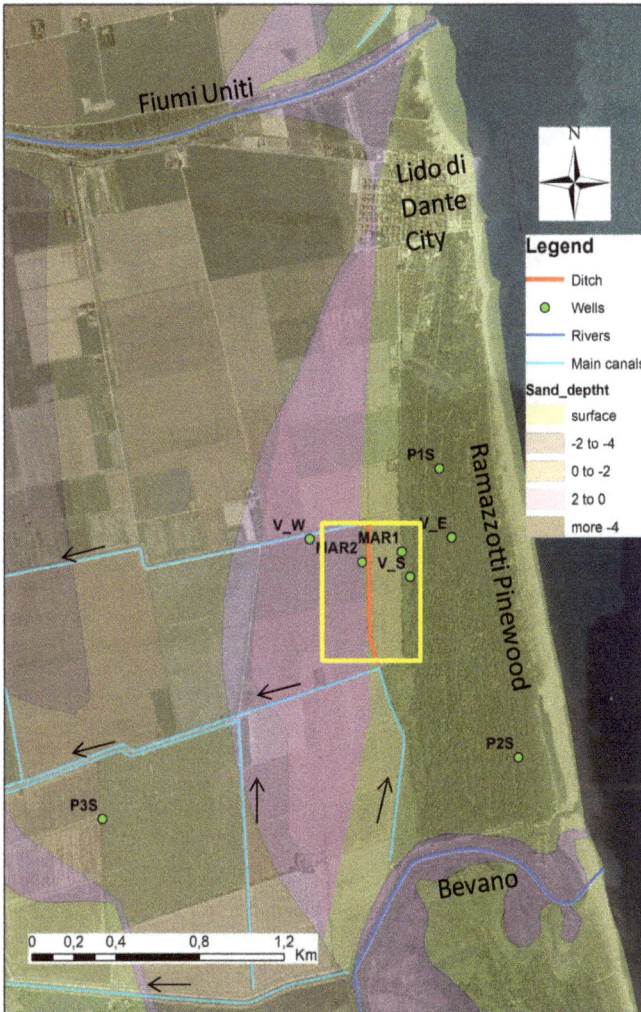

**Figure 2.** Detail of the study area with thickness of sandy top layer, main drainage channels, flux direction and wells used for water table monitoring.

## 3. Soil type and land use

In the coastal area, sediments at the surface are mainly sandy (Figure 3). The sand layers are parallel to the coastline and form paleo and recent dunes.

The sandy soils are called *Cerba* soils, and according to the FAO classification [16] for Soil Taxonomy, they are Calcaric Arenosols and Mesic Aquic Ustipsamments [17].

These soil types are mainly used as arable land or horticulture. Due to their high hydraulic conductivity, these sandy soils are most sensitive for salinization [18],[19]. The *Sant'Omobono* soils (Brown in Figure 3) represents river sand deposits; these soils were deposited along recent or paleo-rivers during flooding events. According to the FAO classification [16] for the Soil Taxonomy they are Calcaric Cambisols, Fine Sandy, Mixed, mesic Udifluventic Ustochrepts, [17]. These kinds of soils are cultivated with vineyards and orchards. Because of their sandy texture, also these soils are at risk of salinization. The *Medicina* (*Cataldi*) sediments (Green colors) are soils with fine texture, mainly created when rivers flowed in presently reclaimed areas. FAO classifies them as Eutric Vertisols [16] and for Soil Taxonomy they are Fine, Mixed, Mesic Ustic Endoaquerts. [17]. They are grown with arable crops only and they are not affected by salinization [18],[19],[20].

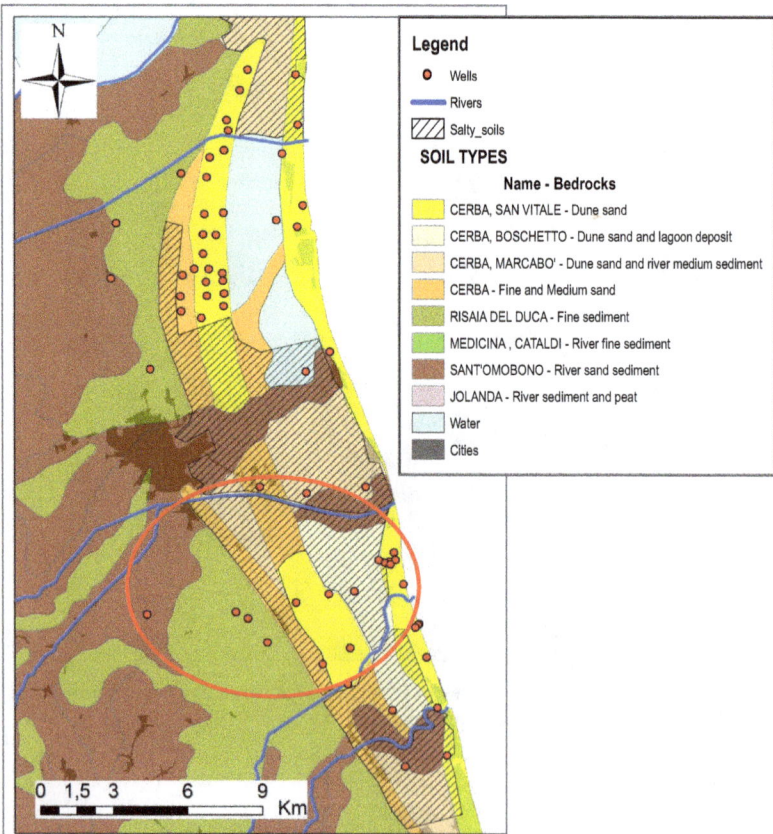

**Figure 3.** Soil types in the Ravenna area and soils at risk of salinization.

The area studied in detail has an extension of 12 ha consisting of two parcels of land of 6 ha on each side of the ditch. The western parcel lies on a 1-m-thick silty sand top layer (*Cerba*

*Boschetto* soil) that confines the sandy aquifer below, whereas the eastern parcel lies on the unconfined part of the aquifer that consists of beach sand (*Cerba San Vitale* soil; Figures 2 and 4).

The differences in soil and lithology, make the land suitable for different kinds of crops: in the last thirty years, on the sandy soil (eastern parcel), the crops consisted of horticulture (strawberries, tomatoes, potatoes, nursery sugarbeet); on the silty soil instead (western parcel), typical extensive arable crops were grown (barley, wheat, maize, alfa-alfa). At our Mediterranean latitude, horticulture crops need irrigation during all the growth phases [13]; arable crops, on the other hand, do not need irrigation during growth. To provide irrigation water for horticulture, the ditch is kept full of freshwater in the summer period. As a consequence, the water level in the ditch is higher than the water table level in the adjacent aquifer. During the winter, the water level in the irrigation ditch matches the adjacent water table level.

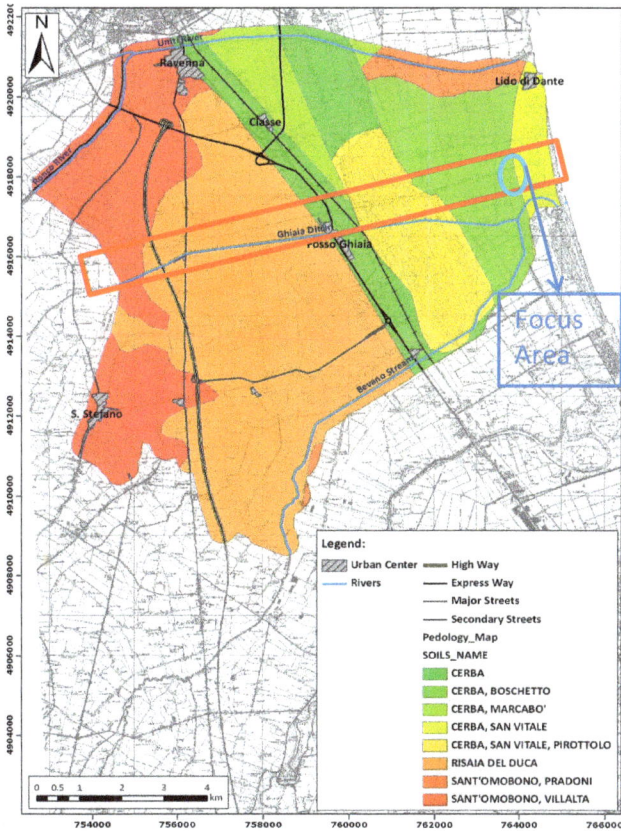

**Figure 4.** *Quinto* Basin soil types, with study section and focus area

Most of the land in the coastal zone near Ravenna is used for agriculture but urban and natural areas (pinewood and water surface body) are also widespread (Figure 1). Most of the natural environments are part of the Po Delta Regional Park. Irrigable crops are the most occurring culture in the province of Ravenna. Vineyards and orchards occupy a small percentage of the territory mostly further inland [12],[13].

The *Quinto* Basin covers a total of 10,355 hectares of which the largest part is used for agriculture (66%), 10% is covered by pine forests, 9% is urban area, and 9% is open surface water or wetlands (Figure 5) [15],[21].

The agricultural land is primarily devoted to arable crops as wheat, barley, alfa alfa, sorghum and corn and only 15% is grown with horticultural and orchards. Horticulture represents less than 5% of the *Quinto* basin and it is practiced on sandy soils (*Cerba* soil) (Figure 4). Orchards are grown in the western part, on *Sant'omobono* soils only that, being the deeper soils in the *Quinto* basin, are the most adapt for fruits and grapes, which are sensitive to water logging (Figures 4 and 5). The most common horticultural products cultivated in the *Quinto* basin are tomatoes, potatoes and nurseries of strawberries and sugar beets. The dominant fruit trees are peaches, apples and vine. For this basin, the data from the last general agricultural census shows that the entire surface cultivated with fruits and horticultural crops are equipped with irrigation systems [12].

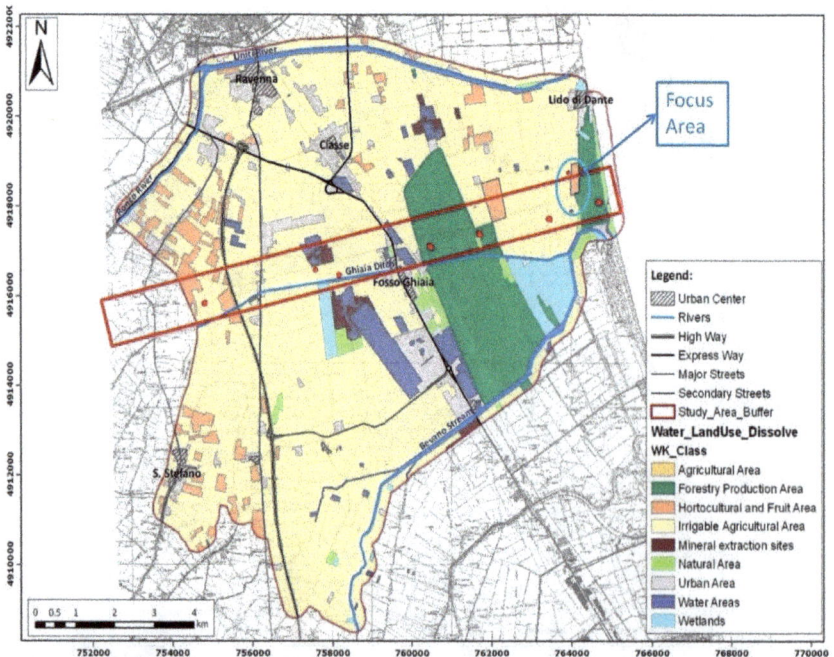

**Figure 5.** *Quinto* Basin land use map.

## 4. Topography, drainage and irrigation

Most of the land near Ravenna is below or just a few centimeters above sea level. Only the recent dune belt and the paleo-dunes are up to 5 m above sea level (Figure 6). These dunes are covered by pine forests (Figure 1). The maximum elevation in the *Quinto* Basin is 5 m and it is measured in the most recent dune belts, whereas the lowest areas are near the gravel quarries 7 km from the coast at 2m below sea level (Figure 6). The farmland around the ditch of our study area is completely flat and lies at a few centimeters above sea level.

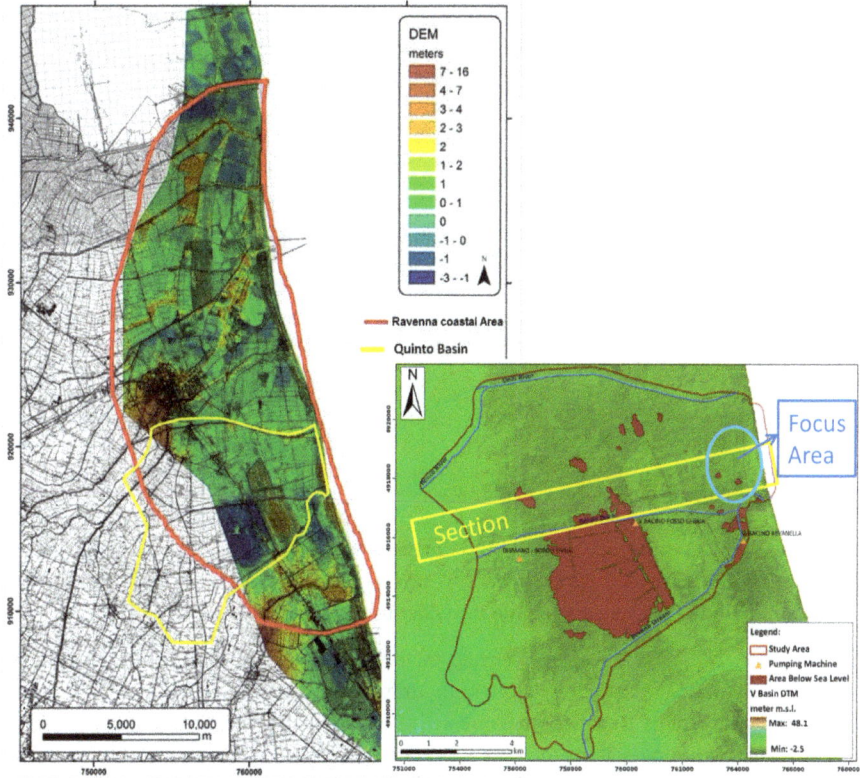

**Figure 6.** Topography of the coastal zone near Ravenna and area below sea level in the *Quinto* basin.

The whole coastal zone of Ravenna is drained mechanically to make agriculture possible despite the low topography. As a consequence, the areas of natural water infiltration are very small and localized along the dune belts. The entire district is divided into small watersheds defined by small topographic features such as rows of dunes, dikes, and levees [22]. Figure 7 shows the drainage basins in the Ravenna Municipality. Only one basin has a natural hydraulic gradient towards the sea (Figure 7).

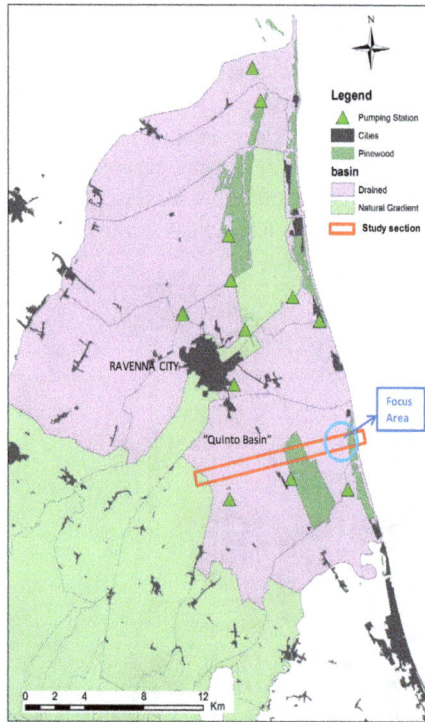

**Figure 7.** Drainage basins in the Ravenna Municipality. The light green color represents natural drainage basins; the pink color shows the mechanically drained basins.

A dense network of pumping stations and channels was built in the last century with the aim to keep the land from flooding and allow farming [23]. In the *Quinto* basin alone, more than 50 km of drainage channels reroute the water to the pumping stations [24].

The pumping stations are typically located inland, several kilometers from the shoreline (Figure 7). The drainage, therefore, creates a hydraulic gradient directed inland [3],[23]. Our study area is drained by two east-west oriented channels that collect the water between the two dune belts and convey it towards the *Fosso Ghiaia* pumping station (Figure 2).

The pumping stations are activated automatically when the water level in the drainage channels reaches a threshold, regardless of the soil water availability or the weather forecasts. As a consequence, the water table is everywhere below sea level with the exception of the dune belts where the water table may be above sea level in the wet periods of the year. The pumping station in the *Quinto* basin, (*Fosso Ghiaia*) causes a deep withdrawal cone down to -2.5 masl in the western part of the *Classe* pinewood (Figure 8). The flux vectors that were calculated based on water table measurements in the wells of Figure 1 are all directed inland showing that sea water flows into the aquifer.

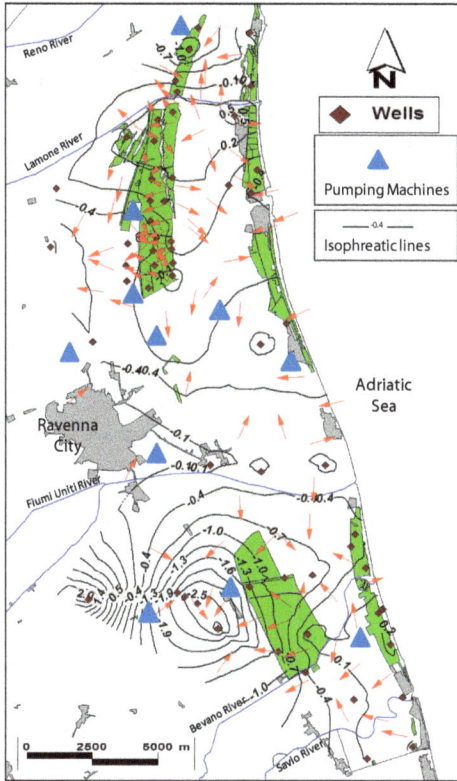

**Figure 8.** Water table contour map and flux vectors indicating the direction of groundwater flow.

The main source of irrigation water is from river water; five rivers are tapped for irrigation, from north to south they are the *Reno, Lamone, Fiumi Uniti, Bevano* and *Savio* rivers. In the summer period, however, there is not enough water in these rivers and additional irrigation water is taken from the *Po* River by means of the *Canale Emiliano Romagnolo* channel (CER). The water from the CER is brought into the irrigation systems via the same channels that are used for drainage in the winter.

The most common irrigation systems are the sprinkler and the drip irrigation system. The first one is used for horticultural crops with a large volume of water distributed twice a week, whereas the second one is typical for orchards and it uses small volumes of water but with a daily treatment. The sprinkler method, therefore, needs a storage basin for the water (ditch, lake or basin). The drip system instead doesn't need a storage basin [25].

In our focus area, water for irrigation was taken for many years from the *Fiumi Uniti* River. Seawater, however, has encroached along the river and, therefore, it was necessary to change the water supply [3]. Recently, irrigation water comes through a pressurized pipeline directly from the CER (Po river source). During the monitoring time, irrigation on tomato parcel started at an early stage of growth in April 2011 and ended at the harvest time in August 2011. The scheduling of irrigation has been rather regular: approximately twice a week. For each irrigation operation, 20 mm/m$^2$ were distributed on the tomato plants (Table 1). No irrigation was given to the alfa alfa in the parcel to the west of the ditch.

In the Ravenna Municipality, irrigation and drainage practices are managed by the same authority: the Land Reclamation Consortium for the *Romagna* Region and for both services the end users costs have recently increased [24],[25].

## 5. Climate

The climate of Ravenna is Mediterranean with generally warm summers and mild winters [26]. The average temperature is about 14°C varying from -5° to 10°C in the winter period and from 20° to 35° C in the summer months [27]. The average annual rainfall calculated on a long term database (1960-2011) is more than 650 mm and it is usually concentrated in the spring and autumn months.

In five of the last six years, total precipitation has been under the average; in particular 2011 was the driest of the last 23 years, with only 385 mm (Figure 9). Compared with the average monthly rainfall of 1960-2011, in fact, most of the autumn rainfall is absent (Figure 9). The same comparison for temperature shows that in 2011 the temperature was 3-4°C higher than the average of the last 51 years, especially in autumn and winter (Figure 10).

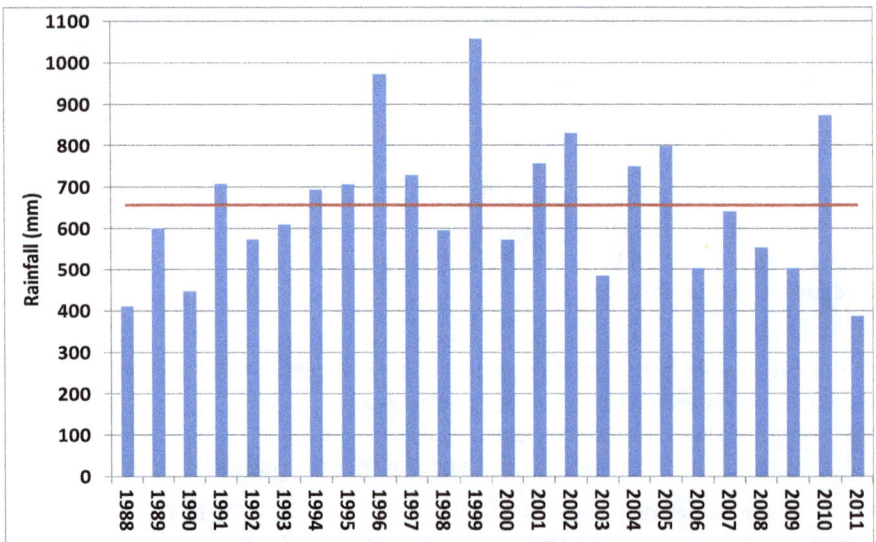

**Figure 9.** Annual precipitation in the city of Ravenna. Red line represents the long term mean value.

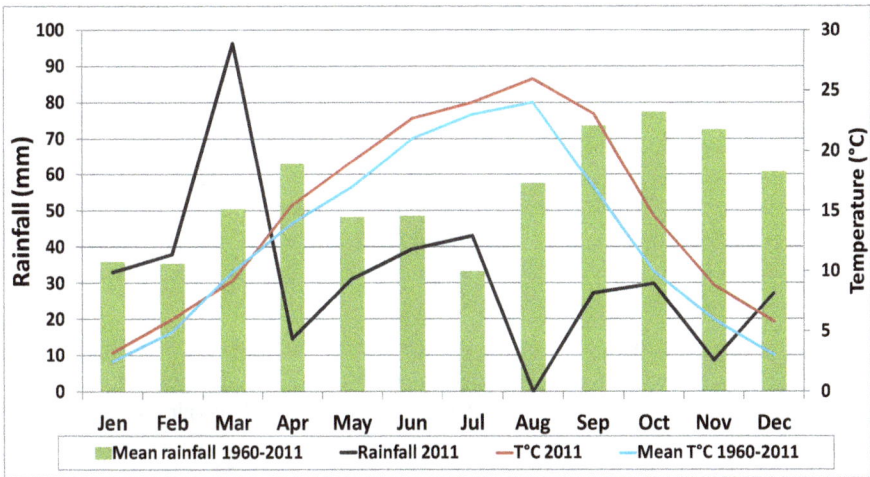

**Figure 10.** Monthly rainfall and temperature for the Ravenna Area.

One of the climate change scenarios proposed for the Mediterranean area [28] indicates a substantial change in the hydrologic budget by the end of this century. According to Antonellini and Mollema 2010 ([7]), evapotranspiration will increase as a result of an increase in temperature; rainfall will be concentrated in short periods, the winter will become more humid and the summer more dry compared to the present, and there will be an increase in extreme events (floods or droughts). Such a forecast for the coastal zone has a particular significance, because a shortage of fresh water in the area next to the sea will promote saltwater intrusion in the ground water and soil salinization. According to Magnan et al. 2009 ([28]), it is possible to mitigate some of these effects, but it is not possible to avoid them completely; it is, therefore, necessary to start the development of mitigation and adaptation strategies now in order to face these issues.

## 6. Geology and geomorphology

The study area is a small portion of the southern part of the wider Po delta plain. The surface aquifer sediments were deposited during the Late Quaternary (Holocene) sea-level fluctuations [29],[30]. During the last glacial maximum (19 kyr ago) the sea-level dropped about 100-120 m in the Adriatic basin and the North Adriatic Sea was became a deep alluvial plain [31][32]. During the optimum climatic period, about 5500 years ago, the paleo-shoreline was located 25-30 km west of its actual position. From the most landward position of the Holocene coast, a rapid migration of a barrier-lagoon-estuarine system was developed to create the Ravenna alluvial plain, especially from Roman age (IV cent. B.C.) to recent times.

The sedimentation in the study area turned from marine to continental only recently (XIX century). Most of the study area emerged during the last shoreline progradational phase

[33], which took place from 1851 to 1894 A.D at an average rate of 20 m/y and in the first half of the past century (4-12 m/y). Nowadays, although the Holocene glacial period with related sea level changes is finished [34], the movements of the coastline still occur: a large part of the coast is under active erosion and only a small part is prograding to sea. This is mostly due to human intervention: the subsidence due to water and gas exploration in combination with a lack of sediments transport by the rivers and the construction of various piers perpendicular to the coast near the harbors and breakwaters parallel to the coast that trap the sediments. These interventions have interrupted the natural process of long shore current and sediment transport [34]. Now beach and dune erosion and shoreline retreat are a major problem along the Adriatic coast [35].

The sandy aquifer consists of two main units: a relatively thick medium-grained sand unit (from 0 to -10 masl) and a lower fine-grained sand unit of small thickness (from -21 to -26 masl). These two sand bodies are separated by a clay-silt and sand-silt continental unit (from -10 m to -21 masl) called *Prodelta* (Figure 11). Lastly, the Flandrian continental silty-clay basement is at a depth varying from -20 m.a.s.l. in the west to -30 m.a.s.l. at the present shoreline [29],[36],[37].

The resulting picture of the aquifer is a wedge shaped dune and beach sand body pinching out in a westerly direction. It is enclosed at the bottom and top by clays and peat deposited in lagoons, marshes and alluvial plains. The only portion where the aquifer is phreatic are the paleodunes and the actual dune belt both covered by pine forests. At the tip of this wedge there are gravel deposits (Figure 11) [29][38].

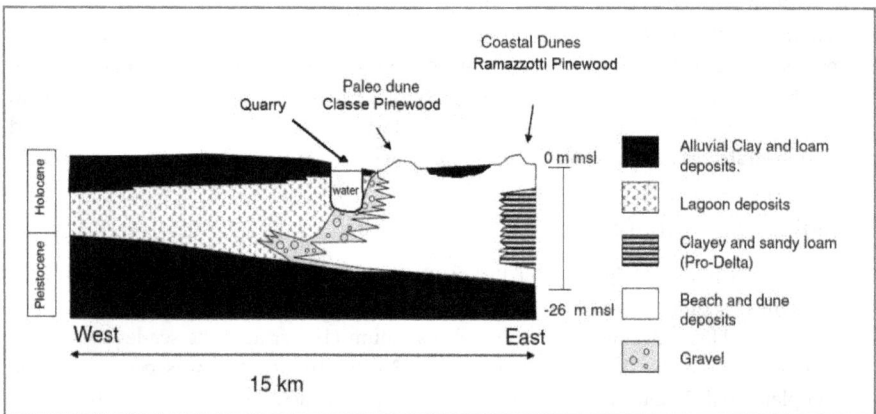

**Figure 11.** Lithological cross section of the Ravenna aquifer in the study area (modified from [29][39])

## 7. Monitoring

A groundwater monitoring campaign has been carried out in 8 wells along the section of Figure 1, in December 2010 and it was repeated in April 2011, with the aim to compare salinity and chemical data in two different seasons. A monthly water table monitoring was performed for the whole transect with increased data density around the ditch, where we installed two shallow wells in February 2011. Overall, 8 wells near the ditch were monitored for water table analysis (Figure 15).

Rainfall was measured with 5 pluviometers located along the section. In the eastern parcel, within the tomato field, we were able to measure both rainfall and irrigation. To assess how much rainfall and irrigation from the surface reaches the groundwater, we buried in the same location a drain-gauge. Geo-electric vertical resistivity surveys (VES) in cross sections normal to the ditch (Figure 12) were carried out in order to determine the depth, shape, and evolution of the fresh water lens [40],[41],[42]. VES surveys were repeated twice: before the tomatoes were planted in February 2011 and during the maximum irrigation period in July 2011. The VES calibration was made using well salinity data, and apparent resistivity data obtained for the area by the process of inversion [43][44].

Groundwater with a salinity ranging from 0 to 3 g/l has been defined as freshwater, based on a study of the relationships between salinity and natural vegetation species richness [7]. Above the salinity threshold of 3 g/l, the species richness decreases dramatically. The brackish water has been defined as water with a salinity ranging from 3 to 15g/l and water with salinity above 15 g/l is considered saline.

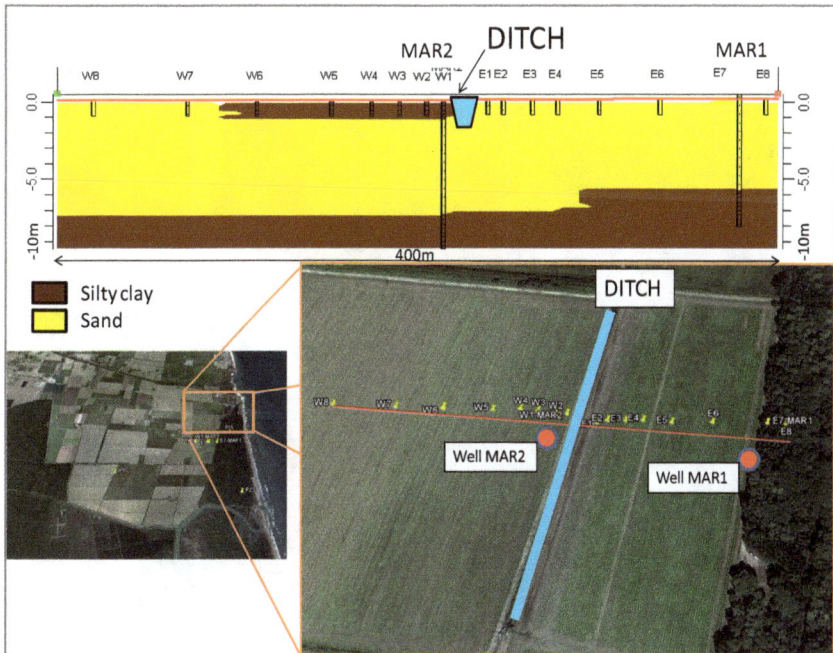

**Figure 12.** Small parcel, studied in detail, with simplified lithology, irrigation ditch (light blue), wells MAR1 and MAR2 (red circles), and 16 V.E.S. locations.

The salinity data show that the size of the freshwater lenses changes throughout the year (Figure 13). Below the coastal pine forest and close to sea, the aquifer is salty and there is a 0.5 m thick fresh-brackish water layer at the top of the aquifer, which disappears in summer. Salinity data indicate a counter-intuitive trend underneath the irrigated agricultural land at 2 km from the sea: in summer the salinity is lower than in the winter. This may be explained by the high irrigation rates during the growing season or by infiltration from the irrigation ditch. Inside the historical *Classe* pine forest, at 4 km from the coast, the groundwater is mostly brackish (10 g/l) with a 1 m thick freshwater layer at the top of the aquifer that, in the coastal pinewood, vanishes rapidly. Fresh groundwater lenses within the quarry belt at 7 km from sea have a larger thickness (2 m) but at the bottom of the aquifer saltwater is also present. At a distance of 10 km from the coast, in agricultural land, there is fresh water throughout the whole thickness of the aquifer (Figure 13).

On the basis of the cations and anions analyses, alkalinity, and salinity, a groundwater type classification used by Stuyfzand [45] was obtained [11]. Seawater is typically of the S4NaClo type meaning a saline water with High Alkalinity, Na being the major cation, Cl the major anion and a zero BEX indicating a stable water without cation exchange taking place (for data see [11]).

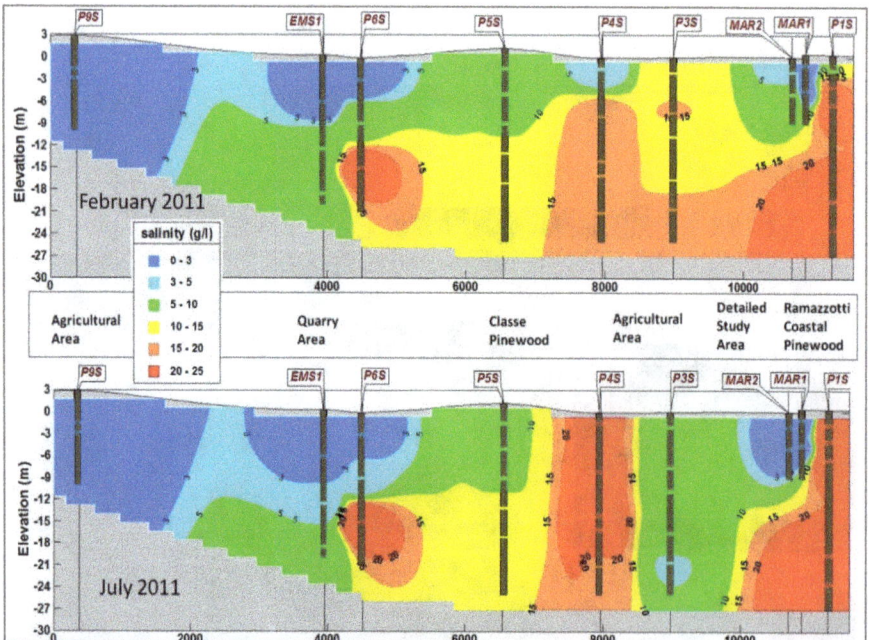

**Figure 13.** Transect 10 km long from the Adriatic coast on the right towards the mountains on the left. Warm colors represent brackish-saline water whereas cold colors are for freshwater.

Apennine river water is typically of the type (g-F)3CaHCO₃ and (f-F)3CaMix, or in other words: fresh water, medium alkalinity and rich in Calcium and varying anions. Po river water coming through the CER is of the type g2CaHCO₃ and slightly less alkaline than Apennine river water [11]. The groundwater water type is diagnostic to evaluate if the river-irrigation water has replaced the saline groundwater. Figure 14 shows that the sodium-chloride water type is dominant in the whole section. It is important to note that there are differences between the winter and spring situations in the area studied in detail. This was caused by the addition of new wells in the second monitoring campaign (MAR1 and MAR2 were monitored in April 2011 only). Excluding this effect, two water type changes were observed in the western part of the transect: from calcium-carbonate in December 2010 to calcium-mix in April 2011 in well P9S and from magnesium to sodium in well EMS. The April water type for the area near the irrigation ditch shows, in MAR1 well, a calcium carbonate water type, which is the same chemical composition of river water. Otherwise, in well MAR2 despite its low salinity (Figure 13), the dominant water type is sodium-chloride (Figure 14).

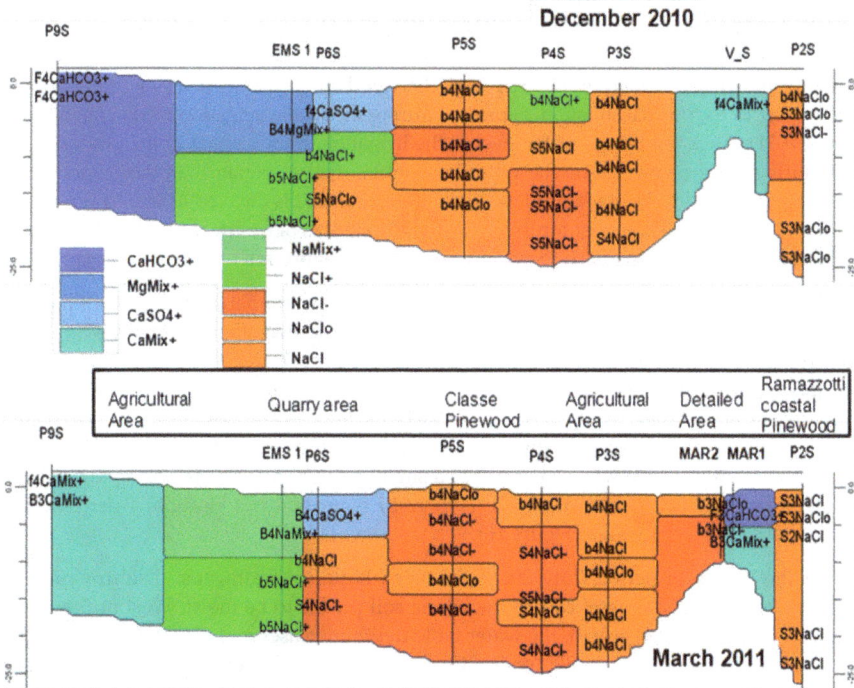

**Figure 14.** 10-km-long transect from the Adriatic coast (on the right) towards the mountains (on the left). The different water types are indicated with different colors. Warm colors represent brackish-saline Na-Cl composition whereas cold colors are for typical freshwater composition.

The BEX index for dolomite-rich aquifers as reported by Stuyfzand [45] is used to study whether the groundwater is becoming more or less saline. The BEX is indicated in Figure 15 with a +, - or 0 character after the water type code. The groundwater samples at the top of the aquifer and in winter show a positive BEX index, which indicates freshening (Figure 14).

Rainfall from January to September 2011 recorded in the five pluviometers along the section has a fairly homogeneous value around 400 mm. Total irrigation of a tomato field in the area studied in detail was 600 mm, applied with a rate of 120 mm per month (April-August). The infiltration rate of water into the aquifer, measured from drain gauges located in the tomato field, was 8 mm/m² in May, 2 mm/m² in June and 0 in the other months (Table 1).

| 2011 (unit mm) | Months | January | February | March | April | May | June | July | August | September |
|---|---|---|---|---|---|---|---|---|---|---|
| | Rainfall | 33 | 38 | 96 | 15 | 31 | 40 | 43 | 0 | 27 |
| | Irrgation | 0 | 0 | 0 | 150 | 120 | 150 | 180 | 90 | 0 |
| | Infiltration | n.a. | n.a. | n.a. | 8 | 2 | 0 | 0 | 0 | 0 |

**Table 1.** Precipitations, irrigation and infiltration measured in tomato parcel from January to September 2011

Water table measurements in April 2011 show that the groundwater is at 0.2 m below sea level in the *Ramazzotti* pinewood and, moving landward, the depth linearly increases up to 1.2 m below sea level (Figure 15). The water level decreases rapidly everywhere during the summer, except for the area surrounding the irrigation channel (Figure 15) where an inverse gradient, with respect to the normal situation, is present during the whole irrigation time (from May to August 2011). As in winter the water level in the irrigation ditch is equal to the groundwater level, during the growing season the water level in the ditch is higher than the groundwater level, creating a hydraulic gradient in the phreatic aquifer that is directed to sea. At the end of the irrigation season (September for the year 2011), the inputs of freshwater end and the water level in the ditch returns to match the groundwater level (Figure 15).

The February 2011 survey results suggest that below the eastern parcel, where the aquifer is phreatic, a 4-m-thick freshwater lens is present before the irrigation season; freshwater is also present west of the ditch but its thickness is only 1,5 m. During both field campaigns in February and July 2011, VES profiles and the depth of the water table and the salinity of the groundwater at the water table were measured by drilling auger holes. The measurements made in February (Figure 16) show that the water level in the ditch is the same of the groundwater and that the whole system has a small gradient towards the pumping station. The freshwater lens vanishes below the *Ramazzotti* pinewood (E8 location) where the the water is salty up to the surface (Figure 16).

During the irrigation period (July 2011), the ditch is full of freshwater. The irrigation of tomatoes in the eastern parcel causes the whole soil profile to be moist. West of the ditch, where the Alfa Alfa was grown, the water table depth increases and also the bottom of the dry soil line becomes deeper (Figure 16). Comparing the VES profilqes made in February 2011 with those made in July 2011 confirms the existence of a well-developed freshwater lens in the eastern part that grows in size during the irrigation season (Figure 16). The largest growth in thickness of the freshwater lens occurs under the ditch and in the western part where the semi-confined aquifer is present.

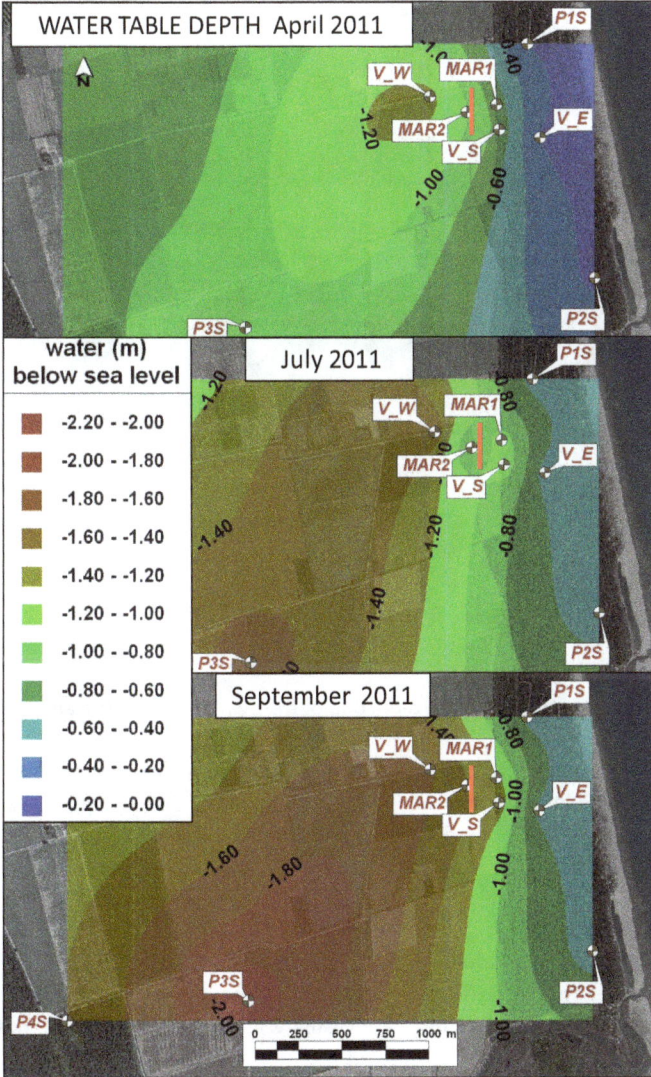

**Figure 15.** Study area water table map for April, July, and September showing also the irrigation ditch (red line).

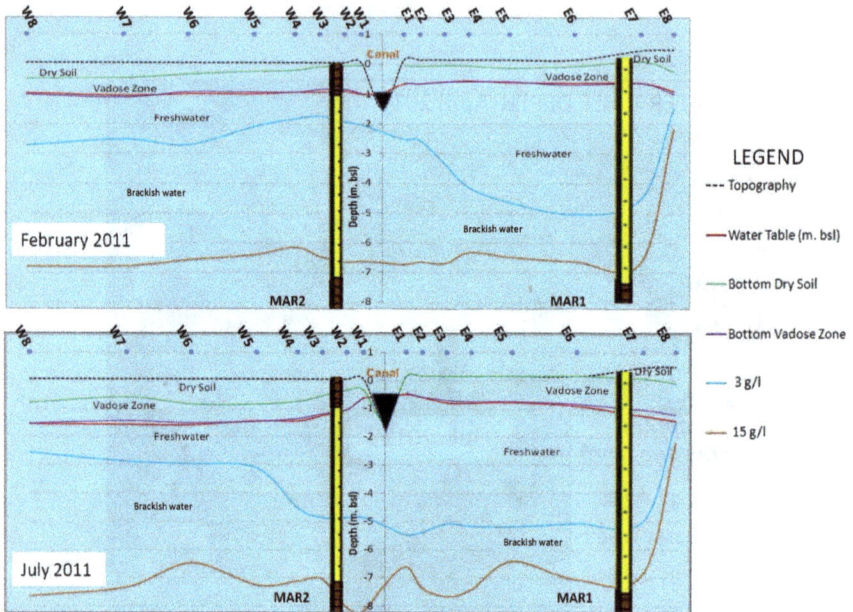

**Figure 16.** Seasonal comparison of VES results for the agricultural land close to the freshwater storage ditch.

## 8. Discussion

As recently reported by IPCC [22] a reduction of freshwater availability and an increase of drought periods length is likely to occur in the Ravenna area during the next 100 years. An increase in temperature with a simultaneous increase of evaporation will cause soil salinization and loss of production. Near Ravenna more than 68 square kilometers are at risk for soil salinization caused by salt water intrusion in the coastal aquifer and along main rivers and channel. A stable isotope analysis shows that the ground and surface water near the city of Ravenna is a mix of sea, rain, and river water. The sea-water is, in part, inherited from the Holocene deepest marine ingression [11]. The aquifer has the shape of a sandy wedge enclosed in clay and this inhibits an efficient flushing of the saline water that is in part left at the bottom of the aquifer from the optimum climate period. Deeper aquifers in the area receive underground recharge from water with a mountain or rainfall sources. Where present, freshwater is located in the recent and paleo-dune belts, because of their high infiltration potential [3],[22],[47]. Where channels and rivers flow on a sandy top layer, there is infiltration of freshwater into the aquifer as along the *Bevano* and *Fiumi Uniti* rivers as well as in the western part of the San Vitale pinewood [11]. There is a small area along the *Bevano* River, where the freshwater lens reaches the bottom of the aquifer but in all other cases the freshwater lenses are floating on top of brackish groundwater. These freshwater lenses are thin (1m maximum), ephemeral (present only during winter) and strongly linked to the climate conditions [3],[22],[48],[49],[50].

Aquifer salinization in our study area is due to the combination of several factors such as low topography, natural and anthropogenic subsidence, mechanical drainage, destruction of the coastal dune belt, and scarcity of freshwater infiltration [11][43][51][52]). As shown in Figure 8, in the whole aquifer, the groundwater head is below sea level in winter also. This situation causes a sea-water encroachment promoted also by the high discharge water volumes of the pumping stations. Based on number of activity hours, we roughly calculated that in the *Quinto* Basin the volume of drainage water was 8 times larger than that of the total rainfall in same basin for the period.

In this contest, with low freshwater availability and poor natural recharge, it is important to look for some technique that is able to counteract soil and groundwater salinization and at the same time encourages water management practices that prevent nitrate leaching, safeguard the economics of agriculture, and save the freshwater resources.

Our attention was focused on a small agricultural area at a distance of 500m from the shoreline, where despite all previous studies [22][51][52][53], the presence of a freshwater lens throughout the year, was found. Although this area is affected by drainage, low topography, and subsidence as the rest of the Ravenna province, the presence of a sandy top layer, the horticulture with irrigation, and a ditch used for freshwater storage, all promoted aquifer recharge. The chemical analysis and electrical surveys showed that the size of the freshwater lens is larger during the summer period (Figure 13, Figure 14 and Figure 16).

In many other countries worldwide, managed aquifer recharge (MAR) is used to store fresh river or rain water in the underground of riverbanks or dunes for later use as drinking water [54].

Managed Aquifer Recharge or MAR is also used for re-pressurization of aquifers subject to falling water levels, declining yields, saline intrusion or land subsidence. Examples of these from around the world include different technologies such as aquifer storage and recovery (ASR), aquifer storage transfer and recovery (ASTR), bank and dune filtration, infiltration ponds, percolation tanks, rainwater harvesting , and soil-aquifer treatment (SAT) [55]-[61]. All above mentioned technologies have a cost that sometimes may be not sustainable by the local economy. In countries such as India, China, and Pakistan artificial recharge from ditches is affordable for the cultivation of rice, but this is usually done far from the coast ([61][62][63][64]). As reported by Gale 2000 ([61]), in many countries, the seepage from ditches or channels is considered a loss of water and they are studying systems to solve this problem. Irrigation ditches, on the other hand, may also play an ecological role: some authors report that ditches support the biodiversity and associated ecosystem services ([65]) and that they promote retention and removal of agrochemicals, decrease soil erosion, and allow the early stage of development for fish, amphibians, and reptiles [66]. Another aspect to consider is the low cost for their construction and management (removal of excess vegetation to restore efficient drainage) [66]. In particular for our area a dense network of ditches is already present and their building costs could be very low or absent.

Irrigation practices tend to become more and more efficient by letting go less and less water into soil-evaporation. The driving forces for that are the scarcity of freshwater and economic incentives. One might say that the water left in the ditch to recharge the aquifer is water that is 'lost'. One of the main challenges in water resources and agricultural management in coastal Mediterranean zones will be to define priorities that may vary with the particular area. In one place it may be important to use as little water as possible for irrigation not leaving anything to waste or end up in the aquifer underneath the top soil. In other places instead, especially in the coastal zone, irrigation water may be left to seep from the ditches into the surface aquifer to create freshwater lenses that sustain the agricultural land as well as the natural areas and at the same time counteract seawater intrusion.

The high salinity of the coastal aquifer near Ravenna is in part inherited from its geological history. Therefore it is unlikely that the aquifer will be ever used as a source for drinking water. By infiltrating (excess) river water on purpose through ditches as the one studied in this paper, however, freshwater lenses can be created that may prevent soil salinization and serve agricultural crops as well as natural areas. Our studied ditch has the same dimension as the drainage channels in the area. The only difference is that it is a closed basin used for storage irrigation water during the summer. By using the existing hydraulic infrastructure for managed aquifer recharge in ditches, more freshwater would flow into the aquifer instead of being drained away by the pumping machines. This system of aquifer artificial recharge would be viable in similar coastal settings throughout the world.

## 9. Conclusions

Despite the fact that there are many technological, agricultural, economic, and water availability factors that are dependent on the local situation and farming practices (crop rotations, volume and timing of irrigation, tillage, economic forcing, etc.) and that the length of the study was of only one year, so that no long term trends can be captured by the analysis presented, some interesting observations have been made:

- The land use in the Ravenna coastal area is complex; natural areas (pine forests, lagoons and wetlands) and industrial activities are placed in a landscape dominated by agriculture.
- The Low topography, the high rates of subsidence, the salty phreatic aquifer require land drainage to allow for any activity. The result is that drainage allows for agriculture

but it decreases the recharge of the aquifer, removes freshwater from the top of the aquifer and increases up-welling of saline water from below.

- The surface hydrology is controlled by a high density network of channels used mainly for drainage but locally for irrigation.
- Due to different geological events, different kinds of sediments were deposited in the area; therefore special soils were developed and crops distribution as well as irrigation water requirements depends strictly from the soil characteristics.
- Irrigation is essential for horticultural crops, which are mainly located on sandy soil where the connection with the aquifer below is guaranteed.
- The water sprayed on the land for irrigation does not infiltrate into the aquifer and this is a good practice to avoid nitrate leaching and contamination of the groundwater. In the area studied in detail, the only irrigation water added to the aquifer is from seepage from the storage ditch.
- The case study shows that the presence of a ditch, full of freshwater for 5 months during the year, generates a permanent freshwater bubble 4 m in thickness, at 500 m distance from the sea. This shows that combined irrigation and drainage management, considering the dense network of existing channels and different types of soil, can be of benefit for the agriculture for soil protection, storage of freshwater resources, and to counteract sea-water intrusion.

## Author details

Nicolas Greggio, Pauline Mollema, Marco Antonellini and Giovanni Gabbianelli
*Interdepartmental Centre for Environmental Science Research (C.I.R.S.A.),*
*University of Bologna, Ravenna, Italy*

## Acknowledgement

We thank the farmer of the *"Cooperativa Braccianti di Campiano"* for their helpfulness and kindness; we thank also University of Bologna, the *Autorità dei Bacini Romagnoli* and the Municipality of Ravenna for funding the several projects that form the basis of this chapter. Special thanks to Murugan Ramasamy and Sathish Sadhasivam for their fundamental support in field work.

## 10. References

[1] Toth G., Montanarella L., Rusco E., 2008 Updated Map Of Salt Affected Soils In The European Union. Threats To Soil Quality In Europe, Office For Official Publications Of The European Communities, Luxembourg (2008), Pp. 61–74

[2] Louw, P.G.B., De, Eeman, S., Siemon, B., Voortman, B.R., Gunnink, J., Baaren, E.S., Van And G.H.P. Oude Essink. 2011. Shallow Rainwater Lenses In Deltaic Areas With Saline Seepage, Hydrol. Earth Syst. Sci. Discuss., 8, 7657-7707.

[3]  Antonellini, M., Mollema, P., Giambastiani, B., Banzola, E., Bishop, K., Caruso, L., Minchio, A., Pellegrini L., Sabia M., Ulazzi E., And Gabbianelli G., 2008. Salt Water Intrusion In The Coastal Aquifer Of The Southern Po-Plain, Italy. Hydrogeology Journal 16,1541-1556. Doi 10.1007/S10040-008-0319-9.

[4]  Custodio, E., 2010. Coastal Aquifers Of Europe: An Overview. Hydrogeology Journal, 18(1): 269 - 280 DOI: 10.1007/S10040-009-0496-1

[5]  FAO 2002 Agricultural Drainage Water Management In Arid And Semi-Arid Areas Ftp://Ftp.Fao.Org/Docrep/Fao/005/Y4263e/Y4263e11.Pdf

[6]  Maas E.V., Hoffman G.J. 1977 Crop Salt Tolerance, Current Assessment J. Irrig. Drain. Div., 103, Pp. 115–134

[7]  Antonellini, M. And Mollema, P,. 2010. Impact Of Groundwater Salinity On Vegetation Species Richness In The Coastal Pine Forests And Wetlands Of Ravenna, Italy. Ecological Engineering. 236 (9), 1201-1211, Doi:10.1016/J.Ecoleng.2009.12. 007.

[8]  Hutchinson, I., 1991. Salinity Tolerances Of Plants Of Estuarine Wetlands And Associated Uplands. Washington State Department Of Ecology Report, Washington, USA.

[9]  IPWEA, 2001. The Local Government Salinity Management Handbook. Institute Of Public Works Engineering Australia (IPWEA), Australia.

[10] Lombardi, T., Lupi, B., 2006. Effect Of Salinity On The Germination And Growth Of Hordeum Secalinum Schreber (Poaceae) In Relation To The Seeds After Ripening Time. Atti Della Società Toscana Di Scienze Naturali, Memorie, Serie B 113, 37–42.

[11] Mollema P, Antonellini A, Dinelli E, Gabbianelli G, Greggio, N, Stuyfzand P J. 2012 Identification Of Fresh And Salt Water Intrusion Types In The Upper Ravenna Coastal Aquifer (Italy) Via Hydrochemical Systems Analysis. Submitted to Applied Geochemistry.

[12] Istat 2000 5° Censimento Generale Dell'agricoltura http://www.census.istat.it

[13] Provincia Di Ravenna 2006 Piano Territoriale Di Coordinamento Provinciale

[14] Marconi V., Antonellini M., Balugani E., Dinelli E.. 2011 Hydrogeochemical Characterization Of Small Coastal Wetlands And Forests In The Southern Po Plain (Northern Italy). Ecohydrol. Vol.4, 597-607

[15] Mollema, P., Antonellini, M., Gabbianelli, G., Laghi, M., Marconi, V., Minchio, A., 2012. Climate And Water Budget Change Of A Mediterranean Coastal Watershed, Ravenna, Italy. Environmental Earth Sciences. 65:257–276 Doi: 10.1007/S12665-011-1088-7.

[16] Fao. 1990. Guidelines For Soil Profile Description. Third Edition (Revised). Soil Resources, Management And Conservation Service, Land And Water Development Division, Fao, Rome.

[17] Soil Survey Staff. 1996. Keys To Soil Taxonomy. Seventh Edition. United States Department Of Agriculture, Washington D.C.

[18] Rose D. A., Konukcu F. & Gowing J. W. 2005 Effect Of Water table Depth On Evaporation And Salt Accumulation From Saline Groundwater. Aust. J. Soil Res. 43, 565-573.

[19] Il'ichev, Andrej T. And Tsypkin, George G. And Pritchard, David And Richardson, Chris N. (2008) Instability Of The Salinity Profile During The Evaporation Of Saline Groundwater. Journal Of Fluid Mechanics, 614. Pp. 87-104. Issn 0022-1120

[20] Filippi N (1994) Carta Pedologica (Soil Map) 1: 250.000. Stampata Dalla Regione Emilia Romagna, Italy

[21] Corticelli S. 2003 Carta Del Uso Del Suolo (Land Use Map). Copertura Vettoriale. Stampata Dalla Regione Emilia Romagna, Italy

[22] Giambastiani, B. M.S., Antonellini, M., Oude Essink, G.H.P., Stuurman, R. J. 2007 Salt Water Intrusion In The Unconfined Coastal Aquifer Of Ravenna (Italy): A Numerical Model. Journal Of Hydrology 340, 94-104.

[23] Giambastiani, B.M.S.,Mollema, P.N., Antonellini, M., 2008. Groundwater Management In The Northern Adriatic Coast (Ravenna, Italy): New Strategies To Protect The Coastal Aquifer From Saltwater Intrusion. In: Groundwater: Modelling, Management And Contamination. Authors: König L.F. And Weiss J.L.; Nova Publishers

[24] Consorzio Di Bonifica Della Romagna 2002 Il Comprensorio E Le Infrastrutture http://www.bonificaromagna.it/Cartografia.ìt/html

[25] Regione Emilia Romagna, Agip, 2009. Riserve Idriche Sotterranee Della Regione Emilia Romagna, Technical Report. Regione Emilia Romagna, Firenze

[26] Giorgi, F., Lionello, P. 2008. Climate Change Projections For The Mediterranean Region. Global Planet Change 63,90–104.

[27] Antolini G., Marletto V., Tomei F., Pavan V., Tomozeiu R. 2008 Atlante Idroclimatico Dell'emilia-Romagna 1961-2008

[28] Magnan, A., Garnaud, B., Billé, R., Gemenne, F., Iddri, Hallegatte, S., Cired-Météo, 2009. The Future Of The Mediterrenean. From Impacts Of Climate Change To Adaptation Issues. Institut Du Développement Durable Et Des Relations Internationals, Paris, France, 43 Pp.

[29] Amorosi A., Colalongo M.L., Pasini G. & Preti D., 1999. Sedimentary Response To Late Quaternary Sea-Level Changes In The Romagna Coastal Plain (Northern Italy). Sedimentology, 46, 99-121.

[30] Catalano, R., Di Stefano, E., Infuso, S., Sulli, A., Vail, P.R., Vitale, F.P. 1995 Sequence And Systems Tracts Calibrations On High Resolution Bio-Chronostratigraphic Scheme: The Central Mediterranean Plio-Pleistocene Record. In: Mesozoic-Cenozoic Sequence Stratigraphy Of Western European Basins (Ed. By P.C. De Graciansky Et Al.). Spec. Publ. Soc. Econ. Paleont. Miner

[31] Rizzini, A., 1974, Holocene Sedimentary Cycle And Heavy Mineral Distribution, Romagna–Marche Coastal Plain, Italy: Sedimentary Geology, V. 11, P. 17–37.

[32] Van Straaten, L.M.J.U. 1970 Holocene And Late Pleistocene Sedimentation In The Adriatic Sea. Geo/. Runsch. 60, 106-131.

[33] Giambastiani B.M.S. 2003. Geomorfologia Del Litorale Ravennate: I Sistemi Dunosi. Unpublished Msc Thesis: University Of Bologna

[34] Stecchi, F., Minchio, A., Del Grande, C.. And Gabbianelli, G., 2003. Historical And Recent Evolution Of The Reno Rover Mouth And Adjacent Areas. In. Gabbianelli G. And Sangiorgi F.(Editors). Volume 4 Of Proceedings Of The Sixth International Conference On The Meditteranean Coastal Environment, Medcoast 03, 7-11 October 2003, Ravenna Italy, P. 55-67.

[35] Bondesan M, Castiglioni Gb, Elmi C, Gabbianelli G, Marocco R, Pirazzoli Pa, Tomasin A 1995b Coastal Areas At Risk From Storm Surges And Sea Level Rise In North-Eastern Italy. J Coast Res 11:1354–1379

[36] Amorosi A., Colalongo M.L., Fiorini F., Fusco F., Pasini G., Vaiani S.C. & Sarti G.,2004 – Palaeogeographic And Paleoclimatic Evolution Of The Po Plain From 150-Ky Core Records. Global And Planetary Change, 40, 55-78.

[37] Correggiari A, Roveri M, Trincardi F. 1996. Late-Pleistocene And Holocene Evolution Of The North Adriatic Sea. Late-Glacial And Early Holocene Climatic And Environmental Changes In Italy: Il Quaternario: Italian Journal Of Quaternary Sciences 9: 697–704.

[38] Amorosi A., Colalongo M.L., Fiorini F., Fusco F., Pasini G., Vaiani S.C. & Sarti G.,2004 – Palaeogeographic And Paleoclimatic Evolution Of The Po Plain From 150-Ky Core Records. Global And Planetary Change, 40, 55-78

[39] Marchesini, L., Amorosi, A., Cibin, U., Zuffa, G.G., Spadafora, E., Preti, D., 2000. Sand Composition And Sedimentary Evolution Of A Late Quaternary Depositional Sequence, Northwestern Adriatic Coast, Italy. Journal Of Sedimentary Research 70 (4), 829–838.

[40] Bulter D. K. 2005 Near-Surface Geophysics, (Seg Investigations In Geophysics Series; N°13), Library Of Congress; Pag. 46

[41] Burger H. Robert 1992, Exploration Geophysics Of The Shallow Subsurface. Prentice Hall Pagg. 291-295

[42] Celico P.. 1998 Prospezioni Idrogelogiche. Vol I, Liguori Editore, V° Ristampa 1998 Pag 243

[43] Capo D. 2011 Phd Thesis "Caratterizzazione Della Falda Freatica Costiera Ravennate" In Press

[44] Loke, M.H. 2001 Electrical Imaging Surveys For Environmental And Engineering Studies – A Practical Guide To 2-D And 3-D Surveys. Residinv Manual, Iris Instruments; 67 Pp.

[45] Stuyfzand, P.J., 1989. A New Hydrochemical Classification Of Watertypes. Iahs Publ. 182, 89–98.

[46] IPCC Technical Paper Vi. 2008 Climate Change And Water - Intergovernmental Panel On Climate Change http://www.ipcc.ch/pdf/Technical-Papers/Climate-Change-Water-En.Pdf

[47] Marconi V., Antonellini M., Balugani E., Dinelli E.. 2011 Hydrogeochemical Characterization Of Small Coastal Wetlands And Forests In The Southern Po Plain (Northern Italy). Ecohydrol. Vol.4, 597-607

[48] Capaccioni B., Didero M., Paletta C., Didero L. 2005 Saline Intrusion And Refreshening In A Multilayer Coastal Aquifer In The Catania Plain (Sicily, Southern Italy): Dynamics

Of Degradation Processes According To The Hydrochemical Characteristics Of Groundwater. Journal of Hydrol 307:1–16

[49] Oude Essink G.H.P., 2001. Improving Fresh Groundwater Supply-Problems And Solutions: Ocean And Coastal Management 44: 429-449.

[50] Chot Pp, Dekker Sc, Poot A. 2004. The Dynamic Form Of Rainwater Lenses In Drained Fens: Journal Of Hydrology 293: 74-84.

[51] Marconi V. Antonellini M. Laghi M, Minchio A., Savelli D.. 2008. A Water Table Fluctuations Model In Sandy Soil Below A Coastal Pine Forest. Geophysical Researches Abstract 10: Egu2010-05761

[52] Marconi V., PhD Thesis 2010 Hydrogeochemical Characterization Of Small Coastal Wetlands And Forests In The Southern Po Plain (Northern Italy).

[53] Marconi V. 2011. Effetti Della Vegetazione E Del Drenaggio, Sull'intrusione Salina Nell'acquifero Freatico Costiero Della Zona Compresa Fra Foce Dei Fiumi Uniti E Foce Bevano (Ravenna). Dottorato Di Ricerca In Scienze Ambientali: Tutela E Gestione Delle Risorse Naturali, Ciclo XXIII. Università Di Bologna

[54] Stuyfzand, P.J. 1999. Patterns In Groundwater Chemistry Resulting From Groundwater Flow. Hydrogeology Journal 7 P 15-27

[55] Böhlke Jk 2002 Groundwater Recharge And Agricultural Contamination Hydrogeology Journal (2002) 10:153–179 Doi 10.1007/S10040-001-0183-3

[56] Boonstra J., Bhutta M.N. 1996 Groundwater Recharge In Irrigated Agriculture: The Theory And Practice Of Inverse Modeling Journal Of Hydrology 174 357 374

[57] Bouwer, H. 1997.Role Of Groundwater Recharge And Water Reuse In Integrated Water Management. Arabian J. Sci. Eng., 22, 123–131

[58] Bouwer, H. 1999. Chapter 24: Artificial Recharge Of Groundwater: Systems, Design, And Management. Hydraulic Design Handbook, L. W. Mays, Ed., Mcgraw–Hill, New York, 24.1–24.44.

[59] Bouwer, H., Fox, P., Westerhoff, P., And Drewes, J. E. 1999 Integrating Water Management And Reuse: Causes For Concern? Water. Qual. Internat., 1999, 19–22

[60] Dillon P.. 2005 Future Management Of Aquifer Recharge Hydrogeol J. 13:313–316

[61] Gale I., 2000 Strategies For Managed Aquifer Recharge (Mar) In Semi-Arid Areas International Association Of Hydrogeologists Commission On Management Of Aquifer Recharge Iah - Mar International Hydrological Programme (Ihp) Unesco Division Of Water Sciences www.iah.org/recharge

[62] Meijer, K.S., 2000. Impacts Of Concrete Lining Of Irrigation Canals, Uda Walawe, Sri Lanka. M.Sc. Thesis, University Of Twente, The Netherlands.

[63] Meijer, K., Boelee, E., Augustijn, D., Van Der Molen, I., 2006. Impacts Of Concrete Lining Of Irrigation Canals On Availability Of Water For Domestic Use In Southern Sri Lanka. Agric. Water Manage. 83, 243–251.

[64] Liqiang Yao, Shaoyuan Feng, Xiaomin Mao, Zailin Huo, Shaozhong Kang, D.A. Barry. 2012 Coupled Effects Of Canal Lining And Multi-Layered Soil Structure On Canal Seepage And Soil Water Dynamics Journal Of Hydrology 430-431, 91–102

[65] Bradbury R.B., Kirby W.B., 2006 Farmland Birds And Resource Protection In The Uk: Cross-Cutting Solutions For Multi-Functional Farming? Biological Conservation, 129, Pp. 530–542

[66] Herzon I., Helenius J. 2008 Agricultural Drainage Ditches, Their Biological Importance And Functioning Biological Conservation 141, 1171–1183

# Irrigation Delivery Performance and Environmental Externalities from a Risk Assessment and Management Perspective

Daniele Zaccaria and Giuseppe Passarella

Additional information is available at the end of the chapter

## 1. Introduction

As a whole the Mediterranean region holds 3 % of the world's freshwater resources and hosts more than 50 % of the "water poor" population, i.e. people with less than 1000 m³ per capita per year. In the Mediterranean countries access to water and irrigation is crucial for land productivity and stability of agricultural yields (Benoit and Comeau 2005). But the balance between water demand and availability in irrigated areas is reaching critical levels (EEA 2012) in parts of the Mediterranean region and is an increasingly difficult task to achieve, both in spatial and temporal terms. Fresh water supplies are in fact mostly limited and the national strategies of many countries are no longer addressed towards developing new water sources and storage infrastructures. On the other hand, water demand is progressively rising up, mainly due to population increase and to policies of agricultural development and farming intensification for food security goals. The European Environment Agency (EEA 2010) reported that climate change is likely to increase the current pressures on water resources and that increasingly much of the Mediterranean countries will face reduced water availability during summer months, while the frequency and intensity of drought is projected to increase in the southern areas.

The recurrent drought periods occurring under Mediterranean climatic conditions thus represent the major water scarcity issue for irrigated agriculture but, besides that, poor irrigation management and inappropriate delivery schedules are often the problems (Clemmens 2006; Hargreaves and Zaccaria 2007). Clemmens and Molden (2007) stressed the importance of flexibility and quality of delivery service on the economic and environmental viability of irrigation projects. Merriam and Freeman (2002) documented that accurate on-farm control of irrigation water deliveries can contribute to reducing drainage and salinity problems on the project scale caused by excess, inadequate and non-uniform applications. Styles (1997) reported that in several areas of the world a significant increase in the number of farmers using irrigation wells has been observed during the last decades, even where less expensive irrigation water was available from the district, in response to the lack of flexible deliveries from the distribution networks. As pointed out by Umali (1993), poor water management by irrigation agencies is one of the leading grounds for irrigation-induced salinity in many agricultural areas. As a matter of fact, salinity problems in irrigated agriculture may often result from seawater intrusion into coastal areas where the water tables have been lowered due to mining of groundwater for irrigation purposes (Kijne et al. 1998). Zaccaria and Scimone (2008) refer that often times, when water distribution by the management authority is unreliable, inadequate in terms of delivery conditions, rigid or not timely matching crop water demand or growers' needs and practices, farmers tend to rely on aquifers as main water source for irrigation.

Sanaee-Jahromi et al. (2001) clarified that the delivery schedule performance relates to how well the water delivery schedule matches the crop irrigation requirements, whereas the operation performance refers to the ability of the system to supply water according to the schedule.

As for soil and aquifer degradation, Paniconi et al. (2001) and Capaccionia et al. (2005) pointed out that in coastal areas periods of intensive groundwater pumping for irrigation purposes can cause a drawdown of water levels in aquifers and give way to seawater intrusion, often leading to salt build-up in the cropped soils.

The present study was conducted on the Sinistra Bradano irrigation system managed by a local Water Users Association (WUA) to supply an irrigated agricultural area located in the western part of the province of Taranto (Apulia region, southern Italy) that stretches along the Ionian coast. Large reductions in the area serviced by the irrigation delivery networks operated by the WUA, and strong increases in the area irrigated by growers through groundwater pumping from farm tube wells occurred during the last 10 years, as documented by Zaccaria et al. (2010) on the basis of records provided by the WUA and by INEA (1999).

Under the perspective of responsible use of natural resources, a simplified Risk Assessment and Management procedure (RA&M) was thus applied to the study area for quantifying the risks of soils and aquifer degradation. Some feasible management options were also appraised for risk mitigation purposes on the basis of specific decision-making criteria.

## 2. Study area description

The "Sinistra Bradano" irrigation scheme (Fig. 1) covers a total command area of 9,651 ha and an irrigable area of 8,636 ha. This area was equipped for irrigation during the period from 1968 to 1974 and extends over an alluvial plain, with land elevation ranging between 24 and 54 m a.s.l. The irrigation system was designed for surface irrigation methods and is subdivided into 10 operational districts, each being composed by sub-units called sectors that consist of a grouped number of farms. The system is managed by a local association of water users, namely the "Consorzio di bonifica Stornara e Tara" that distributes irrigation water to horticultural growers from mid April to late October by rotation delivery schedule. The rotation is fixed for the entire irrigation season with a flow rate of 20 l s$^{-1}$ ha$^{-1}$, 5 hours of delivery duration to each user, and a delivery interval of 10 days.

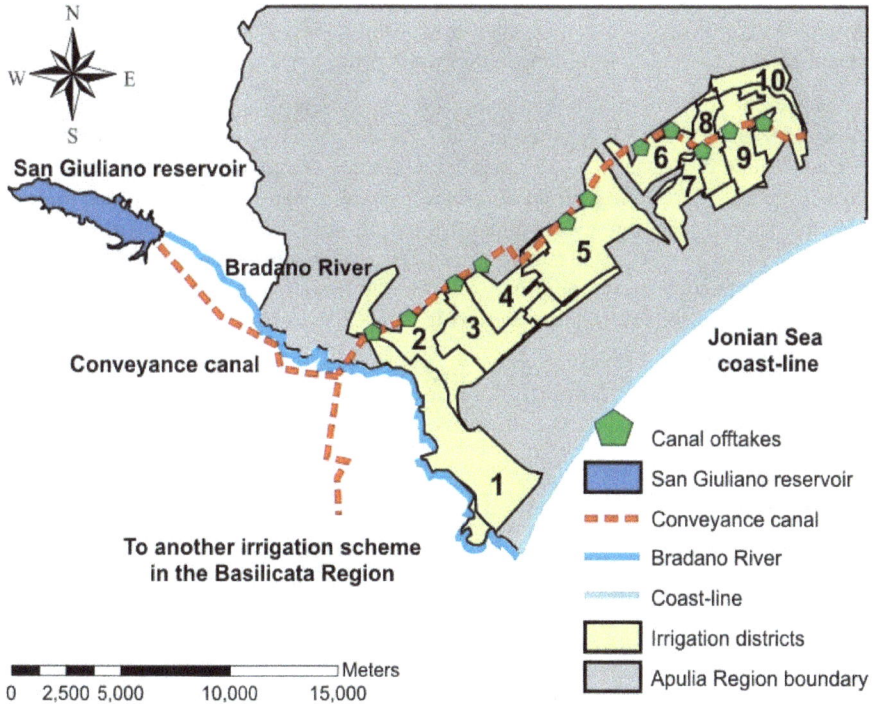

**Figure 1.** Overview of the Sinistra Bradano irrigation system.

The main source is the Bradano River, whose water gets partially diverted and stored in the "San Giuliano" reservoir of a total capacity of 70 Mm$^3$, which is located in in the nearby region of Basilicata. Water is then conveyed from the San Giuliano reservoir to the study area by a main canal along which 10 open-branched district distribution networks originate that divert water to the district distribution networks. Water diversion from the main canal occurs through cross-regulators and undershot gates, which are manually operated by the WUA's staff on a regular basis for implementing the planned delivery schedule. Water is finally distributed to users through gravity-fed branched delivery networks consisting of buried pipelines, and pressure at farm hydrants ranges between 0.3 and 0.6 bars depending on their ground elevation relative to the canal off-takes, thus resulting from the difference in elevation between the inlets of the distribution networks and the lower-elevation irrigated areas.

Climate of the area is semi-arid to sub-humid and referred to as "Maritime-Mediterranean", which is typical of the coastal areas of the Mediterranean region. Precipitation ranges between a minimum of 400 mm, in south-eastern part of the scheme, and a maximum of 730 mm in the northern part of the scheme. The average yearly rainfall is around 550 mm, 35 % of which occurring during the winter months, 32 % during fall and 33 % during spring and summer. There is typically very little summer precipitation, thus summer droughts are frequent, and irrigation is usually needed from April to September. Because of semi-arid climatic conditions, profitable farming in the area depends largely on irrigation.

The main crops grown in the area are citrus, table grapes, olive trees and summer vegetables, whose relative distribution is reported in Table 1 and Figure 2 as referred to the year 2006. Soils are mainly of alluvial type, resulting from deposits onto flat clayey plains that were afterwards subjected to a long period of carbonate leaching. For the purposes of the present study the cropped soils were grouped into five classes, according to the USDA soil textural classification, as shown in the soil map reported in Fig. 3, with most of the cropped areas being on loamy-sand. The electrical conductivity (EC) of soils, measured during a survey campaign in 2006, resulted in a range of values between 0.064 and 0.635 dS m$^{-1}$.

| CROP | AREA (Ha) | Area (%) |
|---|---|---|
| Table-grapes | 3,753 | 43.5 |
| Citrus | 2,208 | 25.6 |
| Vegetables | 2,184 | 25.3 |
| Olives | 432 | 5,0 |
| Almonds | 14 | 0.1 |
| Orchards | 44 | 0.5 |
| **TOTAL** | **8,635** | **100** |

**Table 1.** Cropping pattern and relative distribution in the Study Area

At farm level, micro-irrigation methods are currently used by growers in the majority of cropped areas, whereas sprinkler irrigation covers only 20% of the citrus acreage. Surface irrigation is no longer practiced due to high labour costs. In a few larger farms, small storage reservoirs were constructed by farmers with the aim of buffering the delivery timing and discharge to achieve higher flexibility in crop irrigation management.

As for the service area, even though the cropped area has not changed over the years, the area irrigated with water supplied by the WUA's networks progressively decreased since 1990 and onward, with no significant changes in the cropping distribution. Based on WUA's records reported in Table 2, the area requesting irrigation delivery service from the WUA passed from 2,128 ha in 1997 to only 921 ha in 2007, out of a total cropped and irrigable area of 8,636 ha.

Several farmers and extension officers from the study area were interviewed and reported that the irrigation delivery schedule enforced by the WUA is too restrictive with respect to the prevailing farming conditions, and not often timely to match the actual crop water requirements and farmers' irrigation needs (Zaccaria et al. 2006). The rigid rotation supply may in fact cause wasteful water use due to improper timing, over-irrigation and runoff, and may inhibit good farm management, as documented by some authors (e.g. Merriam et al. 2007).

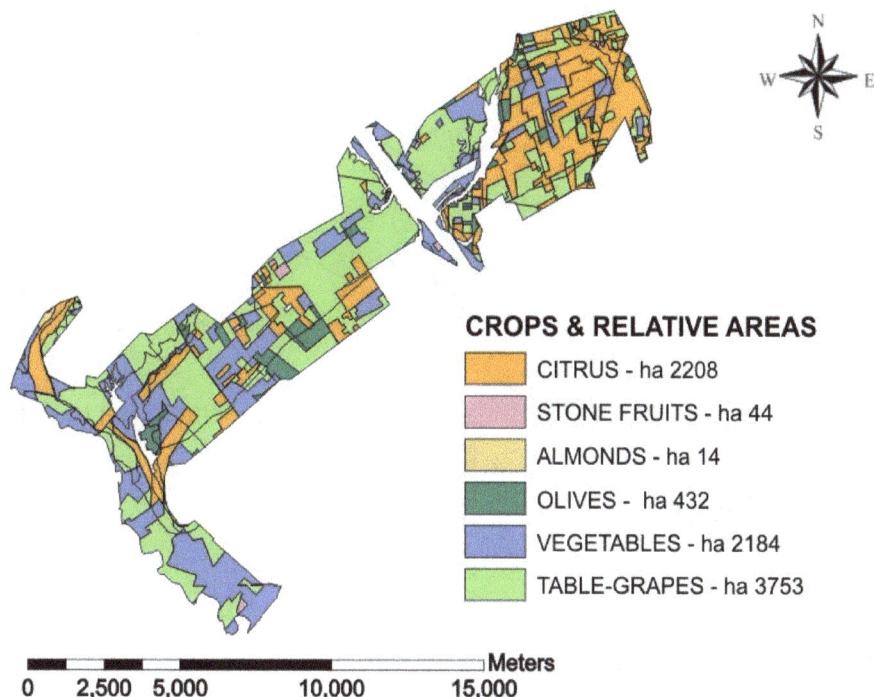

**CROPS & RELATIVE AREAS**

CITRUS - ha 2208

STONE FRUITS - ha 44

ALMONDS - ha 14

OLIVES - ha 432

VEGETABLES - ha 2184

TABLE-GRAPES - ha 3753

**Figure 2.** Cropping pattern of the Sinistra Bradano irrigation systems for 2006

| Year | 1997 | 1998 | 1999 | 2000 | 2001 | 2002 | 2003 | 2004 | 2005 | 2006 | 2007 |
|---|---|---|---|---|---|---|---|---|---|---|---|
| Area (ha) | 2,128 | 2,046 | 2,026 | 2,044 | 1,815 | - | 1,354 | 1,183 | 1,004 | 987 | 921 |
| Area (% of irrigable) | 24.6 | 23.7 | 23.4 | 23.7 | 21.0 | - | 15.7 | 13.7 | 11.6 | 11.4 | 10.7 |

**Table 2.** Areas serviced by the WUA in the years 1997-2007 in the Sinistra Bradano irrigation system (Source: Stornara e Tara Water Users Association, 2008)

The reduction in the area serviced by the WUA indicates that the area irrigated by groundwater pumping has tremendously increased over the years, most likely as a consequence of inadequate water delivery conditions with respect to the actual farmers' requirements. In other words, during the different years farmers irrigated larger areas exclusively relying on groundwater pumping, most likely for avoiding the limitations imposed by the rotation delivery schedule. Major changes, instead, occurred to the farm irrigation methods, as the majority of growers passed from surface methods to pressurized high-frequency irrigation. As reported by the extension agents and farmers' representatives interviewed, when the water supply is flexible and shows no delivery constraints (i.e. storage reservoirs, holding ponds or groundwater pumping), growers usually tend to distribute small amounts of water to cropped fields by means of micro-irrigation systems with high frequency, which also varies during the irrigation season in response to perceived crop water needs.

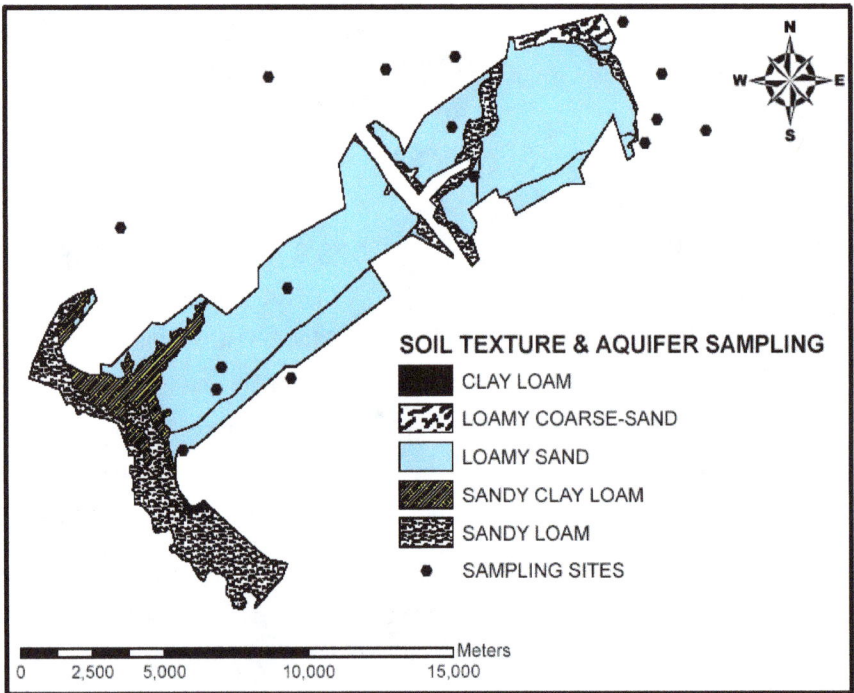

**Figure 3.** Soil map of the Sinistra Bradano irrigation system (textural classification according to the USDA soil classification), and sites of groundwater sampling conducted in 2006

According to extension service agents and growers' representatives, the majority of farmers consider the water distribution conducted by the WUA as not matching the actual crops' needs and farmers' requirements, both in terms of timing and of conditions of delivery. Delivery intervals, flow rates and pressure heads available at hydrants are found to be inadequate by farmers for the prevailing farming practices. As a result, during the last 10 years many growers relied nearly exclusively on groundwater pumping for irrigating their crops for large part of the irrigation season in order to achieve the desired flexibility.

As such, a concentration of groundwater pumping is found to occur during the peak water demand periods (July and August). This has progressively led to high antropogenic pressure on groundwater resources and has started originating aquifer contamination and soil degradation, namely due to seawater intrusion in the groundwater and salt build-up in the agricultural soils, which are considered as the major causes of environmental degradation in the study area.

Some research works conducted in areas bordering the system under study (Polemio and Ricchetti 1991; Polemio and Mitolo 1999; Polemio et al. 2002) revealed that seawater intrusion is progressively increasing in the whole Ionian coastal aquifer. A strong increase in the area subjected to seawater intrusion was also documented by Zaccaria et al. (2010) based on a comparison between two subsequent Regional Water Plans, namely the "Piano Regionale di Risanamento delle Acque" (Regione Puglia 1983) and the "Piano di Tutela delle Acque della Regione Puglia" (Regione Puglia 2007). This increase was found to be consistent with the strong increment in the number of agricultural wells drilled during the last decades throughout the whole area.

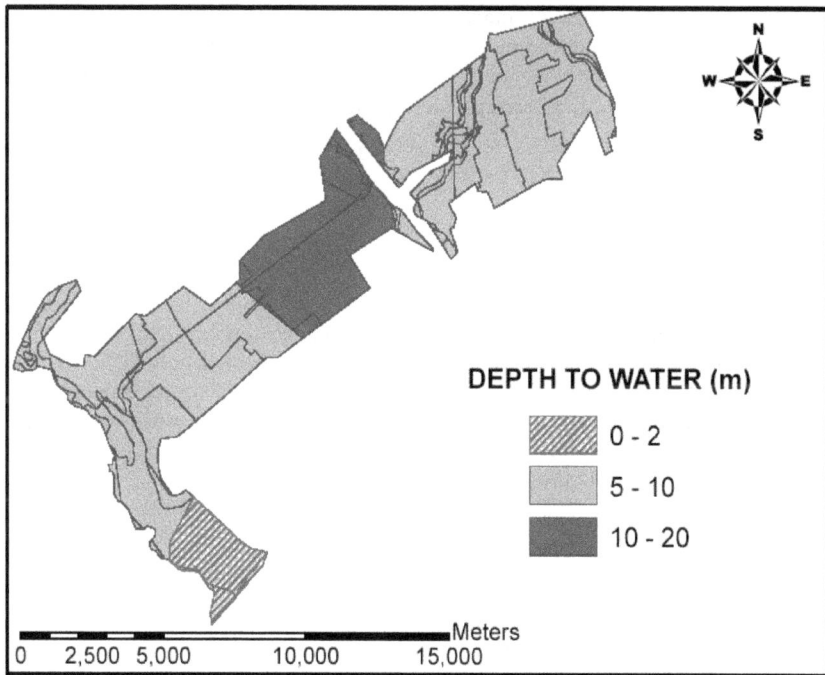

**Figure 4.** Depth-to-water map of the aquifer in the Sinistra Bradano area

The area under study is characterized by abundant groundwater resources coming from both a shallow upper unconfined aquifer and a deeper confined aquifer, whose hydrological set-up was described by Zaccaria et al. (2010) based on the outcomes of previous investigations (Cotecchia and Magri 1967; Cotecchia et al. 1971; Piccirillo 2000). According to Polemio et al. (2002) the shallow aquifer is subjected to heavy utilization and therefore to seawater intrusion. Observations of the water table depth were conducted in 2004 (Regione

Puglia 2007) and led to the development of the depth-to-water map (Fig. 4), which shows that the water table lies at depths ranging from 2 m (in the south-western part) to 20 m (in the north-eastern part) from the ground surface, confirming the easy access to the aquifer by farmers for irrigation purposes.

Seawater intrusion upon coastal groundwater was reported by Polemio et al. (2002) as a real problem for the social and economic development of this area, as results from the analysis of hydro-geological, chemical and physical data collected at boreholes in areas near the study site that revealed quality degradation of coastal plain groundwater, owing to seawater intrusion in the shallow aquifer. This evidence was also supported by data collected in the period 2006-2007 during a research project aiming at monitoring groundwater parameters at regional level (Regione Puglia 2006).

Detailed information on the operational procedures of the distribution networks, on the resulting effects on crop irrigation management by farmers, on the poor performance in water delivery, and on the impending need of system modernization were documented by previous research works and were all described in details by Zaccaria and Lamaddalena (2005), Zaccaria et al. (2010), and Zaccaria and Neale (2012).

## 3. Materials and methods

### 3.1. Soil water balance modeling

Simulations of daily soil water balance in the root zone were performed for forty-two unique crop-soil-climate combinations to compare the amounts of water applied, crop evapotranspiration, delivery schedule performance and the related yield impacts when irrigation is conducted under the current rotational delivery schedule (RDS) or if an alternative flexible delivery schedule is adopted (FDS). The crop-soil-climate combinations were identified by intersecting the cropping pattern map with the soil map and with the areas of influence of three meteorological stations (Ginosa Marina, Castellaneta and Massafra) located within or surrounding the study area, using commercial GIS software (ArcGIS). The procedure, models and data utilized for the above sets of water balance simulations are described in details in Zaccaria et al. (2010) and followed the methodology proposed by Allen et al. (1998). The delivery schedule performance was used as an indicator of potential room for water conservation.

Figure 5 presents the simulation results for the three main crops grown in the study area (vegetables, table-grapes and citrus) under the RDS and FDS scenarios.

For the simulations under rotation delivery scheduling (RDS), fixed irrigation dates and volumes were adopted to reproduce the current deliveries conducted by the WUA, i.e. irrigation intervals of 10 days, flow rate of 20 l s$^{-1}$ha$^{-1}$ with 5 hours of delivery duration. For the simulations under flexible delivery (FDS), the irrigation schedules reproduced those that are commonly used by farmers when they rely on flexible or unconstrained water supply i.e. on-farm storage reservoirs, holding ponds, or groundwater pumping, and according to the irrigation methods and practices commonly utilized in the study area for each crop.

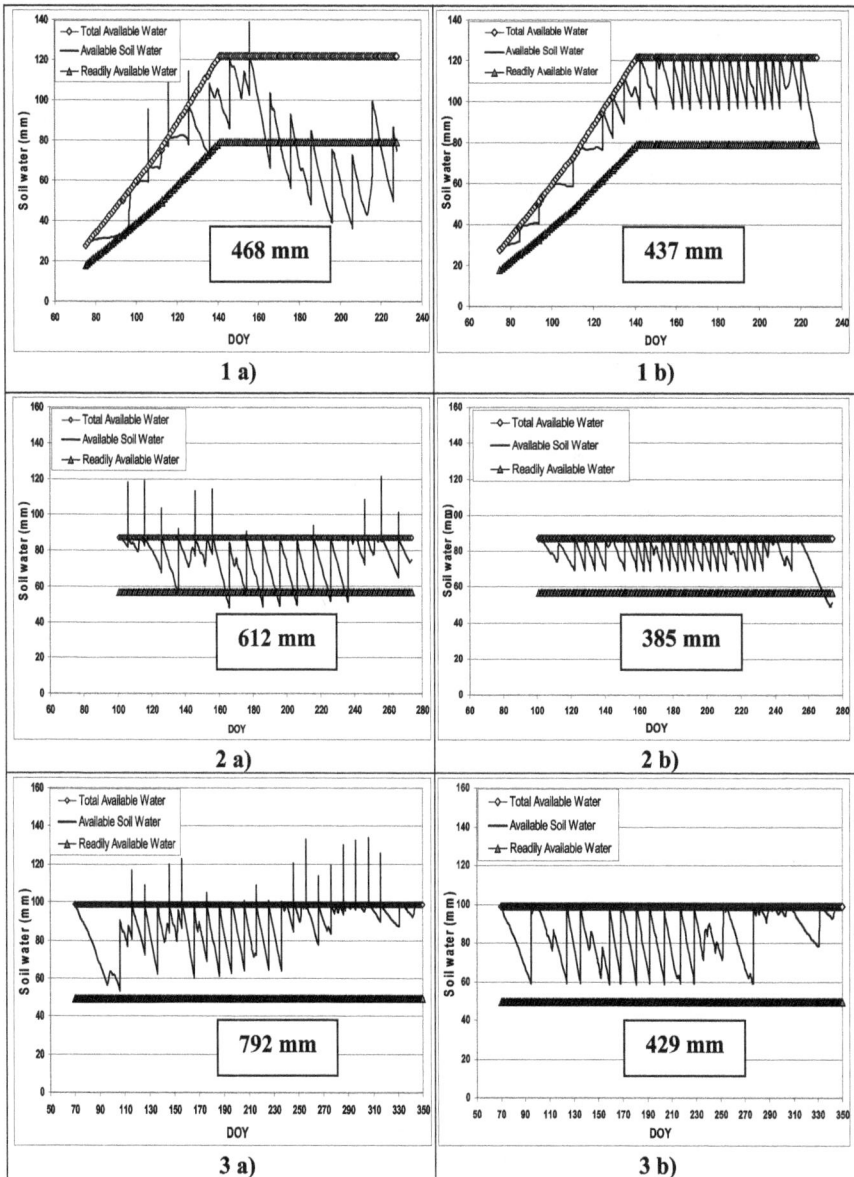

**Figure 5.** Simulated soil water balance for units consisting of 1) vegetables grown on sandy-loam soil in the area of Ginosa Marina, 2) table-grapes grown on loamy-coarse sandy soil in the area of Castellaneta, and 3) citrus grown on loamy-sand soil in the area of Massafra, under the RDS (sections a) and FDS (sections b).

The simulated irrigation scheduling shows that under RDS over-irrigation occurs at different times for the three main crops, whereas soil water deficits take place only for vegetables and table-grapes in the second half of the season. Alternatively, if farmers could rely on FDS, the irrigation management would be more effective at farm level and both water stress and excess applications could be easily avoided.

The results reported in Fig. 5 clearly show that farmers are heavily bounded by the present mode of operation of the water delivery system. If farmers irrigate in compliance with the fixed delivery currently scheduled by the WUA, the crops are likely to experience both situations of water deficit and excess waterings. The comparison between RDS and FDS schedules explains why many growers prefer to irrigate using groundwater pumping rather than rely on deliveries from the irrigation distribution networks. By managing farm irrigation under FDS growers can easily prevent water deficit and water excess to their crops by applying a lower amount of water than that under RDS. The simulation results are supported by information provided by the growers interviewed who reported that, in order to offset the restrictions imposed by rigid rotation delivery and to achieve more effective irrigation timing, many farmers pump water from the aquifer, which in their perception represents an unconstrained and flexible water supply.

## 3.2. Groundwater quality

Groundwater quality was sampled at eighteen sites throughout the Sinistra Bradano area in 2006 (Fig. 3), with two samples collected per each site, the first in February and the second in July. Measurements of total dissolved solids (TDS) and electrical conductivity (EC) were conducted on the groundwater samples, with TDS values determined by means of laboratory measurements using the gravimetric method, whereas EC values were obtained using a conductivity meter (Hanna Instruments, mod. HI 9835). Winter and summer salinity maps were developed based on the spatial interpolation of point-measured values of the TDS and EC, using the inverse weighted distance method embedded in the GIS software package. These maps are presented in Fig. 6 and seasonal changes in groundwater quality were assessed by comparing the aquifer salinity in winter with that of summer. The comparison showed that groundwater salinity increased in 2006 from winter (Fig. 6 – section a) to summer (Fig.6 – section b). The increase in groundwater salinity mainly concerned the eastern part of the study area. From Fig. 6 it can be inferred that the groundwater salinity in winter for the eastern part ranged between TDS values of 1.5 and 1.8 g $l^{-1}$, whereas it reached TDS values between 1.9 and 3.1 g $l^{-1}$ in summer, which is most likely related to the intensive groundwater pumping during period of peak demand, specifically from May to August. The western-most part of the study area showed no significant increment of groundwater salinity. This can be reported as the main consequence of the inadequate delivery schedule enforced in the area.

## 3.3. Crop evapotranspiration and crop performance under saline irrigation

Salts brought into the soil water solution through irrigation with saline water can reduce crop evapotranspiration by making soil water less available to root extraction by plants, thus creating low osmotic potential in the root zone. In other words, the total potential energy of soil water solution can be reduced due to the presence of salts. Some salts can even have toxic effects on plants or induce nutrient deficiencies, thus reducing plants metabolism and growth. Many plants can make physiologic adjustments and reduce the negative effects of low osmotic potential of soil water by adsorbing ions from soil solution and by synthesizing organic osmolytes. Both processes involve the use of metabolic energy by plants that often results in reducing growth and canopy development under saline conditions.

The response of different crops to salinity may vary, according to their different tolerances and to the physiologic capability to make the required osmotic adjustments, with some crops being able to yield acceptable productions at higher soil salinity than others. Keller and Bliesner (2000) developed a widely practiced approach for predicting the crop yield reductions due to salinity based on a yield-salinity equation adapted from Ayers and Westcott (1985), which is reported hereafter.

$$Y_r = \frac{Y_a}{Y_m} = \frac{\max EC_e - EC_w}{\max EC_e - \min EC_e}$$

where:
$Y_r$ = relative yield
$Y_a$ = actual crop yield
$Y_m$ = maximum expected crop yield when $EC_e$ < min $EC_e$
max $EC_e$ = electrical conductivity of the saturated soil extract that will reduce the yield to zero (dS m⁻¹)
min $EC_e$ = electrical conductivity of the saturated soil extract that will not decrease crop yield (dS m⁻¹)
$EC_w$ = electrical conductivity of the irrigation water (dS m⁻¹)

Values for min $EC_e$ and max $EC_e$ for the main crops grown in the study area were taken from Keller and Bliesner, as adapted from Ayers and Westcott, and are listed in the Table 3.

| CROP | min $EC_e$ (dS m⁻¹) | max $EC_e$ (dS m⁻¹) | Sensitivity to salinity |
|---|---|---|---|
| Table-grapes | 1.5 | 12.0 | Medium Sensitive |
| Citrus | 1.7 | 8.0 | Sensitive |
| Vegetables | 1.5-2.5 | 10.0-14.0 | Medium Sensitive |
| Olives | 2.7 | 14.0 | Medium Tolerant |
| Almonds | 1.5 | 7.0 | Sensitive |
| Orchards | 1.5 | 6.5 | Sensitive |

**Table 3.** Salt tolerance of agricultural crops commonly grown in the study area

Provided that the impact of salinity on plants is a time-integrated process, generally only the seasonal effects are considered to predict the reduction in crops evapotraspiration, growth and yield as occurring over an extended period of time. The above equation is thus not expected to be accurate for predicting salinity effects on crop evapotranspiration and yield for short periods.

Within the present research the likely crop yield reductions due to the use of saline irrigation water were not estimated, as this process requires the collection of multi-annual data on soil and aquifer salinity at short intervals with the aim of assessing the time of crop exposure to different levels of salinity in the soil water and to determine the evolution of soil water salinity along the year as resulting from seasonal rainfall leaching salts from the root zones.

### 3.4. The ERA&M procedure

The local climatic conditions, as well as the intensive farming of agricultural areas together with the inadequate distribution of water supplies make "business-as-usual" not environmentally-viable in the area on the long run. In view of a strategic change to the existing situation, a simplified Risk Assessment and Management (RA&M) procedure was applied to the study area through a new framework to identify viable counter-measures and mitigation of the existing environmental concerns and risks.

The applied ERA&M procedure (Fig. 7) was developed within the STRiM project (www.strim.eu) funded by the EU under the INTERREG IIIB CADSES Programme. It is a simplified framework for conducting environmental risk assessment and management, predominantly based on the Environmental Risk Management guidelines issued by the Department of Environment Food and Rural Affairs (DEFRA, 2002) of United Kingdom, which focus on risk management and applicability to any type of environmental risk. The STRiM RA&M framework consists of 5 iterative steps and is linked to other key environmental protection decision-making procedures such as the Environmental Impact Assessment (EIA), the Strategic Environmental Assessment (SEA) and the framework conceived by the European Environmental Agency (EEA) on Driving Forces, Pressures, State, Impacts and Responses (DPSIR). Both the Risk Assessment (RA) and Risk Management (RM) phases require datasets to support decision-making, often in the form of indicators. In order to harmonize environmental protection management, the STRiM framework has the novelty of linking the DPSIR indicators and monitoring framework with RA and RM, something that was not attempted before. The framework embeds risk assessment into the risk management process and, as such, includes a number of key aspects emerging throughout the various steps of the process. Among these issues, the most relevant are: a) the importance of accurately defining the actual hazards or environmental problems; b) the need to prioritize all relevant risks prior to proceeding with their quantification through the data collection; c) the need to consider the risks while taking into account feasible management solutions through the use of option-appraisal from the initial stages; d) the iterative nature of the process.

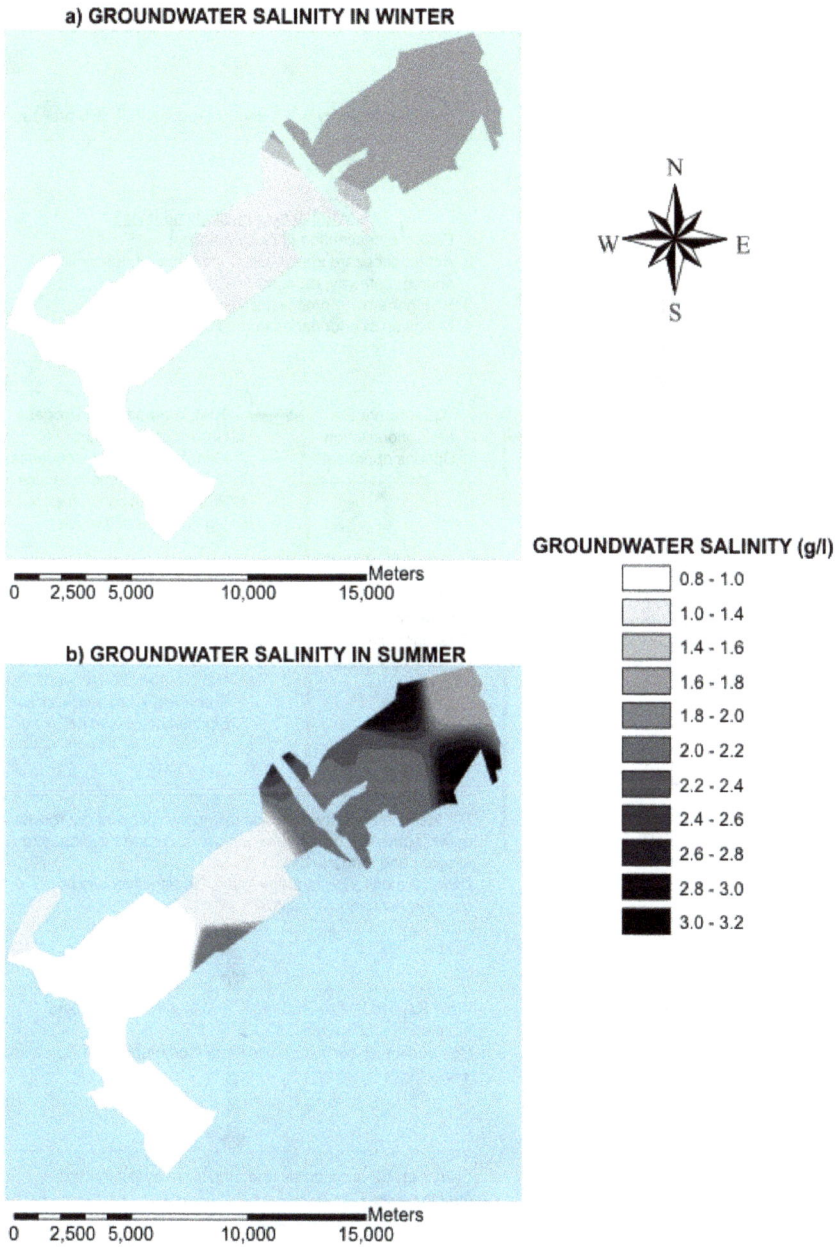

**Figure 6.** Map of groundwater salinity in the study area during winter (February) (a) and summer (July) (b) for the year 2006

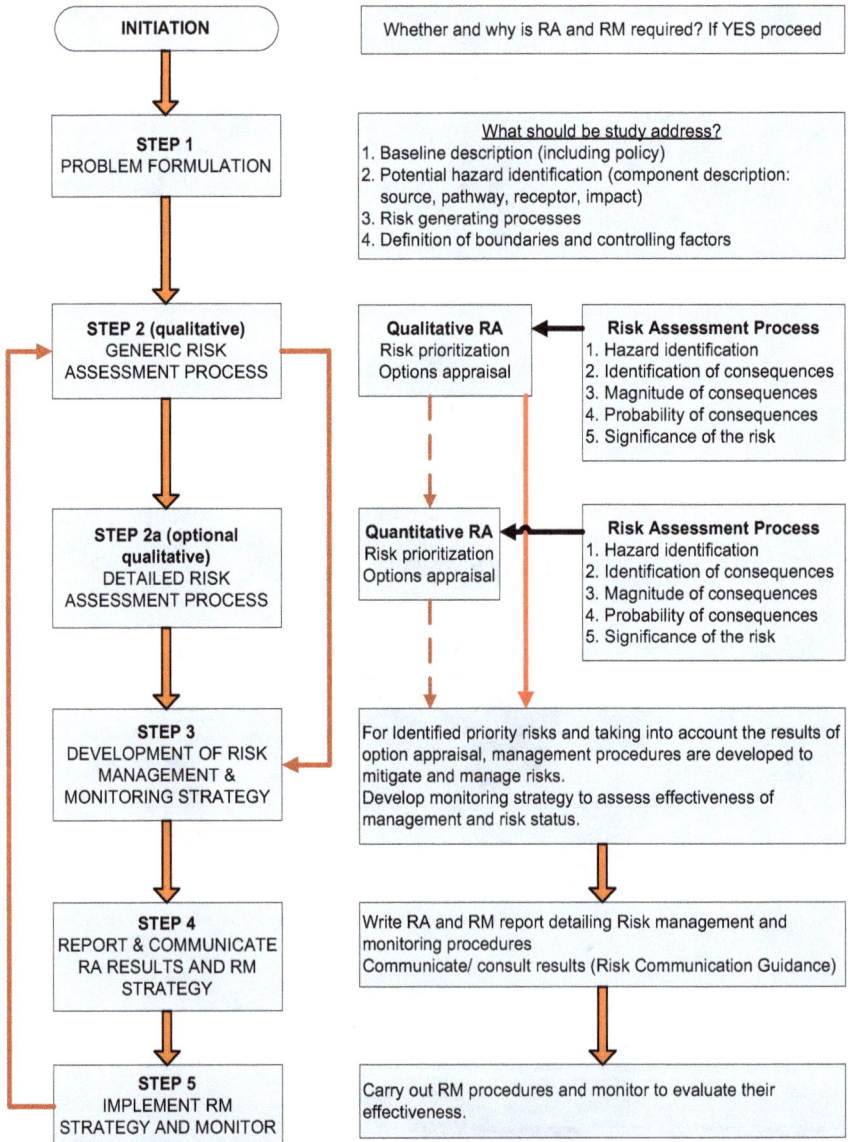

**Figure 7.** The STRiM Risk Assessment and Management Framework (modified from DEFRA, 2002)

## 4. Hazard identification and Risk-generating processes

The aquifer over-exploitation is the primary environmental hazard impending in the study area. In view of this hazard, managing "business-as-usual" represents the intention for which the RA&M is required, the intention being defined as *"any course of action, intentional or otherwise, which by its nature may pose a risk to the environment - natural or built - and the life it sustains"*. The "business-as-usual" or baseline scenario in the study area consists in maintaining the intensive farming practices along with the irrigation delivery schedule enforced by the water management authority. The secondary hazards resulting from the aquifer over-exploitation are those indicated in Fig. 8, whereas the sources, pathways, receptors and impacts are indicated in the Table 4 for the primary hazard.

The potential causes concurring to aquifer over-exploitation are the intensive groundwater pumping (S1) by farmers during peak irrigation demand periods (July and August) and the inadequate water distribution through the irrigation networks (S2). This situation is driven by the existing market-oriented agriculture that is based on water-demanding crops, and by the current operation of the irrigation distribution system that does not match with crops and farmers' water requirements. The primary pathway (P1) goes through groundwater pumping, which in some periods may occur beyond the safe yield of aquifer due to concentration of withdrawals. This has the effect of depressurizing the aquifer, giving way to seawater intrusion and to aquifer contamination by saline water. The receptor of salination by seawater intrusion is thus the aquifer itself.

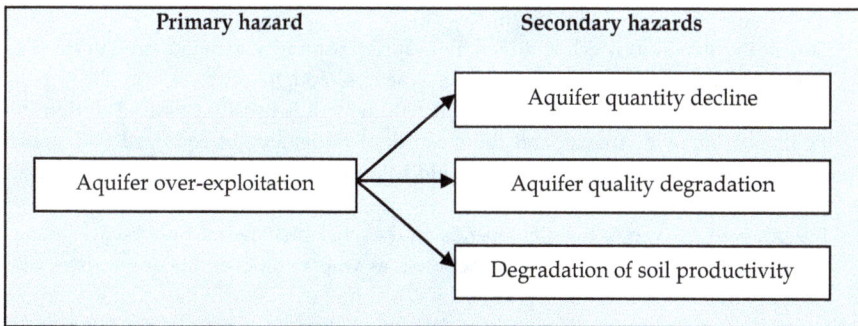

**Figure 8.** Primary and secondary hazards identified in the study area

The secondary pathway (P2) is again through groundwater pumping by farmers and through the distribution of saline water onto irrigated fields. The major potential impact is the salts build-up in the irrigated soil (I2.1) resulting from the distribution of saline irrigation water and from the water evaporation and transpiration processes.

As for the identification of risk-generating processes, the current water distribution and the conditions of water delivery (discharge and pressure head at hydrants) being not adequate for proper farm irrigation management both concur to the environmental hazard. The extensive use of groundwater pumping throughout the study area results in drawdown and qualitative deterioration of aquifer, as well as in salts loads being progressively brought onto cropped plots through saline irrigation water. If leaching is not properly conducted on a regular basis, or in case salts are not flushed away from the root zones by the action of seasonal rainfalls, soils are progressively subjected to salts build-up, which may negatively affect their productivity. The soil and aquifer salination thus represent the main risk-generating processes.

| Hazard | Source | Pathway | Receptor | Impact |
|--------|--------|---------|----------|--------|
| H1 Aquifer over-exploitation | S1 - Intensive pumping by farmers during peak demand periods | P1-Aquifer | R1-Aquifer | I1.1 - Aquifer Depletion |
| | | | | I1.2 - Salination by seawater intrusion |
| | S2 - Poor water distribution through the irrigation networks | P2-Aquifer | R2– Soils | I 2.1 - Salt build-up in the soils |

**Table 4.** Hazards sources pathways receptors and impacts

## 5. Controlling factors of hazards and magnitude of impacts

The aquifer over-exploitation is tightly dependent upon the following factors:

- **Crop water demand**, which is driven by evapo-transpirative demand, growth stage of crops, prevailing farming and irrigation practices, and by effective rainfall. A peak concentration of crop water demand in the study area is usually observed during the months of July and August, and the majority of farms are not equipped with water storage facilities (holding ponds) that could help them buffering the irrigation demand with the water delivery by WUA.

- **The adopted delivery schedule** depends on the available flow rate, on the design and capacity of the existing distribution network, as well as on operational resources and skills provided by the technical staff of the WUA. In the study area the rotation delivery is not agreed upon with farmers, but is instead dictated by the WUA following a supply-driven approach. More flexible arranged deliveries would allow partially overcoming the rigid water distribution.

- **On-farm irrigation practices**, can range from full replenishment of soil water depletion from the root zone to different levels of deficit irrigation, on the basis of the crops grown, the specific sensitivity of the different growth stages to water deficits, the target yields, and the farmers' skills and capability in field water management. Full irrigation is the most common irrigation practice in the study area. Micro-irrigation methods allow maximizing crop yields even when using saline water. Leaching of salts from top soil layers is usually not carried out by the majority of farmers, but flushing of salts mainly occurs due to fall and winter rains.

- **Natural leaching and aquifer recharge** mainly depends on rainfall intensity and distribution, vegetation cover, soils' hydraulic features, and slope. In the study area natural leaching and partial aquifer recharge usually occur during fall and winter months but, as pointed out by previous investigations, those are not sufficient to avoid aquifer salinity increase and salts build-up in the soils on the long run.

The overall magnitude of impacts was estimated based on three criteria, namely a) the spatial distribution of impacts, b) their time-duration, and c) the time necessary to onset the impacts. These impacts were assigned a partial score for each criteria based on a scale ranging from 1 to 4. For instance, the scale related to the spatial distribution of impacts assigned scores according to the following ratings:

Nowhere (0%): score = 0;

Localized (< 5%): score = 1;

Scattered (5-15%): score = 2;

Widespread (15-50%): score = 3;

Throughout (> 50%): score = 4.

The overall magnitude of impacts resulted by multiplying the partial scores assigned for the three criteria, thus on a scoring scale ranging from 0 to 64, then classified from "negligible" (score 0) to "mild" (score 1-22) to "moderate" (score 23-43) to "severe (score 44-64). The calculated values for the magnitude of impacts are reported in Table 5.

| Hazard | Receptor | Impact | Spatial scale | Temporal scale | Time of onset to impact | Overall magnitude |
|---|---|---|---|---|---|---|
| H1 Aquifer over-exploitation | R1 Aquifer | I1.1.1 Aquifer depletion | Throughout (> 50%) 4 | Medium term (5-20 years) 2 | Medium (1-10 years) 3 | Moderate 24 |
| | | I1.1.2 Salination by seawater intrusion | Throughout (> 50%) 4 | Medium term (5-20 years) 2 | Medium (1-10 years) 3 | Moderate 24 |
| | R2 Agricultural soils | I1.2.1 Salts build-up | Throughout (> 50%) 4 | Medium term (5-20 years) 2 | Immediate (0-1 year) 4 | Moderate 32 |

**Table 5.** Estimated magnitude of impacts

## 6. Estimation of risk probabilities

The estimation of the overall probability of hazards is also based upon three criteria, respectively the probabilities of hazard occurring, of the receptors being exposed, and of harm resulting to the receptor. Within each criteria, the probabilities were assessed and classified on a High (score 3) to Negligible (score 0) scale. The overall probabilities of hazards were finally obtained by combining the partial scores assigned in each criterion and afterwards classifying the overall scores based on the following probability scale:

Negligible (when score ~ 0);

Low (when score = 1-9);

Medium (when score = 10-18);

High (when score = 19-27).

The overall probabilities for the study area are those reported in Table 6

|  |  | H1 | |
|---|---|---|---|
| Probability of hazard occurring | receptor independent | High (3) | |
| Probability of receptors being exposed | R1 | High (3) | |
| | R2 | High (3) | |
| Probability of harm occurring to receptor | R1 | High (3) | |
| | R2 | High (3) | |
| Overall probability | | H1.R1=27 (high) | H1.R2=27 (high) |

**Table 6.** Probability estimation

## 7. Risk significance

Risk significance is assessed considering the magnitude of consequences and the probability of effects occurring. In case of qualitative risk assessment, a simple two-ways entry matrix that considers simultaneously the probability and magnitude of consequences, such as the one reported in Table 7 can provide a consistent basis for decision-making.

Evaluation of the risk significances for the 3 impacts that were analyzed in the present case study led to results reported in Table 8. The results from the evaluation were then used to prioritize the most relevant risks and conduct options appraisal to identify viable and consistent management solutions.

As for risk communication process, the results from the risk prioritization should be communicated to the technical staff and to the decision-makers of the WUA through thematic meetings. Also, outcomes from the evaluation of magnitude and probability and from the risk prioritization stage should be disseminated to farmers' groups and to opinion leaders by means of extension service activities and through specific field focus meetings.

| Increasing acceptability | Consequences | | | |
|---|---|---|---|---|
| | Severe | Moderate | Mild | Negligible |
| **Probability** | | | | |
| High | high | high | medium/low | near zero |
| Medium | high | medium | low | near zero |
| Low | high/medium | medium/low | low | near zero |
| Negligible | high/medium/low | medium/low | low | near zero |

**Table 7.** Risk significance evaluation matrix.

| Risk | Significance score |
|---|---|
| Risk (H1. R1.I1.1) | Moderate x High = **High** |
| Risk (H1. R1. I1.2) | Moderate x High = **High** |
| Risk (H1. R2. I2.1) | Moderate x High = **High** |

**Table 8.** Risk Significance for the study area

## 8. Appraisal of risk management options

Options appraisal consists in the identification of the most suitable risk-management techniques. This entails scoring, weighting and reporting the different risk management options, and comparing alternatives prior to selection. Viable options can be appraised on the basis of various criteria. For the present study, alternative risk management techniques were evaluated according to: a) social risk acceptability by stakeholders; b) technical feasibility; c) effectiveness in risk alleviation; d) duration of effects; e) costs for implementing the risk management options. The results from options appraisal for the three major risks, namely aquifer quantitative depletion, aquifer degradation, and salts build-up in the agricultural soils are shown in Table 9a, 9b and 9c.

| Risk 1 Aquifer depletion | Timing Instant result to progressive | Social acceptability (-- to ++) | Feasibility (-- to ++) | Effectiveness in risk alleviation (-- to ++) | Duration | Cost Low to high |
|---|---|---|---|---|---|---|
| Business as usual (zero option) | Never -- | Acceptable + | Very feasible ++ | Very ineffective -- | Never -- | Very affordable ++ |
| Limit water pumping from Groundwater | Immediate + | Unacceptable - | Feasible + | Very effective ++ | Short term - | Unaffordable - |
| Improved rotation in water delivery | Medium +/- | Acceptable + | Feasible + | Effective + | Medium term +/- | Very affordable ++ |
| Decrease water tariffs by WUO to compensate for pumping costs | Long term - | Very Acceptable ++ | Feasible + | Effective + | Short term - | Unaffordable - |
| Water delivery on-demand | Medium +/- | Acceptable + | Feasible + | Very Effective ++ | Medium term +/- | Affordable + |

a)

| Risk II Aquifer salination | Timing Short term to permanent solution | Social acceptability (-- to ++) | Feasibility (-- to ++) | Effectiveness in risk alleviation (-- to ++) | Duration Instant result to progressive | Cost Low to high |
|---|---|---|---|---|---|---|
| Business as usual (zero option) | Never -- | Acceptable + | Very feasible ++ | Very ineffective -- | Never -- | Very affordable ++ |
| Stop groundwater pumping | Medium +/- | Very unacceptable -- | Feasible + | Very effective ++ | Medium +/- | Unaffordable - |
| Limit groundwater pumping to safe yield of aquifer | Medium +/- | Unacceptable - | Feasible + | Effective + | Medium +/- | Unaffordable - |

| | | | | | | |
|---|---|---|---|---|---|---|
| Rotation irrigation delivery + conjunctive use | Medium +/- | Acceptable +/- | Feasible + | Effective + | Medium +/- | Affordable + |
| Irrigation delivery on-demand | Medium +/- | Acceptable + | Feasible + | Effective + | Medium +/- | Affordable + |
| Artificial aquifer recharge | Immediate ++ | Neither unacceptable nor acceptable +/- | Feasible + | Very effective ++ | Medium +/- | Affrodable + |

b)

| Risk III Salts build-up in the agricultural soils | Timing Instant result to progressive | Social acceptability (-- to ++) | Feasibility (-- to ++) | Effectiveness in risk alleviation (-- to ++) | Duration | Cost Low to high |
|---|---|---|---|---|---|---|
| Business as usual (zero option) | Never - - | Acceptable + | Very feasible ++ | Very ineffective - - | Never - - | Very affordable ++ |
| Improved rotation delivery | Long term - | Acceptable ++ | Very feasible ++ | Effective + | Medium Term + | Affordable + |
| Improved rotation delivery + conjunctive use | Medium +/- | Very acceptable ++ | Feasible + | Effective + | Medium Term + | Affordable + |
| Irrigation delivery on-demand | Medium +/- | Very acceptable ++ | Feasible + | Very effective ++ | Long term ++ | Affordable + |
| Improved on-farm irrigation practices (leaching) | Medium +/- | Neither unacceptable nor acceptable +/- | Feasible + | Very effective ++ | Long term + | Very affordable ++ |
| On-demand delivery + leaching | Immediate + | Acceptable + | Feasible + | Very effective ++ | Long term + | Affordable + |

c)

**Table 9.** Risk Management option selection matrices for: a) aquifer depletion (Risk 1), b) aquifer salination (Risk 2), c) salts build-up in the agricultural soils (Risk 3)

## 9. Conclusive remarks

Selecting a suitable risk management option strongly depends on the weights attributed by the evaluator to the decision criteria for the different options with respect to the zero-alternative (business-as-usual). Some of the identified management options pertain to alternative operation of the large-scale distribution network, whereas some others entail improved water management practices at the farm scale or mixed options.

As for the risk related to aquifer quantitative depletion, the preferred option could be to operate the distribution network by an improved rotation delivery, which could better match crop water requirements in terms of timing of delivery. This would require some accurate estimation of irrigation requirements and improved irrigation scheduling plans, as well as some extension service activities to assist farmers in the effective use of available water.

As for the risk of aquifer salination, since it is tightly linked to the amount and concentration of groundwater pumping during the irrigation season, conducting artificial aquifer recharge would be very effective in reducing the pressure over the groundwater. For mitigating the existing effects on aquifer salinity, a strong reduction in groundwater pumping should also be enforced along with artificial aquifer recharge. These two measures in conjunction would most likely allow decreasing the existing level of salinity and inverting the trend of progressive salinity increase in the whole study area.

As for the risk of salts build-up in the agricultural soils, the on-demand delivery in conjunction with improved irrigation practices (leaching) at the farm level would result as the best management options. These techniques would entail some modernization works to the irrigation distribution network as well as extension service activities to train farmers on aspects related to soil-water balance and salinity balance for the major crops grown in the area, and for the prevailing farming practices and irrigation methods.

Overall, selecting the most suitable and viable risk management option would be a matter of strategic planning by the Regional Administration and by the WUA, as well as of the available financial resources, human resources and skills available and required for implementing the options.

Combining the risk management options for the above three risks would result in bringing together conflicting objectives for different stakeholders that may be involved in the land planning and land use. Land users may in fact primarily or exclusively be interested in mitigating the risk of salts build-up in the cropped soils, whereas land planners, and the actors responsible for sustainable use of natural resources, would be inclined to address broader objectives with high priority, such as the reduction of aquifer depletion and salination.

## Author details

Daniele Zaccaria*
*Division of Land and Water Resources Management, Mediterranean Agronomic Institute of Bari (CIHEAM-IAMB), Valenzano, Bari Italy*

* Corresponding Author

Giuseppe Passarella
*Water Research Institute (IRSA), National Research Council (CNR), Bari, Italy*

# 10. References

Allen RG, Pereira LS, Raes D, Smith M (1998) Crop Evapotranspiration: Guidelines for computing Crop Water Requirements. Irrig. Drain. Paper. 56, FAO, Rome, p 300.

Ayers RS, Westcott DW (1985) Water Quality for Agriculture. Food and Agricultural Organization of the United Nations, Irrigation and Drainage Paper n. 29.

Benoit G, Comeau A (eds) (2005) *A sustainable future for the Mediterranean : the Blue Plan's environment and development outlook.* London : Earthscan.

Capaccionia B, Diderob M, Palettab C, Diderob L (2005) Saline intrusion and refreshing in a multi-layer coastal aquifer in the Catania Plain (Sicily, Southern Italy): Dynamics of degradation processes according to the hydro-chemical characteristics of ground waters. J. Hydrology 307:1–16.

Clemmens AJ (2006) Improving irrigated agriculture performance through an understanding of the water delivery process. Irrig and Drain 55(3):223–234.

Clemmens AJ, Molden DJ (2007) Water uses and productivity of irrigation systems. Irrig Sci 25(3):247-261.

Cotecchia V, Magri G (1967) Gli spostamenti delle linee di costa quaternarie del Mar Ionio fra Capo Spulico e Taranto. Geologia Applicata e Idrogeologia, 2:3-28.

Cotecchia V, Dai Pra G, Magri G (1971) Morfogenesi litorale olocenica tra Capo Spulico e Taranto nella prospettiva della protezione costiera. Geologia Applicata e Idrogeologia, 6:65-78.

DEFRA (2002) Department for Environment, Food and Rural Affaires, Guidelines for Environmental Risk Assessment and Management, HMSO, 2002.

EEA (2010) Adapting to climate change — SOER 2010 thematic assessment' (http://www.eea.europa. eu/soer/europe/adapting-to-climate-change), in: The European environment — state and outlook report 2010, State of the environment report No 1/2010.

EEA (2012) Towards efficient use of water resources in Europe, EEA Report No. 1/2012, ISSN 1725-9177, DOI: 10.2800/95096. © EEA, Copenhagen, 2012.

Hargreaves GH, Zaccaria D (2007) Better management of renewable resources can avert a world chrisis. J Irrig Drain Engng 133(3):201-205.

INEA (1999) Quadro di riferimento per lo studio è il monitoraggio dello stato dell'irrigazione in Puglia: Consorzio di Bonifica Stornara e Tara, Taranto.

Keller J, Bliesner RD (2000) Sprinkle and Trickle Irrigation, ISBN: 1-930665-19-9, The Blackburn Press, Caldwell, New Jersey 07006, 2012.

Kijne JW, Prathapar SA, Wopereis MCS and Sahrawat KL (1998) How to Manage Salinity in Irrigated Lands: a Selective Review with Particular Reference to Irrigation in Developing Countries. SWIM Paper 2, International Irrigation Management Institute, Colombo, Sri Lanka.

Merriam JL, Freeman BJ (2002) Irrigation water supplies to not inhibit improved water management. ICID 18th Intl. Congress, Montreal, Trans. Question 50, Rep. No. R3.13.

Merriam JL, Styles SW, Freeman BJ (2007) Flexible irrigation systems: concept, design, and application. J Irrig Drain Engng 133(1):2-11.

Paniconi C, Khlaifi I, Lecca G, Giacomelli A and Tarhouni J (2001) Modeling and analysis of seawater intrusion in the coastal aquifer of eastern Cap- Bon, Tunisia. National Agronomic Institute of Tunisia (INAT) and Center of Advanced Studies, Research and Development in Sardinia (CRS4), Italy, Kluwer Academic publishers.Transport in porous media 43: 3 – 28.

Piccirillo M (2000) Studio idro-geochimico delle acque sotterranee nella zona di Palagiano (Taranto). Universita' degli Studi di Bari, Bari.

Polemio M, Ricchetti E (1991) Caratteri idrogeologici dell'aquifero della piana costiera di Metaponto. In: Proc. "Il rischio idrogeologico e la difesa del suolo", Accademia Nazionale dei Lincei, pp 423-428.

Polemio M, Mitolo D (1999) La vulnerabilita' dell'acquifero nella piana costiera di Metaponto. Ricerca Scientifica e Istruzione Permanente 93:417-426.

Polemio M, Limoni PP, Mitolo D, Santaloia F (2002). Characterization of Ionian-Lucanian Coastal aquifer and Seawater intrusion hazard. Proc. 17th SWIM, Delft, The Netherlands, May 6-10, pp 422-434.

Regione Puglia (1983) Piano regionale di risanamento delle acque.

Regione Puglia (2006) Progetto Tiziano - Sistema di monitoraggio acque sotterranee.

Regione Puglia (2007) Piano di Tutela dell Acque della Regione Puglia.

Sanaee-Jahromi S, Feyen J, Wyseure G, Javan M (2001) Approach to the evaluation of undependable delivery of water in irrigation schemes. Irrigation Drain Syst 15(3):197-213.

Styles SW (1997) Alleviation of surface and subsurface drainage problems by flexible delivery schedules. 27th Congress Proc., IAHR/ASCE Water Res. Engrg. Div. Managing Water: Coping with Scarcity and Abundance, San Francisco, pp 717-722.

Umali DL (1993) Irrigation-induced salinity: a growing problem for development and the environment. World Bank Technical Paper Number 215. Washington, D.C., USA: The World Bank.

Zaccaria D, Lamaddalena N (2005) Reliability criteria for re-engineering of large-scale pressurized irrigation systems. In: Proceedings of the USCID 3rd International Conference on Irrigation and Drainage (San Diego, CA, March 30 – April 2, 2005). U.S. Com. on Irrigation and Drainage, Denver, pp. 547-556.

Zaccaria D, Inversi M, Lamaddalena N (2006) Assessment of the environmental sustainability of irrigated agriculture in a large-scale scheme – A case study. In: Ground water and surface water under stress: competition, interaction, solutions (Proc. USCID 2006 Water Management Conference, Boise, ID, October 25-28, 2006). U. S. Com. on Irrigation and Drainage, Denver, pp 413-424.

Zaccaria D, Scimone M (2008) Assessment and Management of Environmental Risks Resulting from Operation of Large-scale Irrigation Delivery Systems: a Case Study in Southern Italy. In: Proceedings of the International Congress on Environmental Modeling and Software (iEMSs 2008). Barcelona (Spain), July 6-10, 2008.

Zaccaria D, Oueslati I, Neale CMU, Lamaddalena N, Vurro M, Pereira LS (2010) Flexible Delivery Schedules to Improve Farm Irrigation and Reduce Pressure on Groundwater: a Case Study in Southern Italy. Irrig Sci 28:257–270, doi 10.1007/s00271-009-0186-8.

Zaccaria D, Neale CMU (2012) Modeling delivery performance in pressurized irrigation systems from simulated peak-demand flow configurations. Manuscript submitted to Irrigation Science – Manuscript ID IrrSci-2012-0025, under publication.

# Chapter 5

# Nonpoint Pollution Caused by the Agriculture and Livestock Activities on Surface Water in the Highlands of Jalisco, Mexico

Hugo Ernesto Flores López, Celia De La Mora Orozco,
Álvaro Agustín Chávez Durán, José Ariel Ruiz Corral,
Humberto Ramírez Vega and Víctor Octavio FuentesHernández

Additional information is available at the end of the chapter

## 1. Introduction

The agriculture and livestock industry in Mexico and particularly in the State of Jalisco, Mexico, has been moved forward technologically, resulting in important productivity increment. However, the use of these technologies involves the application of large amount of fertilizers and pesticides for prolonged periods of time. Furthermore, the fertilizers are often applied incorrectly, increasing the inefficiency and resulting in high environmental impact on soil and water in the lower watershed area.

The current agriculture and livestock lands management practices, with the excessive agrochemical use, have created ecological, environmental and economic problems such as: 1) the soil erosion, which is detrimental to the organic matter availability and soil cover, 2) the surface water and groundwater pollution caused by manure and sediments, 3) the disruption of wild life habitats, and 4) negative effects on the rural landscape. When environmental agrochemical products such as nutrients, pesticides, compost, gases (nitrogen oxide and methane), are used in combination with incorrect agricultural management practices, the result is the nonpoint source pollution (NSP). When this problem is visualized in large scale such as region, lakes or rivers, the result is a high contamination problem in the lowest watershed area, which is both difficult and costly to solve. However, when the pollution sources are in a reduced area, like a small agriculture farm or grassland area, the magnitude decreases and the problem is more easily solved.

On a small scale farm, the combination of management practices for fertilizers and pesticide applications, in accordance with soil and climate characteristics, create a complex processes, particularly when they occur in different times and spaces. This situation is particularly difficult to understand and consequently, the preventive or corrective actions that limit the spread of pollution in agricultural/livestock lands and water bodies are difficult to avoid (Flores *et al.*, 2009). The authors stated that a feasible option to solve the NSP problem of surface water is control or prevention in the initial stages, along with reducing the amount of fertilizers and pesticides by the adoption of new technologies or improving the efficiency of agrochemical products. Apart from, the implementation of adequate conservation practices, it is essential to limit the transportation of sediments and nutrients to outside of agriculture and livestock lands. However, before any technology application is implemented, it is necessary to know the processes associated with NSP from the small farm scale to the watershed scale, first, involving the producer efficiency criteria of their land and then, selecting and including the best management practices to prevent and/or control the NSP.

## 2. The non-point source pollution problem

Contamination is the introduction of substances into the environment indirectly or directly by humans, provoking not only negative impacts on the environment, but also, putting at risk human health or other living organisms by interfering with legitimate environment use. Loehr (1984) mentioned that the pollution sources may be classified as point sources (direct or localized) and non-point sources (diffuse or not localized). The point sources discharge pollutants though piping, channels, or ditches--much like those that come from city wastewater treatment plants, the food processing industry, and runoff from large pig farms. Discharges from these sources are usually constant and also related to the municipal industrial activities.

The non-point source pollution (NSP) is caused by inadequate agricultural and livestock practices, where pollutants move with the runoff, dragging their sediments, and putting at serious risk both the surface water and groundwater. The agricultural sources of NSP includes; the loss of the superficial soil where fertilizer, pesticides and animal manure were applied, and were transported to water sources such as stream and rivers through runoff. The amount of pollutants transported is a function of variables that include the type of soil, the land slope, land use, and the natural route that water follows through the natural drainage networks (Deliman and Leigh, 1990). Some of the variables involved in this kind of contamination are uncontrollable by humans, such as the land topography and the rain, compared to others like cultivation covering, and the time and location of agrochemical application and management, which s can easily be controlled (Loehr, 1984). Nevertheless, combining these variables cannot relate them to the same and unique origin of discharge, which creates the problem of source identification, as well as potential impact evaluation over the transport route and the dams where water is stored (Deliman and Leigh, 1990).

The eutrophication is a natural process that can be accelerated by the water enrichment with excessive inorganic nutrients such as nitrogen (N) and phosphorous (P), which are considered responsible for the excessive growth of algae and aquatic plants in water bodies (Schnoor, 1996). The eutrophication holds a close relation to the dissolved inorganic N (DIN) and the dissolved inorganic phosphorous (DIP), in the proportion DIN:DIP; from a stoichiometric viewpoint of algae and aquatic plants, if this proportion is greater than 7:1 (Gold and Oviatt, 2005), 12:1 (Pietilainen, 1997) or 14:1 (Schnoor, 1996), P is the limiting nutrient, but if the proportion is lower than 5:1 (Pietilainen, 1997) or 7:1 (Gold and Oviatt, 2005), N is the limiting nutrient. Concentrations in water of 0.3 ppm of inorganic N and 0.015 ppm of inorganic P are the levels in which eutrophication could become troublesome (McCool and Renard, 1990). In the United States of America approximately 90% of lakes studied demonstrated a DIN and DIP higher than 14:1; as a result, P was defined as the limiting factor (Carpenter, 2005; Sharpley *et al.*, 2003).

In México, the total P has been used as an indicator of the trophic state of water bodies. Thus, a level of over 0.118 mg $L^{-1}$ in tropical lakes (Sobrino-Figueroa, 2007), or water bodies in warm environments with more than 0.035 mg $L^{-1}$ (Díaz-Zavaleta, 2007b), is considered eutrophic. With this criterion for the characterization of a trophic state, Díaz-Zavaleta (2007a) found that many bodies of water in México have eutrophication problems. In Tepatitlán, Jalisco, Ramírez *et al.* (1996) it was determined that the N and P contents in water samples at two points and at two depths of the El Jihuite dam, where the proportion of DIN:DIP was 14:1 and 13:1, for the surface and the bottom of the reservoir, and a total P concentration greater than 0.1 ppm, making the dam eutrophized, and P the limiting nutrient. The eutrophication in surface water is one of the principal problems caused by the NSP which comes from agricultural and livestock lands.

## 2.1. The agricultural and livestock activity in the highland of Jalisco, Mexico

In the Highlands of Jalisco, Mexico, there are more than 241,000 hectares are utilized for agriculture activities with corn being the most common product at about 90%. However, of this, 40% of the corn is forage which is used as feed for dairy cattle (SIAP-SAGARPA, 2011). On the other hand, the cattle inventory in the Highlands of Jalisco is about 226,000 cattle, 1.715 millions of pigs and 82 millions of chickens (SIAP-SAGARPA, 2011); the cattle alone generate about 4.9 millions of tons of manure (Flores *et al.*, 2012). Even though, the manure produced by the Highland region of Jalisco is an important source of nutrients for corn and grassland, it also contains other components such as fecal coliforms bacteria, total coliforms bacteria, and enteric pathogens (Flores *at al.*, 2012). These components may deteriorate the water quality in lakes, dams and dikes (Torres and Calva, 2007; Soupir *et al.*, 2006; Ramírez *et al.*, 1996).

In order to maintain the productivity of the agricultural and grazing lands and to improve the water quality in dams and dikes, a feasible solution to this problem is to use best management practices in agricultural systems where fertilizer and manure were applied; however, before this happens, the transport process of nutrients and coliforms along with other organisms in water must be identified (Mishra *et al.*, 2008; Pachepsky *et al.*, 2006).

## 2.2. Surface water quality

Rainfall, surface runoff and erosion by water have been identified as some of the means by which agricultural lands, nutrients and micro-organisms are lost; this in conjunction with inadequate management of superficial water bodies. (Flores *et al.*, 2012; Flores *et al.*, 2009; Mishra *et al.*, 2008; Soupir *et al.*, 2006; Oliver *et al.*, 2005; Jamieson *et al.*, 2004; Ferguson *et al.*, 2003).

The negative impact caused by the inadequate management of the natural resources is well known. In surface water bodies, specifically, the effects of this improper management may cause water quality degradation through the excessive nutrient input, resulting in eutrophication problems. Furthermore, the pollutants can reach the groundwater and contaminate the principal water source for domestic use. The region that composes the Highlands of Jalisco, Mexico is not exempt from this problem; the agricultural activities that take place in the area contribute to the contamination of surface water through runoff. Due to the semiarid climate, this region is also more dependent on dam water for domestic use. The Jihuite dam which is located at 5 km north of Tepatitlán de Morelos in Jalisco, Mexico, is supplying approximately 30% of water for domestic use for about the 120,000 inhabitants of Tepatitlán de Morelos. For its location, the Jihuite dam is the source of many pollutants such as pesticides and excessive fertilizers, applied during agricultural activities along the watershed. The pollutants are carried by the runoff and reach the dam, causing a negative impact. Some studies evaluated the water quality characteristics of the dam using nutrient concentration by sampling eight different sites of said dam; the parameters were the following; pH, temperature, total hardness, color, electric conductivity (EC), salinity, total dissolved solids (TDS), dissolved oxygen (DO), nitrates, nitrites, and chlorides (De La Mora, 2010). Some parameters evaluated by De La Mora *et al.* (2010) have shown an increase with respect to the results obtained by Ramirez *et al.* (1996) such as nitrates, nitrites and TP. In 1996 the superficial nitrates, nitrites and TP concentrations were about 0.018, <0.02, and <0.05 mg $L^{-1}$, respectively. In 2010 the nitrates, nitrites and TP concentrations were as follows; 1.3875, 0.00775 and 0.54875 mg $L^{-1}$, respectively.

De La Mora *et al.* (2011) also evaluated the dissolved oxygen concentration at different water depths in three selected sampling sites along the dam; the results were as follows: the DO at sampling site one was from 5.27 mg $L^{-1}$ in the surface to 1.19 mg $L^{-1}$ at six meters of depth. Sampling site two was from 9.5 mg $L^{-1}$ to 1.5 mg $L^{-1}$ at eight meters of depth and sampling site three 7.47 mg $L^{-1}$ to 1.05 mg $L^{-1}$ at eight meters of depth. Results showed that the oxygen reduction at every meter indicates anoxic conditions.

These results demonstrated a significant increment of nitrites, nitrates and TP concentration, which suggests the deterioration of water quality in el Jihuite dam. Results also suggest that the increment of agricultural industries in the watershed is the principal cause of the water deterioration. Moreover, additional water analyses have to be performed in order to evaluate the toxicological characteristics of the water for concerned citizens and policy makers.

## 2.3. The rainfall effect on the non-source pollution

The rainfall plays an important role in the NSP process in the agricultural lands, particularly the characteristics related to rainfall distribution and to the intensity and quantity of rain;

the latter is very much associated with the time of the fertilizer and manure application (Oliver *et al.*, 2005; Ferguson *et al.*, 2003; Saini *et al.*, 2003).

The NSP initiates with the impact of raindrops on the soil or on the manure that is left on grazing lands. The process of soil wetness by rainfall provokes saturation of the superficial soil level, removal of soil particles, and the destruction of cow manure. This process becomes even more critical when the soil is continuously wetted and dried when combined with soil tillage. Alcalá (2011) demonstrated that the first storms of the rainy season that produced runoff in corn crops with the application of cattle and chicken manure generated the highest levels of fecal coliforms. But furthermore, Flores *et al.* (2012) showed that storms with a smaller quantity of rain have a larger quantity of coliforms when the manure is applied to corn; when the quantity of rain increases the level of coliforms, the water diminishes, causing dilution of coliforms in rainwater. However, the authors found the opposite effect in grassland where the cattle graze, which requires a larger quantity of rain to increases the amount of fecal coliforms in the water (Figure 1).

On the other hand, the threshold of rain to produce superficial runoff depends on the type of crop and the kind of covering that is present on the topsoil. It can be noticed in Table 1, that the corn and grassland areas showed the lowest rainfall value when compared to bare soil and tequila-making agave. The value of the rain threshold that has been used to define storms with runoff is about 12 mm (Xie *et al.*, 2002) or 12.7 mm (Wischmeier and Smith, 1978).

| Soil covering types | Threshold rain (mm) |
|---|---|
| Bare soil | 6.2 |
| Agave tequilero | 6.7 |
| Native grass | 12.4 |
| Corn | 9.5 |

**Table 1.** Rain thresholds that produce surface runoff in four different kinds of soil covering (Flores *et al.*, 2009).

## 2.4. The soil erosion effect on the non-source pollution

The soil erosion is defined as the process of detachment, transport and deposition of particles by erosive agents, like the rainfall, surface runoff or wind (Meyer and Wishmeier, 1969; Hairsine and Rose, 1992). The detachment and transport of particles is related to flow mechanisms in rill and interrill erosion. The rill erosion is considered a function of the superficial flow capacity as a means of detaching sediment, the sediment transport capacity, and the present sediment load; the interrill erosion is described as the detachment process of particles caused by the impact of rain drops followed by transportation in the wide and shallow superficial flow transport and their delivery to furrows and channels (Flanagan *et al.*, 1995).

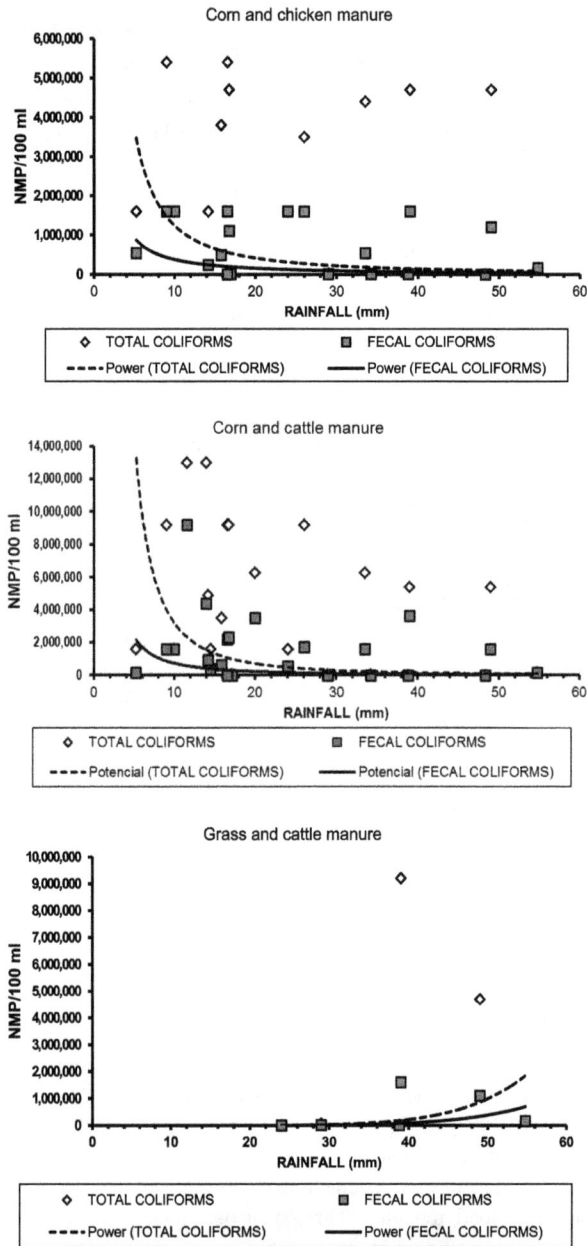

**Figure 1.** Tendency of fecal and total coliforms correlation with rainfall and corn treated with cattle and chicken manure, and grass treated with cattle manure (Flores *et al.*, 2012).

The soil loss in corn, agave tequilero and grass land with coverage in fields of Highland Jalisco are shown in Table 2 (Flores *et al.*, 2009). It can be observed that the highest soil loss was in the agave with respect to the pasture coverage and intermediate maize coverage. The largest loss of the tequila-agave soil, with respect to soil with respect to soil coverage, was attributed to morphology of the agave leaves; this plant has leaves designed to capture water and lead it to the plant base so that it is absorbed through its root system.

| Crop | Soil Loss (t/ha) | | Superficial Runoff (mm) | |
|---|---|---|---|---|
| | 2002 | 2003 | 2002 | 2003 |
| Corn | 13.35 | 11.62 | 159.4 | 180.2 |
| Grass | 0.51 | 0.38 | 78.6 | 68.1 |
| Agave tequilero | 27.04 | 36.59 | 237.6 | 278.1 |
| Bare soil | 24.93 | 35.13 | 222.5 | 248.6 |

**Table 2.** Soil loss and superficial surface runoff in corn and agave crops, grassland bare soil, observed in an Alfisol, of the Highlands of Jalisco, Mexico.

However, the type of flow created plays an important role in the loss of soil generated by the agave tequilero. The transport of particles removed by splashing along the slope rapidly decreases with the depth and velocity of flow, along with other important factors that influence this process such as the raindrop size and the final velocity of rain fall (Kinnell, 1993). The transport capacity inter rills result highly depends highly on the rain intensity and the length and slope of the land, even though momentum of rain *momentum* and the kinetic flows of energy also explain the increment in the transport capacity (Nearing *et al.*, 1991; Guy *et al.*, 1987; Gilley *et al.*, 1987). Also, an increment in the rate of removed particles has been observed under turbulent flow conditions (Nearing and Parker, 1994), such as what is generated after water flows out from agave plant.

Figure 2 shows the accumulation of the loss of soil over time and the kinetic energy of rain for corn, agave tequilero, grass and bare soil. It can be observed that grass has very low soil erosion, indicating the protective effect that it has on soil during the rainy season. Corn has a high response to soil erosion up until the crop has complete coverage over the soil, except when there are storms with a highly intensive erosive level. The agave tequilero has a slightly different behavior in bare soil, an effect caused by the generation of flows concentrated in the base of the agave tequilero.

In the context of soil erosion, the size and the type of particles that are transported also play an important role in the dispersion of contaminants, particularly those of phosphorous and pesticides. The content of lime and silt particles of soil is considered as the medium of transportation of nutrients to the bodies of water. (Sharpley and Menzel, 1987). Gabriels and Moldenhauer (1978) concluded that the distribution of the size of sediment particles (PSD) has an important implication for the carrying capacity and for the chemical deposition mechanism which can result in contaminated materials. Miller y Baharudd (1987) found that in Alfisol and Ultisol soils of Georgia, USA, that when the runoff started the fine layer of sediments was the predominant, but after 15 minutes, the layer from 0.15 to 0.05 mm became dominant, having lower amounts of sand and. However, Braskerud (2005) mentioned that the fine sediment particles are maintained in suspension for a longer period of time when runoff increases, which allows for its transportation at greater distances.

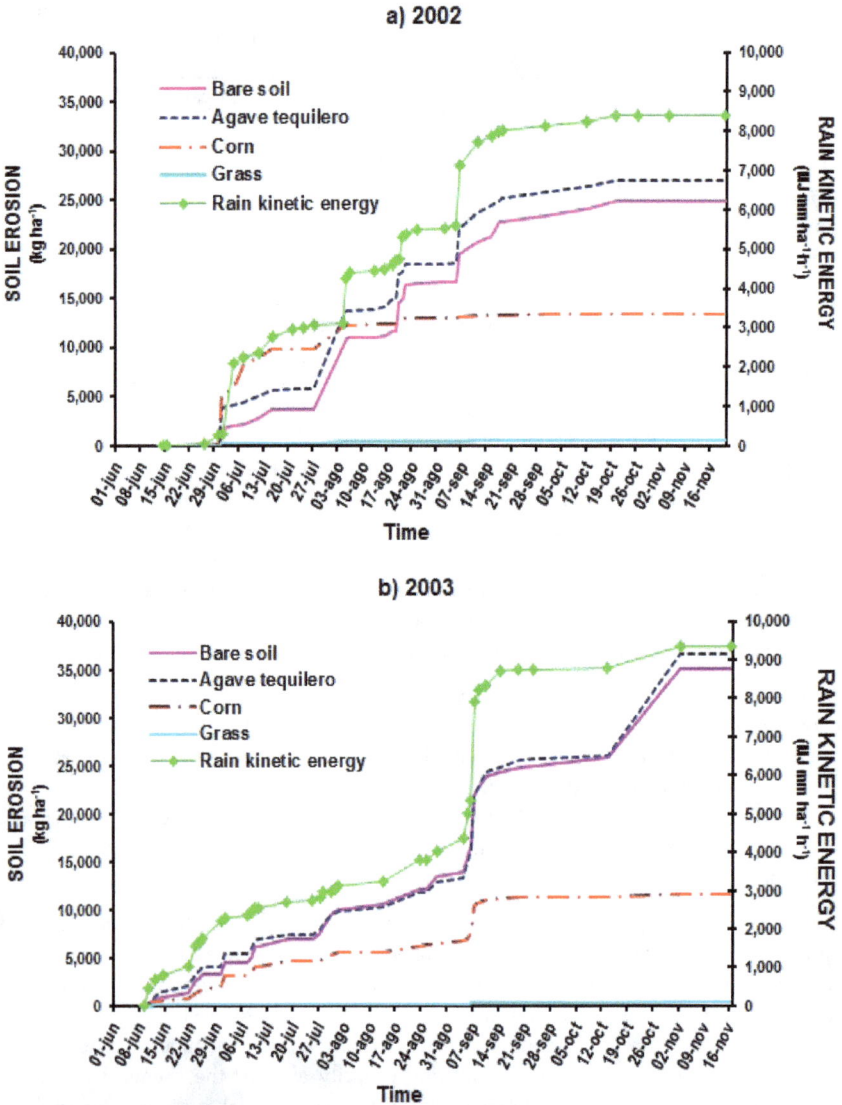

**Figure 2.** The soil erosion accumulated and the rain erosivity, observed during 2002-2003 cycle, in corn, grass, agave tequilero and bare soil, in the Highlands of Jalisco, Mexico.

Flores *et al.* (2009) carried out the PSD soil analysis for Alfisol in the Highlands of Jalisco. The sediment in suspension and that, which precipitated from the runoffs of fields where corn, agave tequilero and bare soil are present, are showed in Figure 3.

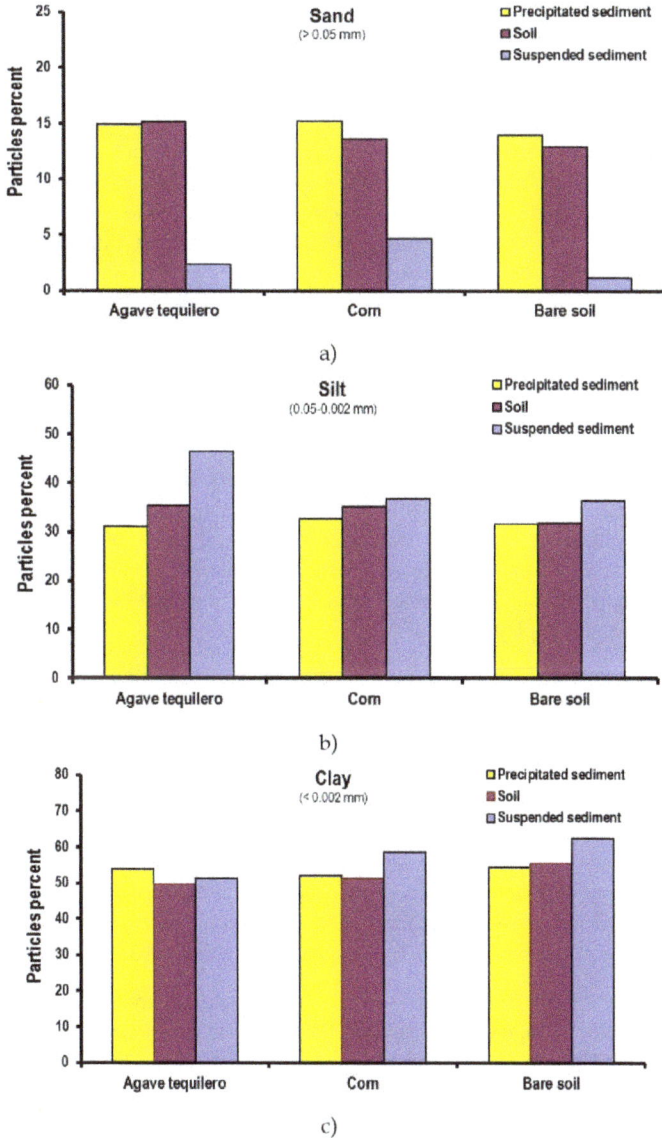

a)

b)

c)

**Figure 3.** Percent of particles: a) sand, b) silt and c) clay, proceeding from soil and suspended and precipitated sediment, of site runoff of the agave tequilero, corn and bare soil.

The sand content in soil and the precipitated sediment is similar; however the suspended sediment is reduced by important proportions, particularly in bare soil. The silt content in soil and the precipitated and suspended material from the agave tequilero, corn and bare soil are similar, but the suspended material increased for the three crops, although the silt amount was more considerable in agave tequilero; this was probably due to the increase in runoff observed for agave tequilero with respect to bare soil and corn. In general the silt layer of soil is considered more easily transported (Young, 1980) and is also the one maintains its proportion through a storm (Gabriels and Moldenhauer, 1978; Miller and Baharudd (1987). The percentage of clay in soil, precipitated material, and suspended material in the agave tequilero, corn, and bare soil are very similar. However, the clay in suspended sediment of corn and bare soil increases slightly, while the agave tequilero is reduced with respect to the precipitated sediment. It's reported that the particles size depends on the texture of the layer from the origin (Young, 1980), as in the present case where the clay layer in soil predominates; also important, is the duration of the storm (Miller and Baharudd, 1987) and the soil cover (Gilley *et al.*, 1986).

## 2.5. The surface runoff effect on the non-source pollution

Part of rainfall which is converted into runoff is grouped in three routes: superficial flow, subsurface flow, and subterranean flow (Linsley *et al.*, 1988). The runoff is a hydrologic term that describes the lateral movement of water which flows from the ground in surface and subsurface from causing a short term increase in the outflow of the drainage area, Meanwhile, the surface runoff only considers the movement of water over the soil surface occurring during heavy rain and flowing until it reaches a channel in the direction of the slope (Haygarth and Sharpley, 2000).

The surface runoff occurs in a laminar flow over the soil surfaces without an existing, precise concentration clearly moving towards the draining channel as such, the flow is visualized as one-dimensional process, proportional to the potential of storage per unit area (Mays, 2001; Wanielista *et al.*, 1997). On the other hand, the surface flow is considered as laminar in its initial condition, but when the depth and flow velocity increase to a critical value, it is converted into a turbulent flow, defined by the Reynolds Number (adimensionless index that express the product of medium-flow velocity in a channel and its effective hydraulic radius divided by the cinematic viscosity) higher than 500. The turbulence created by this surface flow delays the particles' sedimentation and maintains them in suspension, apart from increasing the flow capacity to separate new soil particles (Hillel, 1998).

Table 2 shows the surface runoff of corn, agave tequilero, grass and bare soil. The highest runoff was observed in agave tequilero, and the lowest in grass, corn and bare soil, had intermediate rate. The runoff observed in agave tequilero is attributable to the foliar plant structure; leaves reduce direct impact capture the water which falls over its leaves and is transported to the plant base, creating a concentrated flow in the soil. Table 2 shows the highest amount of runoff occurs in the agave tequilero, followed by the bare soil and corn, with grass being the least amount.

The surface runoff also plays an important role in corn and pasture when it interacts with the application of manure particularly from a point of view of water quality with fecal matter and total coliforms. Figure 4 shows the surface runoff with the content of fecal and total coliforms which increases with overland flow; however dilution of coliforms in water also occurs (Flores *et al.*, 2012).

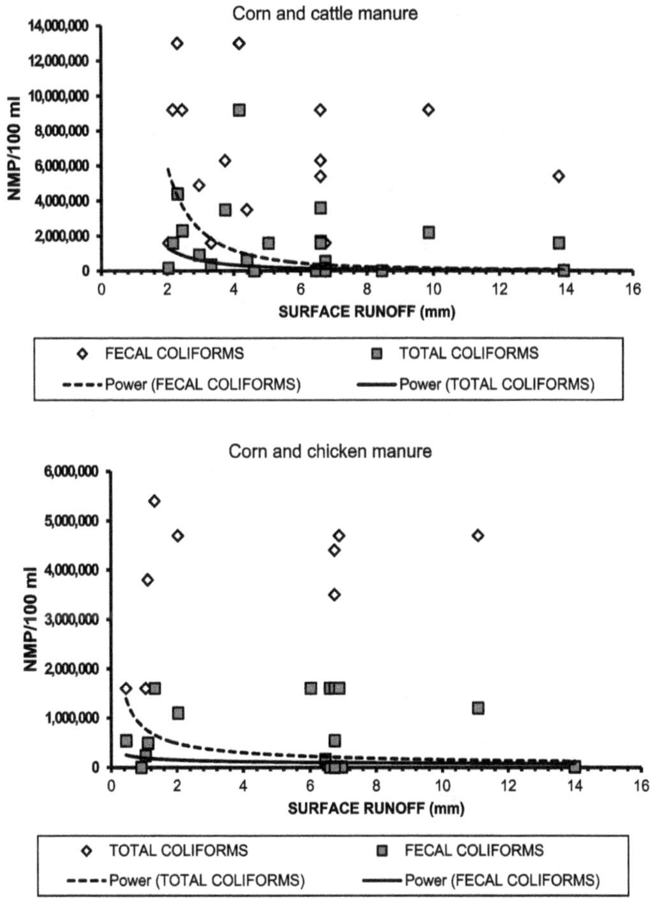

**Figure 4.** Relation of surface runoff with the total and fecal coliforms content in corn where treated with cattle and chicken manure.

## 2.6. Nitrogen losses from agricultural system

Water and nutrient deficiencies (nitrogen, phosphorous, calcium, etc.) are the most important factors that limit the productivity in any crop. One ton of agricultural soil in the first centimeters of the superficial layer can contain 4 kg of nitrogen, 1 kg of phosphorous, 20 kg of potassium and 10 kg of calcium. The soil erosion with a soil loss of about 18 $\text{tha}^{-1}\text{year}^{-1}$, would represent a total of 72 kg $\text{ha}^{-1}$ of nitrogen, which is almost half of the average of nitrogen fertilizer (152 kg $\text{ha}^{-1}\text{year}^{-1}$) applied to corn crops in the United States (Pimentel *et al.*, 1989).

Haygarth and Jarvis (1999) make reference that the practice of tilling on agriculture lands may have direct effects on the loss of phosphorous , also favoring vulnerability of the soil to erosion as a result of the removal of the vegetation cover . It was also indicated that the comparison between conventional tillage against conservation tillage resulted in 67% more runoff a, 90% loss of sediment, and a 90% increase in the loss of phosphorous.

From the combination of nutrient management practices in crops and soil, and rain characteristics, processes of such complexity arise, that when visualizing them as system, with different scales and occurrence times, understanding becomes difficult. This situation makes it difficult to define the preventive or corrective actions that should be taken to limit the transportations of nutrients from agricultural and pastoral lands to the bodies of water. One option is to consider the transportation of nutrients from land to surface water bodies as a system that requires knowledge of the components and an understanding of the interactions that exist between them.

The surface runoff is the means of nitrogen transportation, while the mechanisms that regulate it are the characteristics of storms, soil and the very own crop (Follett and Delgado, 2002; Haygarth and Jarvis, 1999). In Mexico, several studies demonstrate that organic and inorganic nitrogen, applied to the soil, is directly related to the amount of nitrates and ammonium in surface and groundwater during the NSP processes.

Estrada-Botello *et al.* (2002), in a study about the balance of inorganic nitrogen in the humid tropics of Mexico, found that the ammonium nitrogen loss was about 6.8 mg $\text{N-NH}_4$ $\text{L}^{-1}$ in the surface runoff, while the underground drainage was about 2.7 mg $\text{N-NO}_3$ $\text{L}^{-1}$; however, when some type of drainage is introduced in farming lands, the values of nitrogen leaching can be about 16.53 kg $\text{ha}^{-1}$ $\text{año}^{-1}$ without applying surface drainage and with maximum nitrate and ammonium concentrations at about 26.4 mg $\text{L}^{-1}$ and 22.0 mg $\text{L}^{-1}$, respectively. (Estrada-Botello *et al.*, 2007). In the region of Tuxtla in Veracruz, Mexico, Uribe-Gómez *et al.* (2002) found that the average loss of nitrogen in living wall terraces was around 23 kg $\text{ha}^{-1}$, probably due to the high levels of nitrogen from the decomposition pruning refuse that was placed on the surface soil.

In the Highlands of Jalisco, Mexico, the loss of nitrogen is mainly associated with hydrologic and edaphic factors along with the management of the production systems for agave tequilero, corn and pasture. As a result, the nitrates may be leached and contaminate the groundwater or transport the runoff toward the surface water together with the ammonic nitrogen and nitrate nitrogen (Goulding, 2004). However, it very difficult that they adsorb to iron oxides, like the goethite or hematite which are presented in the Luvisol soils in the Highlands of Jalisco (INEGI, 1994), due to the fact that the electrostatic attraction forces acting in this kind of mineral are weak (Parfitt, 1989).

In agreement with nitrogen measurements in runoff from field plots with agave tequilero runoff, pasture and corn, Flores *et al.* (2009), found tendencies of the related nitrogen loss with surface runoff, shown in Figure 5. The highest rate of nitrogen loss due to surface runoff was found in native pasture due to the contributions of refuse. Meanwhile the agave tequilero and corn had a medium rate, while the lowest rate of nitrogen loss was found in bare soil.

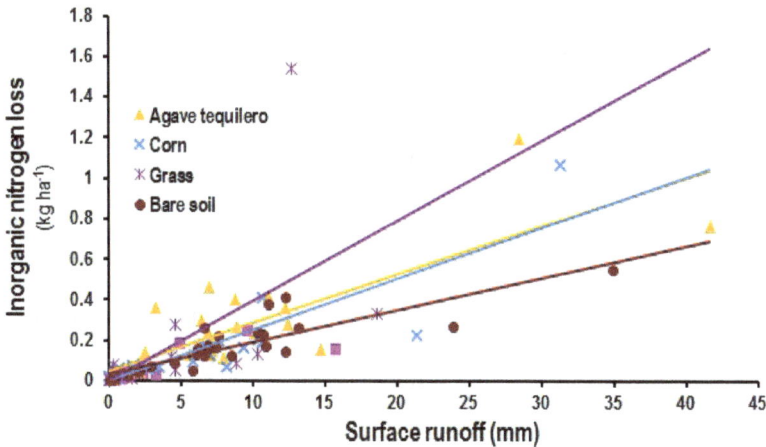

**Figure 5.** Relation between the loss of inorganic nitrogen and the surface runoff, in runoff plots that have bare soil, agave tequilero, corn and native grass.

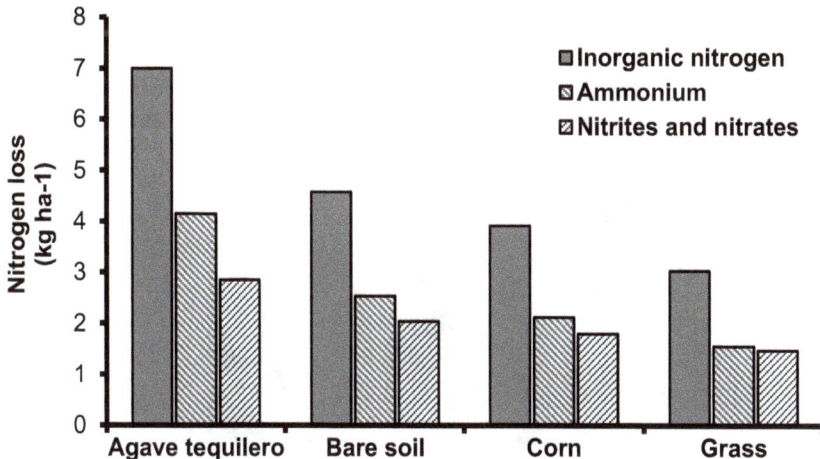

**Figure 6.** The loss of inorganic nitrogen in the form of ammonium, nitrites and nitrates s in four runoff plots the Highland of Jalisco, Mexico.

Figure 6 shows that the highest loss of inorganic nitrogen occurs in the agave tequilero with 7 kg ha⁻¹ and 4.15 kg ha⁻¹ of ammonic nitrogen at 2.85 kg ha⁻¹, of nitrate, and lower values for native grass lands at about 3.03 kg ha⁻¹, with nitrates at 1.56 kg ha⁻¹, and about 1.48 kg ha⁻¹ of ammonia nitrogen.

## 2.7. The loss of phosphorous in agricultural and livestock systems

In the south of the Jalisco Highlands soils are rich in iron oxides (INEGI, 1994), a characteristic that makes the loss of phosphorous utilized as the principal means of transporting mainly sediments to finer layers (Flores, 2004; Sharpley *et al.*, 2003; Sei *et al.*, 2002).

Figure 7, shows the content of total phosphorous (TP), organic phosphorous (OT) and inorganic phosphorous (IP), in the exported, precipitated and suspended sediments, which was measured in the runoff fields with agave tequilero, pasture, bare soil and corn. The concentration of IP loss was higher in the suspended sediment with respect to the precipitated sediment, due to the fact that the size of sediments is finer in the agave tequilero, bare soil and corn. For pasture, the IP in precipitated sediments was higher when compared with the others crops, but due to the fact that suspended sediment was not present in adequate amounts for the analysis, the IP was considered inestimable. This response is associated with a high clay content in suspended sediment, as can be observed in the distribution of sediment particle sizes for corn, agave tequilero and bare soil, results are shown in Figure 7.

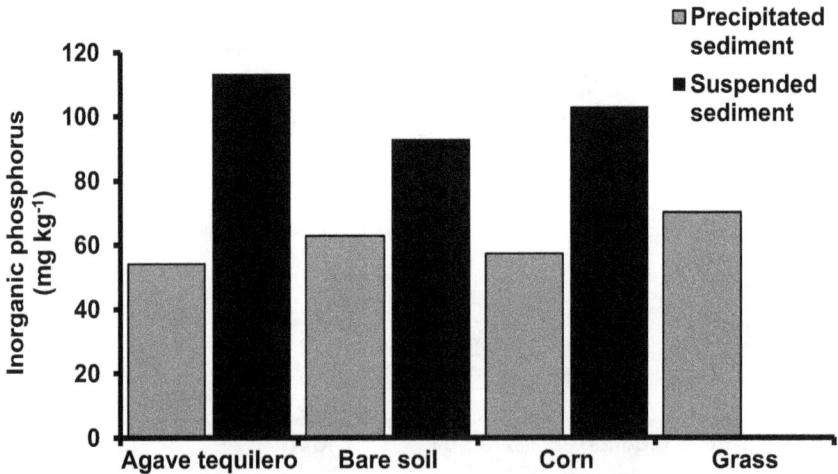

**Figure 7.** Content of inorganic phosphorous, in the exported, precipitated and suspended sediment, which was measured in the field plots of agave tequilero, grass, bare soil and corn.

Figure 7 shows the loss of TP, OP and IP, measured in fields for the agave tequilero, corn, grass and bare soil. In all of the cases, the loss of OP was about 71 to 80% of TP. Due to the fact that phosphorous particles are dependent on the exported material size, the agave tequilero showed the highest hydric erosion and consequently the highest TP loss. In contrast, grassland had the lowest soil erosion resulting in a minimum loss of TP.

Even though, the P loss is mainly associated with the finest layer of sediment (suspended sediment), high amounts of precipitated sediment have a similar granulometric composition to the original soil (Table 3). For that reason, this nutrient that is removed from soil and deposited along the land surface and in the drainage network can in the future, be considered as phosphorous source when it once again enters into the surface runoff. (Haygarth et al., 2000; Haygarth and Jarvis, 1999). Table 3 shows the soil and precipitated sediment amount; clay layers (< 2 μm), silt layers (2 – 50 μm) and sand layers (> 50 μm), with the P retention percent in the precipitated sediment. The precipitated sediment and the soil presented a similar granulometric composition; clay is around 55%, silt is about 32% and sand is approximately 13%. Table 3 also shows that the clay layer has a higher P retention (about 57.5%), and the sand has the lowest P retention (20%). Significant differences were found between layers.

| Name | *Layer | Precipitated sediment ( % ) | Soil ( % ) | **P retained ( % ) |
|------|--------|------------------------------|------------|--------------------|
| Clay | < 2 μm | 54.40 | 55.40 | 57.52 a |
| Silt | 2 - 50 μm | 31.65 | 31.71 | 39.59 b |
| Sand | > 50 μm | 13.95 | 12.89 | 20.07 c |

* From pipette method.
** Includes around 50% of layer< 2 μm.

**Table 3.** Percentage of Granulometric layers in the suspended and precipitated exported material of the soil and the P_retained in the precipitated sediment layers.

On the other hand, the suspended sediment principally constitutes for silt and clay (Table 3). This condition allows the particles be transported for greater distances to reach the superficial water bodies and transport with them, an important quantity of P (Figure 8). Much of the P contained in the exported material is of organic origin, but from the point of view of eutrophication, the inorganic P is more important; furthermore, a large amount is presented in suspended sediment. Pimentel et al. (1989) suggested that the soil erosion does not remove all the soil components in the same way; in fact, many studies indicate that the exported material is generally 1.3 to 5 times richer in organic material than the soil that remains. Organic matter is important to soil quality because of it's the positive effects on the water retention, the soil structure and the cationic exchange; furthermore, it is the highest nutrient source for plants. For this reason, 95% of nitrogen and the 15% to 80% of phosphorous is located in the organic matter.

From the results shown here, the transportation of P in the agricultural and pasture lands to the aquatic environments occurs by water erosion through surface runoff, in such a way that P is absorbed by soil particles and organic matter that is transported during runoff.

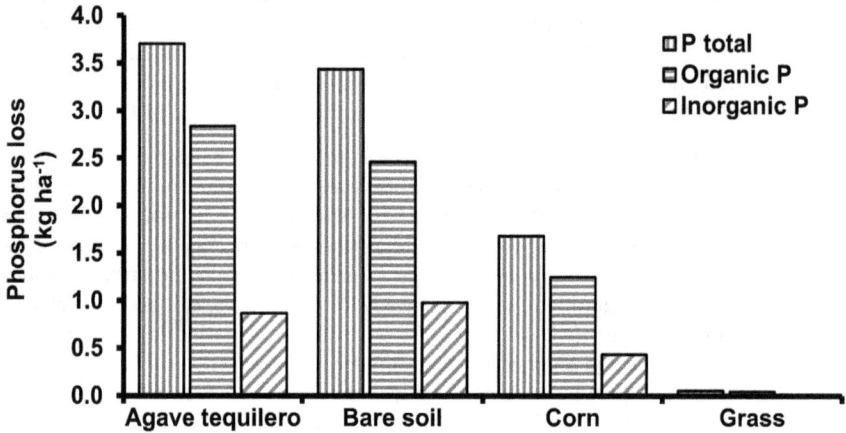

**Figure 8.** Loss of total phosphorous, organic and inorganic agave tequilero, corn, pasture lands and bare soil measured in field plots.

Golterman and Oude (1991) indicated that the surface freshwater reservoirs (lakes, dams, or dikes), requires a reduction of P and N before entering to the system, even though this task requires a large amount of money to improve reduction from 40% to 70% of the nutrients that cause the eutrophication. Loehr (1984) mentioned that using the Best Management Practices is a cost effective method to prevent or control the NSP from the agricultural and livestock industries.

## 2.8. Effects of Management Practices on pollution diffusion

The Best Management Practices (BMP) is conceptualized as the combination of cultural recommendations of management, structure and norms that in an effective and economic manner avoids, minimizes or mitigates the environmental impact on the productive processes of the agricultural and livestock systems (Minnesota Pollution Control Agency, 2000). However, giving sustainable solutions to NSP problems through the BMP's should stem from the farmers' decision making, identifying the causes and quantifying the effect. The simulation model is a tool that can be used for such purposes. Furthermore, this tool considers the farmers' decisions related to the agricultural management practices, that are based on processes associated with crop growth and its productivity, the soil erosion, and surface runoff from agricultural lands. Such is the case of the EPIC model (Erosion-Productivity Impact Calculator) which has been utilized to carry out the task at the field scale (Flores *et al.*, 2011; Flores *et al.*, 2009; Semaan *et al.*, 2007; 2006; Wang *et al.*, 2006; Guerra *et al.*, 2005; Villar-Sánchez *et al.*, 2003; Lacewell *et al.*, 1993), However, in combination with geographic information systems (GIS), it is possible to evaluate the impact of the agricultural systems at the watershed scale (Liu *et al.*, 2007).

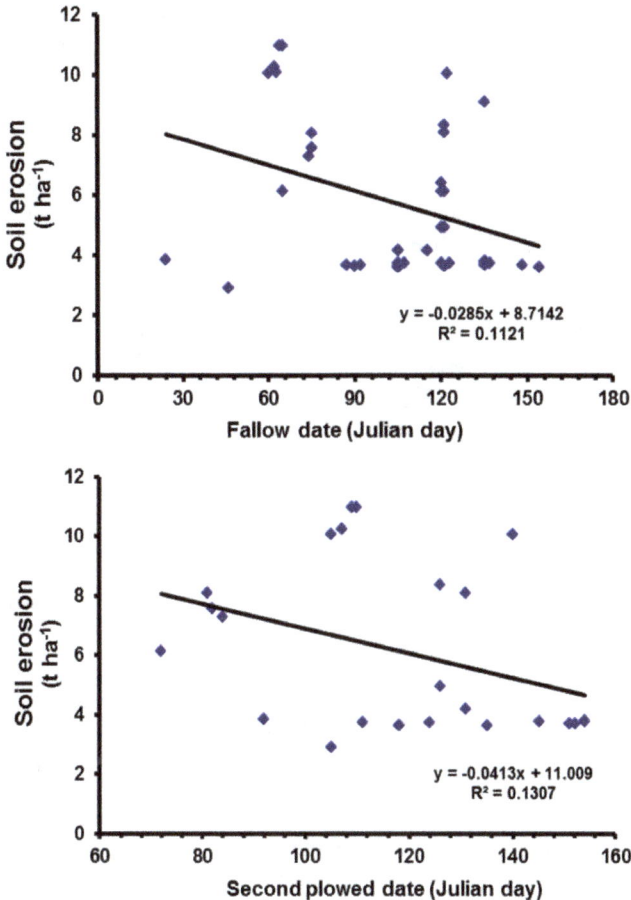

**Figure 9.** Relation between soil erosion in the corn field with the dates of fallowing and tilling during soil preparation.

Flores *et al.* (2009) used EPIC to estimate soil erosion to corn crops in the Highlands of Jalisco. The authors identified that the corn management practices correlated to soil loss at probability level ≤10 % significance, as follows: the fallow date (-0.335), the second plow date (-0.362), the sowing date(0.896), the amount of seed utilized (-0.250), the organic fertilization date (-0.418), the date of the first chemical fertilization (0.851), the date of the first weed control (0.390), the emergence date of corn plants (0.933), the second weed control date (0.939),the date of pest control for soil (0.835), the amount of products utilized in the soil to control pests (0.399), the silage date (-0.405), the corn harvest date (0.546). In the soil preparation, two practices are associated with soil erosion: fallowing and tilling of the ground as Figures 9a and b show. From these figures it can be seen that when these practices are carried out near to the sowing date, the tendency is to reduce the loss of soil, an effect associated with soil rugosity, which, when there is high soil erosion, reduces to produces a decrease in the runoff velocity, favoring the water infiltration (Kirkby, 1988; Podmore and Hugins, 1980).

Other management practices that resulted in important effects for soil loss were the sowing date (SD) at a rate of 429.5 kg for day with a delay in planting as shown in Figure 10. However, when the corn SD was separated from the corn systems at the date before start of growing season (GS),a loss of soil at a rate of about 30.4 kg per day was observed, while the later SD at the beginning of GS, had soil erosion increases at rate of 362.9 kg per day that was delayed by the SD.

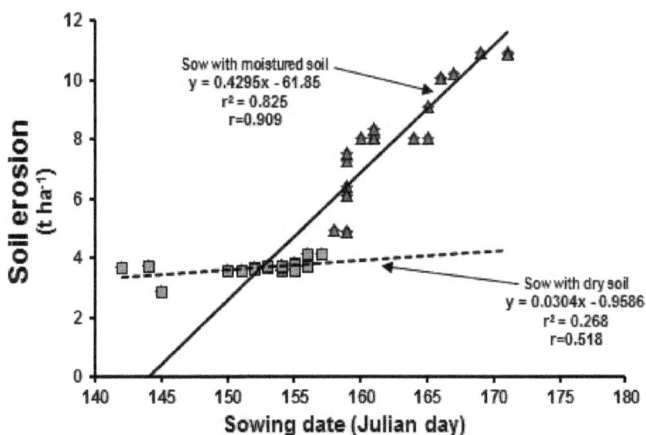

**Figure 10.** Relation between the losses of soil by erosion with the planting date of seasonal corn in the Highlands of Jalisco, Mexico.

On the other hand, the average accumulation of soil loss (SL) at field scale, cultivating corn was about 6.12 t ha⁻¹. But at the time when the accumulation of soil losses was separated from the SL in the crops with ST before the beginning of the GS, and the sowing was started after the GS, the average soil erosion was about 3.67 kg ha⁻¹ y 7.89 kg ha⁻¹, respectively (Figure 11). Figure 11, shows that the protective effect of foliage against soil erosion; the highest loss of soil occurs before the corn foliage covers the soil, which occurs around 40 days after the emergence of the crop. Afterwards, only storms with high erosive qualities can generate SL in the corn.

## 2.9. Scaling effects on the soil erosion, surface runoff and nutrient loss

Sims and Wolf (1994) mentioned that the processes of soil erosion, surface runoff, and nutrients loss, can be visualized through their management at three scales; 1) field scale, 2) farm scale, and 3) watershed scale, region and/or State level. At the field scale, the management practices are developed efficiently, favoring or limiting the impact of practices

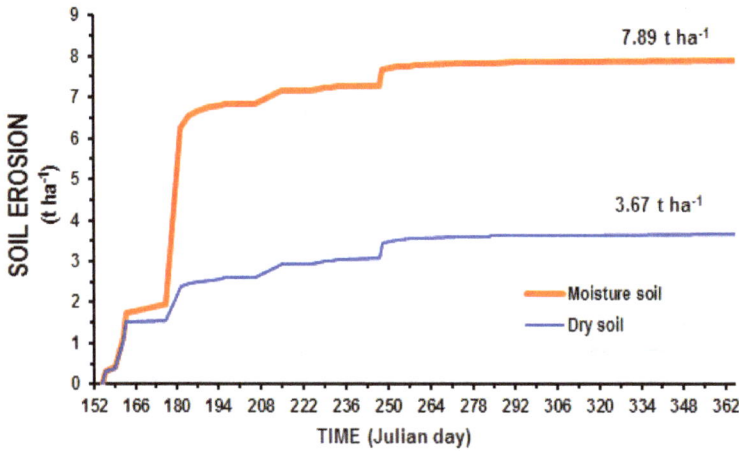

**Figure 11.** The effect of humid conditions on planting over the soil loss in farmlands of the watershed El Jihuite.

on soil erosion, surface runoff, nutrient loss and crop productivity, as an effect of physical processes such as volatilization, leaching, erosion, runoff and etc. At the farming scale, the farmer's decision is considered over many farms where practices are applied and the effect can be reduced or amplified depending of the environmental characteristics. At the watershed scale, regionally or stately, it is a similar scenario to the farm scale;, however at this level, the effects occur at different time and spaces of origin, with a high magnitude and an elevated cost to remediate the impact.

Flores (2004) showed the scale effect on soil erosion, surface runoff and the loss of inorganic P and N which have a non-lineal relationship with an area increase (Figure 12). Canto et al. (2011) also showed a non-lineal effect similar to that of Flores et al. (2004) between a runoff area with the production of sediments, even though this non lineal effect could be modified by soil coverage and other soil and drainage area characteristics (Liu et al., 2012; Delmas et al., 2012; Descroix et al., 2008). With a precipitation of about 1019.3 mm, a substantial reduction was observed in the amount of runoff in accordance with the measurement scale, such as show in Figure 12.

At a small scale of field plots (about 50 m² in area), a higher surface flow was generated, however, at drainage scale (about 22 hectares), the runoff was significantly lower. The runoff is generated during the period of flooding and excessive rain (Stomph et al., 2002); the influence of these mechanisms may reduce when the contributing area increases (Critchley and Siegert, 1991).

The soil erosion the same scale effect can also be observed where the great part of soil loss occurs during runoff in corn field, less in farm scale and even less in the drainage scale: furthermore, this exhibits a non-lineal correlation (Figure 12). At the watershed scales, Martínez et al. (2001) a non lineal response was found with the production of sediments in function to contributing area.

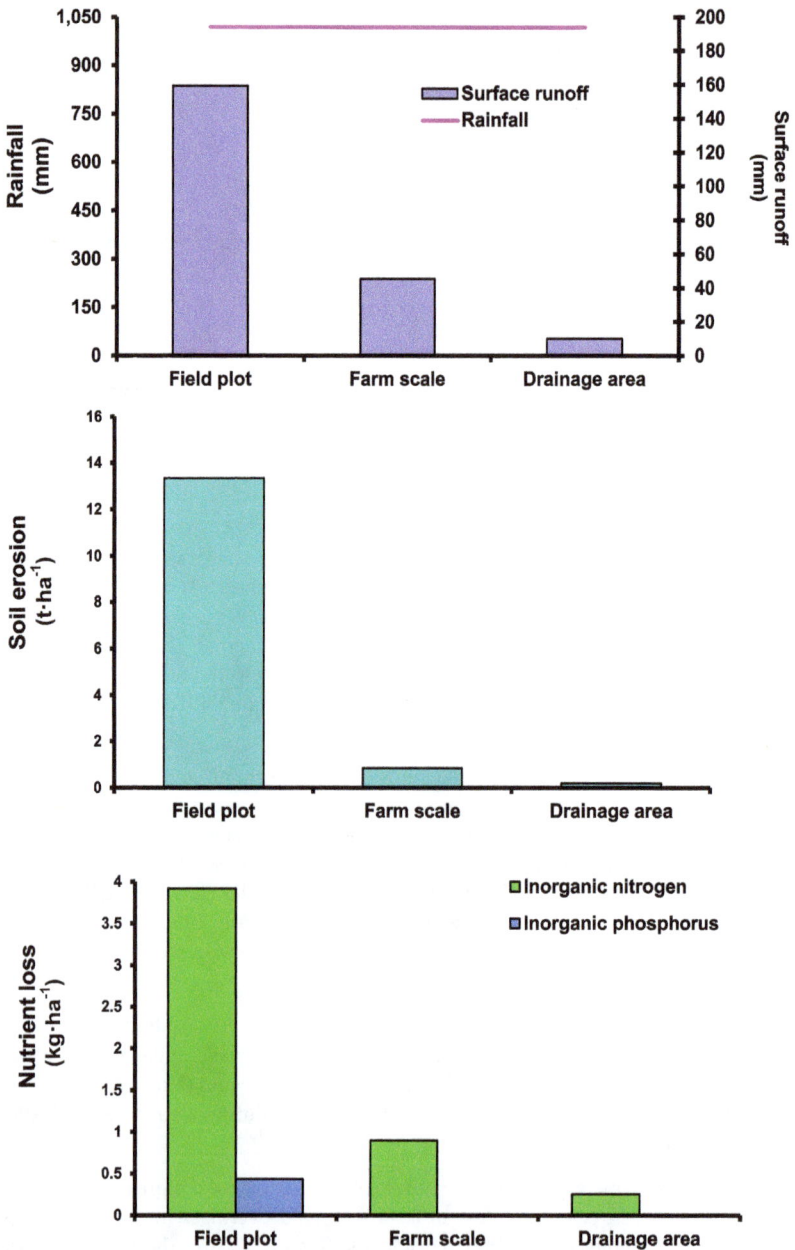

**Figure 12.** The surface runoff, soil erosion and loss of nutrients losses in measurement scales of corn field plots (50 m²), farm scale denominated watershed (5000 m²) and drainage area (22 ha).

The loss of nitrogen and phosphorous showed a non lineal response similar to that of the runoff and soil loss greater in the corn field plot and the lower in the drainage area, even though, the results were more contrasting with the nitrogen. The nutrient input depends on the characteristics of the drainage area the nutrient concentration, in a way so that the organic nitrogen has been correlated with arable land and the stream density, while the phosphorous with the soil texture (Arheimer and Lidén, 2000).

## 3. Conclusion

The agriculture and livestock industry in the Highlands of Jalisco, Mexico, has been an important source of economic growth. However, the use of technologies involving the application of large amounts of fertilizers and pesticides for prolonged periods of time, and often applied incorrectly, increases inefficiency and results in high environmental impact on the soil and water of the lower watershed area. This is known as non-point source pollution, which is considered to be the principal cause of the water deterioration in this region of México.

The first storms that had a smaller quantity of rain during the growing season and produced runoffs in the corn crops where cattle and chicken manure had been applied generated the highest levels of fecal coliforms. But the opposite effect was noticed in grassland where cattle graze, which requires a larger quantity of rain to increases the amount of fecal coliforms in the water.

The surface runoff for corn, agave tequilero, grass and bare soil exhibited different behavior. The highest runoff was observed in agave tequilero, and the lowest in grass, corn and bare soil, had intermediate rate. The runoff observed in agave tequilero is attributable to a concentrated flow created in the soil. The surface runoff plays an important role in corn and grass when it interacts with the application of manure, thereby reducing water quality with fecal matter and total coliforms, and adding nutrients.

With respect to the loss of soil, it was observed that the grassland has very low soil erosion, which indicates the protective effect that it has on soil during the rainy season; corn, however, has a high propensity to soil erosion up until it has complete soil coverage; the exception is when there are storms with a highly intensive erosive level. The agave tequilero has the highest soil erosion rate, and finally, the soil bare exhibited the slightly different behavior, an effect caused by the generation of flows concentrated in the base of the agave tequilero.

Nutrient loss in the aforementioned crops is mainly associated with hydrologic and edaphic factors. The highest rate of nitrogen loss due to surface runoff was found in native grass due to the introduction of refuse; meanwhile the agave tequilero and corn had a moderate rate of nutrient loss, while the lowest rate of nitrogen loss was found in bare soil.

The amount of exported phosphorus is dependent on the exported material size. The agave tequilero showed the highest level of soil erosion by water and consequently the highest total phosphorus loss. In contrast, grassland had the lowest soil erosion resulting in a minimum loss of total phosphorus. The loss of organic phosphorous was about 71% to 80% of total phosphorous. The inorganic phosphorous loss was higher in the suspended sediment than in the precipitated sediment, due to the fact that the agave tequilero and bare soil had high phosphorus loss, while corn and grass had less.

The management practices had an important impact on the process of soil erosion, runoff and nutrient loss. The most important effect that the sowing date had on the soil erosion rate may reduce when the planting date is delay. However, when corn was sowing before the growing season, the soil erosion had a rate of 30.4 kg per day. Nevertheless, sowing later of the growing season, the soil erosion increases at 362.9 kg per day.

The scale effect was observed in soil erosion, runoff, and nutrient loss. The loss of all of these showed a non-lineal response with the runoff and soil loss greater in the corn field plot and lower in the drainage area; this occurred even though the results were more contrasting with the nitrogen.

Due to, the highest surface runoff, nutrient loss and soil erosion effect occurs in the agricultural field scale (depending on the type of crops), the modification of current agricultural management practices toward sustainable practices is necessary. To achieve this goal, it is necessary to encourage farmers to use sustainable management practices, to reduce the rate of soil loss and runoff, and to improve efficiency in the use of nutrients in farmland and grazing. The results of these actions would be observed with sustained soil productivity and an improvement in the quality of water in dams, dikes or rivers of the region. At the watershed scale, legislative and socio-economic aspects that favor the improvement of agricultural systems, contributing not only to its development but also to the quality of life of the region's inhabitants, needs to be taken into consideration.

## Author details

Hugo Ernesto Flores López, Celia De La Mora Orozco,
Álvaro Agustín Chávez Durán and José Ariel Ruiz Corral[1]
*Instituto Nacional de Investigaciones Forestales Agrícolas y Pecuarias, México*

Humberto Ramírez Vega and Víctor Octavio FuentesHernández
*Universidad de Guadalajara, México*

## Acknowledgement

The results presented in this document came from research supported partially by the Consejo Nacional de Ciencia y Tecnología (CONACYT), the Instituto Nacional de Investigaciones Forestales Agrícolas y Pecuarias (INIFAP), the Fundación Produce Jalisco and the Centro Universitario de los Altos of the Universidad de Guadalajara. All of these institutions are acknowledged.

# 4. References

Alcalá G., J. 2011. Study of the dynamics of water quality in the Carts dam in Tepatitlán, Jalisco. Microbiological and physical - chemical analysis. Thesis engineer in livestock systems. Universidad de Guadalajara, Tepatitlán de Morelos, Jalisco, México. 72 pag.

Arheimer, B. & R. Lidén. 2000. Nitrogen and phosphorus concentrations from agricultural catchments-influence of spatial and temporal variables. Journal of Hydrology., vol. (227), No. 1–4, (January 2000), pp. 140-159, ISSN: 0022-1694

Braskerud, B. C. 2005. Retention of soil particles and phosphorus in small constructed wetlands in agricultural watersheds. *In*: Nutrient Management in Agricultural Watersheds: A Wetlands Solution, Dunne, E. J., K. R. Reddy, and O. T. Carton (Eds). Wageningen Academic Publishers. 121-131. ISBN: 9076998612, Wageningen, The Netherlands.

Cantón, Y., A. Solé-Benet, J. de Vente, C. Boix-Fayos, A. Calvo-Cases, C. Asensio, & J. Puigdefábregas. 2011. A review of runoff generation and soil erosion across scales in semiarid south-eastern Spain. Journal of Arid Environments, Vol. (75), pp. 1254-1261, ISSN: 0140-1963

Critchley, W. & K. Siegert. 1991. A Manual for the Design and Construction of Water Harvesting Schemes for Plant Production. Food And Agriculture Organization Of The United Nations. Rome, Italy.

De La Mora-Orozco, C., H.E. Flores López, J. García Velasco, Á.A. Durán Chávez & J.A. Ruiz Corral. 2011. Taxonomic characterization of phytoplankton in the Jihuite dam in Tepatitlán de Morelos, Jalisco. Technical brochure No. 8. Research station Center-Highland of Jalisco. ISBN: 978-607-425-669-7. Tepatitlán de Morelos, Jalisco, México.

Deliman, P.N. & M.W. Leigh. 1990. Assessing nonpoint pollution potential of surface waters using a Geographical Information System, Proceedings of the Symposia by the Committee on Watershed Management of the Irrigation and Drainage Division of the American Society of Civil Engineers. ISBN: 0-87262-767-5, Durango, Colorado, USA, July 9-11, 1990

Delmas, M., Pak, L.T., Cerdan, O., Souchère, V., Le Bissonnais, Y., Couturier, A., & Sorel, L. 2012. Erosion and sediment budget across scale: A case study in a catchment of the European loess belt. Journal of Hydrology, Vol. (420–421), pp. 255–263, ISSN: 0022-1694

Descroix, L., González Barrios, J.L., Viramontes, D., Poulenard, J., Anaya, E., Esteves, M., and Estrada, J. 2008. Gully and sheet erosion on subtropical mountain slopes: Their respective roles and the scale effect. Catena, Vol. (72), pp.325–339. ISSN: 0341-8162

Díaz-Zavaleta, G. 2007a. Analysis of some Limnological studies made in Mexican bodies of water. *In*: Limnology of Mexican Dams. Theoretical and practice aspects, Arredondo-Figueroa, J. L., G. Díaz-Zavaleta, y J. T. Ponce Palafox (Ed.), AGT Editor, S. A., pp. 348-368. ISBN: 978-968-463-136-6, México, D. F.

Díaz-Zavaleta, G. 2007b. A simplified model for the evaluation of trophic Lakes and dams of Mexico. *In*: Limnology of Mexican Dams. Theoretical and practice aspects, Arredondo-Figueroa, J. L., G. Díaz-Zavaleta, y J. T. Ponce Palafox (Ed.), AGT Editor, S. A., pp. 469-483. ISBN: 978-968-463-136-6, México, D. F.

Estrada-Botello, M.A., I. Nikolskii-Gavrilov, F. Gavi-Reyes, J.D. Etchevers-Barra, & O.L. Palacios-Velez. 2002. Inorganic nitrogen balance in a plot with subsurface drainage in the humid tropics. Terra, Vol. (20), No. (2), (Abril-Junio, 2002), pp. 189-198. ISSN: 0187-5779.

Estrada-Botello, M.A., Nikolskii-Gavrilov, I ., Mendoza-Palacios, J.D., Cristóbal-Acevedo, D., de La Cruz-Lázaro, E ., Brito-Manzano, N.P., Gómez-Vázquez, A. & Bakhlaeva-Egorova, O. 2007. Inorganic nitrogen leaching in agricultural soil with different types of drainage in humid tropic. Trópico húmedo, Vol. (23), No. (1), pp.1-14, ISSN: 0186-2979.

Ferguson, CH., De Roda-Husman, A.M., Altavilla, N., Deere, D., & Ashbolt, N. 2003. Fate and transport of surface water pathogens in watersheds. *Critical Reviews in Environmental Science and Technology.* Vol. 3, No. 3, pp. 299-361. ISSN: 1547-6537

Flanagan, D.C., J.C. Ascough H, A.D. Nicks, M.A. Nearing & J.M. Laflen. 1995. Overview of the WEPP erosion prediction model. Technical documentation WEPP. Report No. 10. National Soil Erosion Research Laboratory.USDA-ARS-MWA. West Lafayette, Indiana, USA.

Flanagan, DC, JCA II, AD Nicks, MA Nearing & JM Laflen. 1995. Chapter 1: Overview of the WEPP Erosion Prediction Model, In: *USDA- Water Erosion Prediction Project : Hillslope Profile and Watershed Model Documentation,* Vol. NSERL Report No. 10, Flanagan, D. C. and M. A. Nearing (Eds). USDA-ARS National Soil Erosion Research Laboratory. West Lafayette, Indiana, USA.

Flores-López., H.E., Paredes M., R., Ruvalcaba G., J.M., De La Mora O., C., Pérez D., J.F. & Ireta M., J. 2011. Methodology for the assessment of the value-added form the Programme of Corn with High Performance (PROEMAR) 2010 in Jalisco and Guanajuato. ISBN: 978 - 607-425-746-5, Technical brochure No. 2. Research Station Centre-Highland of Jalisco. Tepatitlán de Morelos, Jalisco, México.

Flores-López, H.E. 1994. Agroclimatic analysis of the Northeast of Jalisco, Mexico, for management in the production of maize (*Zea mays* L.). Master of science thesis. Colegio de posgraduados, Texcoco, Estado de México, México.

Flores-López, H.E. 2004. Routes of surface transport of nitrogen and phosphorus in a drainage area of Jalisco, Mexico. Doctor of science thesis. Natural Resources Institute. Colegio de posgraduados, Texcoco, Estado de México, México

Flores-López, H.E., A.L. Hernández-Jáuregui, U. Figueroa-Viramontes, & A.A. Castañeda Villanueva. 2012. Microbiological water quality resulting from nonpoint pollution from the application of manure on maize and grass. Water Technology and Sciences (in Spanish). Vol. III, Special Number TyCA-RETAC, February-March, 2012, pp. 127-141, ISSN: 2007-2422

Flores-López., H.E., R. Carrillo-González., N. Francisco-Nicolás., C. Hidalgo-Moreno., J.A Ruíz-Corral., H., A. A. Castañeda-Villanueva., & R. Velazco-Nuño. 2009. Contributions of nitrogen and phosphorus from three agricultural systems of "el Jihuite" watershed in Jalisco, Mexico. Agrociencia, Vol. (43), No. (7), 1 de octubre - 15 de noviembre, (2009), pp. 659-699. ISSN: 1405-3195

Follett, R.F. & J.A. Delgado.2002. Nitrogen fate and transport in agricultural systems. Journal of Soil and Water Conservation, Vol. (57), No.(6), pp.402-408, ISSN: 1941-3300

Gabriels, D. & W.C. Moldenhauer. 1978. Size distribution of eroded material from simulated rainfall: effect over a range of texture. Soil Science Society of America Journal, Vol. (42), No. (6), pp.954-958, ISSN: 1435-0661

Gilley, J.E., S.C. Finkner & G.E. Varvel. 1986. Runoff and erosion as affected by sorghum and soybean residue. Transaction of the ASAE, Vol. (29), pp. 1605-1610, ISSN: 0001-2351

Gold, A. J., & C. A. Oviatt. 2005. Nitrate in fresh water and nitrous oxide in the atmosphere. *In*: Nitrate, Agriculture and the Environment, Addiscott, T. M. (ed), 110-126, CABI PUBLISHING. ISBN: 0-85199-913-1, Cambridge, Massachusetts, USA.

Golterman, H.L. & N.T. de Oude. 1991. Eutrophication of lakes, rivers and coastal seas. *In*: The handbook of environmental chemistry: water pollution, O. Hutzinger (Ed.). 80-124. Springer-Verlag, ISBN: 0-387-51599-2, Berlin, Germany

Goulding, K. 2004. Pathways and losses of fertilizer nitrogen at different scales. *In*: Agriculture and the Nitrogen Cycle, Mosiet, A. R., J. K. Syers, and J. R. Freney (eds), 209-217, SCOPE 65. ISBN: 1-55963-710-2, Washington, D. C. USA.

Guerra, L.C., Hoogenboom, G., Hook, J.E., Thomas, D.L., Boken, V.K., & Harrison, K.A. 2005.Evaluation of on-farm irrigation applications using the simulation model EPIC. Irrigation Science, Vol. (23), pp.171–181, ISSN: 1432-1319

Guy, B.T., W.T. Dickinson & R.P. Rudra. 1987. The roles of rainfall and runoff in the sediment transport capacity of interrill flow. Transaction of the ASAE, Vol. (30), pp. 1378-1386. ISSN: 0001-2351

Hairsine, P.B. & C.W. Rose. 1992. Modeling water erosion due to overland flow using physical principles 1. Sheet flow. Water Resources Research, Vol. (28), pp.237-253, ISSN: 1944–7973

Haygarth, P.M. & A.N. Sharpley 2000.Terminology for phosphorus transfer. Journal of Environmental Quality, Vol. (29), No. (1), pp. 10-15, ISSN: 1537-2537

Haygarth, P.M. & S.C. Jarvis. 1999. Transfer of phosphorus from agricultural soils. Advances in Agronomy, Vol. (66), pp.195-249, ISSN: 0065-2113

Haygarth, P.M., A.L. Heathwaite, S.C. Jarvis & T.R. Harrod. 2000. Hydrological factors for phosphorus transfer from agricultural soils. Advance in Agronomy, Vol. (69), pp. 153-178, ISSN: 0065-2113

Hillel, D. 1998. Environmental soil physics. Academic Press. ISBN: 978-0-12-505103-3. San Diego, California, USA.

INEGI. 1994. Tour description. Section Guadalajara, Jalisco – Zamora, Michoacán. *In*: Guide for technical tours "1" and "10": Guadalajara – México city. 15° Congreso Mundial de la Ciencia del Suelo. 10–16 de Julio. INEGI. ISBN: 970-13-0362-B, Aguascalientes, México.

Jamieson, R., R. Gordon, D. Joy, D., & H. Lee. 2004. Assessing microbial pollution of rural surface waters. A review of current watershed scale modeling approaches. *Agricultural Water Management*. Vol. (70), pp. 1-17, ISSN: 0378-3774

Kinnell, P.I.A. 1993. Sediment concentrations resulting from flow depth/drop size interactions in shallow overland flow. Transaction of the ASAE. Vol. (36), No. (4), pp. 1099-1103, ISSN: 0001-2351

Kirkby, M. 1988. Hillslope runoff processes and models. Journal of Hydrology, Vol. (100), No. (1-3), (July 1988), pp. 315-339, ISSN: 0022-1694

Lacewell, R.D., M. E. Chowdhury, K. J. Bryantz, J. R. Williams, &V. W. Bensony. 1993. Estimated Effect of Alternative Production Practices on Profit and Ground Water Quality: Texas Seymour Aquifer. Great Plains Research, Vol. (3), pp. 189-213, ISSN: 1052-5165

Linsley, R.K., M.L.H. Kohler & J.L.H. Paulhus. 1988. Hydrology for engineers. SI metric edition. McGraw-Hill Company Book. ISBN: 0-07-100599-4, United Kingdom.

Liu, L., J.R. Williams, A. J.B. Zehnder, & H. Yang. 2007. GEPIC – modelling wheat yield and crop water productivity with high resolution on a global scale. Agricultural Systems, Vol. (94), No. (2), (May 2007) pp. 478–493, ISSN: 0308-521X

Liu, Y., B. Fu, Y. Lü, Z. Wang & G. Gao. 2012. Hydrological responses and soil erosion potential of abandoned cropland in the Loess Plateau, China. Geomorphology Vol. (138), pp. 404–414, ISSN: 0169-555X

Loehr, R.C. 1984. Pollution control for agriculture. Academic Press, Inc. ISBN: 9780124552708, Orlando, Florida, USA.

Martínez M., M.R., R. López M. & E. Hernández F. 2001. Relationship sediment - erosion in the upper basin of Papaloapan. Abstracts of the 11th National Congress of irrigation. Symposium 5. Integrated watershed management ANEI-S50109. Guanajuato, Guanajuato, México.

Mays, L.W. 2001. Water resources engineering. John Wiley & Sons, Inc. ISBN: 0-471-29783-6, New York, USA.

McCool, D.K. & K.G. Renard. 1990. Water erosion and water quality. Advance in Soil Science, Vol. (13), pp.175-185, ISSN: 0176-9340

Meyer, L.D. & W.H. Wishmeier. 1969. Mathematical simulation of the process of soil erosion by water. Transaction of the ASAE, Vol. (12), pp. 754-758, 762. ISSN: 0001-2351

Miller, W.P. & M.K. Baharuddin. 1987. Particle size of interrill eroded sediments from highly weathered soils. Soil Science Society of America Journal, Vol. (51), pp. 1610-1615, ISSN: 1435-0661

Minnesota Pollution Control Agency. 2000. Protecting water quality in urban areas. 15.03.2012. Available from http://www.pca.state.mn.us/water/pubs/sw-bmpmanual.html.

Mishra, A., B.L. BEnham, & S. Mostaghimi. 2008. Bacterial transport from agricultural lands fertilized with animal manure. Water Air Soil Pollution. Vol. (189), 2008, pp. 127-134. ISSN: 1573-2932

Nearing, M.A. & S.C. Parker. 1994. Detachment of soil by flowing water under turbulent and laminar conditions. Soil Science Society of America Journal, Vol. (58), pp. 1612-1614, ISSN: 1435-0661

Nearing, M.A., J.M. Bradford & S.C. Parker. 1991. Soil detachment by shallow flow at low slopes. Soil Science Society of America Journal, Vol. (55), pp.339-344, ISSN: 1435-0661

Oliver, D.M., L. Heathwaite, P.M. Haygarth, & C.D. Clegg, 2005. Transfer of Escherichia coli to Water from Drained and Undrained Grassland after Grazing. Journal of Environment Quality. Vol. (34), pp. 918-925, ISSN: 1537-2537

Pachepsky, Y.A., A.M. Sadeghi, S.A. Bradford, D.R. Shelton, A.K. Guber, & T. Dao. 2006. Transport and fate of manure-borne pathogens: Modeling perspective. *Agricultural Water Management*. Vol. (86), pp. 81-92, ISSN: 0378-3774

Parfitt, R.L. 1989. Phosphate reactions with natural allophone, ferrihydrite and goethite. Journal of Soil Science, Vol. (40), pp.359-369., ISSN: 2231-6833

Pietilainen, O.P. 1997. Agricultural phosphorus load and phosphorus as a limiting factor for algal growth in Finnish lakes and rivers. *In*: Phosphorus loss from soil to water, Tunney, H., O.T. Carton, P.C. Brooks y A.E. Johnston (Ed.), CAB INTERNACTIONAL. ISBN: 0-85199-156-4. New York, USA.

Pimentel, D., T.W. Culliney, I.W. Buttler, D.J. Reinemann & K.B. Beckman. 1989. Low input sustainable agriculture using ecological management practice. Agriculture, Ecosystems & Environment, Vol, ( 27), No. (1), pp.3-24, ISSN: 0167-8809

Podmore, T.H. & L.F. Huggins. 1980. Surface roughness effects on overland flow. Transaction of the ASAE, Vol. (23), No. (6), pp. 1434-1439, 1445. ISSN: 0001-2351

Ramírez V., H., Martínez S., J.A., Flores L., H.E., Díaz M., P. Alemán M., V. & Guzmán A., J.M. 1996. Study of the water quality of the Jihuite dam. *In*: Executive summary of the project "The Jihuite agricultural and forestry watershed management". SAGAR-INIFAP-CIPAC-CEAJAL.

Saini, R., L.J. H. Alverson, & J.C Lorimor. 2003. Rainfall timing and frequency influence on leaching of Escherichia coli RS2G through soil following manure application. Journal of Environmental Quality. Vol. (32), No.(5), pp. 1865-1872. ISSN: 1537-2537

Schnoor, J.L. 1996. Environmental modeling: fate and transport of pollutants in water, air and soil. John Wiley & Sons, Inc. ISBN: 0-471-12436-2, New York, USA

Sei, J., J.C. Jumas, J. Oliver-Fourcade, H. Quiquampoix & S. Stauton. 2002. Role of iron oxides in the phosphate adsorption properties of kaolinites from the Ivory coast. Clay and Clay Minerals, Vol. (50), No.(2), pp.217-222, ISSN: 1552-8367

Semaan, J, Flichman, G., Scardigno, A., & Steduto, P. 2007. Analysis of nitrate pollution control policies in the irrigated agriculture of Apulia Region (Southern Italy): A bio-economic modelling approach. Agricultural Systems, Vol. (94), pp.357–367, ISSN: 1405-3195

Sharpley, A.N. & Menzel, R.G. 1987.The impact of soils and fertilizer phosphorus on the environment. Advances in agronomy, Vol. (41), pp.297-324, ISSN: 0065-2113

Sharpley, A.N., T. Daniel, T. Sims, J. Lemunion, R. Stenvens & R. Parry. 2003. Agricultural phosphorus and eutrophication. United States Department of Agriculture. Agricultural Research Service.ARS-149. 04.02.2012. Available from: http://www.ars.usda.gov/is/np/Phos&Eutro2/agphoseutro2ed.pdf

SIAP-SAGARPA. 2011. Agriculture Yearbook 2010. 15.12.2011. Available from: http://www.siap.gob.mx/index.php?option=com_content&view=article&id=181&Itemid=426.

Sims, J.T. & Wolf, D.C. 1994. Poultry waste management agricultural and environmental issues. Advances in agronomy, Vol. (52), pp.1-83, ISSN: 0065-2113

Sobrino-Figueroa, A. 2007. Biological indicators in aquatic systems. *In*: Limnology of Mexican Dams. Theoretical and practice aspects, Arredondo-Figueroa, J. L., G. Díaz-

Zavaleta, y J. T. Ponce Palafox (Ed.). 677-719, AGT Editor, S. A. ISBN: 978-968-463-136-6, México, D. F.

Soupir, M.L., S. Mostaghimi, E.R. Yagow, C. Hagedorn, & D.H. Vaughan. 2006. Transport of fecal bacteria from poultry litter and cattle manures applied to pastureland. Water, Air, and Soil Pollution. Vol. (169), pp. 125-136, ISSN: 1573-2932

Stomph, T.J., N. de Ridder, T.S. Steenhuis & N.C. Van De Giesen. 2002. Scale effects of hortonian overland flow and rainfall-runoff dynamics: laboratory validation of a process-based model. Earth Surface Process Landforms, Vol. (27), pp.847-855, ISSN: 1096-9837

Torres, R. & L.G. Calva. 2007. Importance of bacteria in dams. *In:* Limnology of Mexican Dams. Theoretical and practice aspects, Arredondo-Figueroa, J.L., Díaz-Zavaleta, G. y Ponce Palafox, J.T. (Ed.). 371-412, AGT Editor, S.A. ISBN: 978-968-463-136-6, México, D.F.

Uribe-Gómez, S., N. Francisco-Nicolás, & A. Turrent-Fernández. 2002. Soil loss in an Entisol with practices of conservation in Los Tuxtlas, Veracruz, Mexico. Agrociencia, Vol. (36), pp.161-168, ISSN: 1405-3195

Villar-Sánchez, B., López-Martínez, J., Pérez-Nieto, J., y Camas-Gómez, R. 2003. Simulation EPIC Model Applied in the Prediction of Soil Tillage Systems Effects. TERRA, Vol. (21), No. (3), pp. 381-388. ISSN: 0187-5779

Wang, X., Harmel, R.D., Williams, J.R., & Harman, W.L. 2006. Evaluation of EPIC for assessing crop yield, runoff, sediment and nutrient losses from watersheds with poultry litter fertilization. Transactions of the ASABE, Vol. (49), No.(1), pp.47–59, ISSN: 0001-2351

Wanielista, M., R. Kersten & R. Eaglin. 1997. Hydrology: water quantity and quality control. John Wiley & Sons, Inc. ISBN: 0-471-07259-1, New York, USA.

Wischmeier, W. H. & Smith, D.D. 1978. Predicting rainfall erosion losses-a guide to conservation planning.U.S.Department of Agriculture.Agriculture Handbook no. 537. USA. 26.02.2012. Available from: http://topsoil.nserl.purdue.edu/usle/AH_537.pdf

Xie, Y., B. Liu & M.A. Nearing. 2002. Practical thresholds for separating erosive and non-erosive storms. Transaction of the ASAE, Vol. (45), pp. 1843-1847, ISSN: 0001-2351

# Climate Change and Carbon Sequestration in Dryland Agriculture

# Chapter 6

# Climate Change and Carbon Sequestration in Dryland Soils

Peeyush Sharma, Vikas Abrol, Shrdha Abrol and Ravinder Kumar

Additional information is available at the end of the chapter

## 1. Introduction

Climate change is the biggest threat to humanity with implications for food production, natural ecosystems, health etc. The primary greenhouse gases are carbon dioxide ($CO_2$), methane ($CH_4$) and nitrous oxide ($N_2O$). Although carbon dioxide is the most prevalent greenhouse gas in the atmosphere, nitrous oxide and methane have longer durations in the atmosphere and absorb more long-wave radiations. Therefore, small quantities of methane and nitrous oxide can have significant effects on climate change. The mean global level of greenhouse gases in the atmosphere is increasing to a level that can generate serious climate changes in air temperature, aggressive weather cycles and greater frequency of storms (Osborn et al., 2000). The primary sources of greenhouse gases in agriculture are the production of nitrogen based fertilizers; the combustion of fossil fuels such as coal, gasoline, diesel fuel, natural gas; and waste management. Livestock enteric fermentation results in methane emissions. Increased levels of greenhouse gases enhance the naturally occurring greenhouse effect by trapping even more of the sun's heat, resulting in a global warming effect. The average surface temperature of the earth is likely to increase by 2 to 11.5°F (1.1-6.4°C) by the end of the 21st century, relative to 1980-1990, with a best estimate of 3.2 to 7.2°F (1.8-4.0°C) (Fig. 1). The average rate of warming over each inhabited continent is very likely to be at least twice as large as that experienced during the 20th century.

These changes in greenhouse gas emissions generally are linked to human activities. Scientists have concluded that warming of the climate system is "equivocal" and there is a "very high confidence that the globally averaged net effect of human activity since 1750 has been one of warming" (IPCC, 2007). The concentration of carbon dioxide ($CO_2$) in the atmosphere increased from 285 ppm at the end of the nineteenth century, before the industrial revolution, to about 366 ppm in 1998 (equivalent to a 28-percent increase) as a consequence of anthropogenic emissions of about 405 gigatonnes of carbon (C) ($\pm$ 60 gigatonnes C) into the atmosphere (IPCC, 2001). This increase was the result of fossil-fuel combustion and cement production (67 percent) and land-use changes (33 percent). Acting as carbon sinks, the marine and terrestrial ecosystems have absorbed 60 percent of these emissions while the remaining 40 percent has resulted in the observed increase in atmospheric $CO_2$ concentration. Agricultural ecosystems represent 11% of the earth's land surface and include some of the most productive and carbon-rich soils. Agriculture accounts for approximately 13% of total global anthropogenic emissions and is responsible for about 47% and 58% of total anthropogenic emissions of methane ($CH_4$) and nitrous oxide ($N_2O$). Besides $CH_4$ from enteric fermentation (32%), $N_2O$ emissions from soils due to fertilization constitute the largest sources (38%) from agriculture (US-EPA, 2006; IPCC, 2007; Stern, 2006). The annual greenhouse gas emissions from agriculture are expected to increase in coming decades due to increased demand for food and shifts in diet. Conservation tillage, nutrient management, cover cropping and crop rotation can drastically increase the amount of carbon stored in soils. Now scientists use carbon dioxide equivalents to calculate a universal measurement of greenhouse gas emissions as greenhouse gases have varying global warming potentials Table (1).

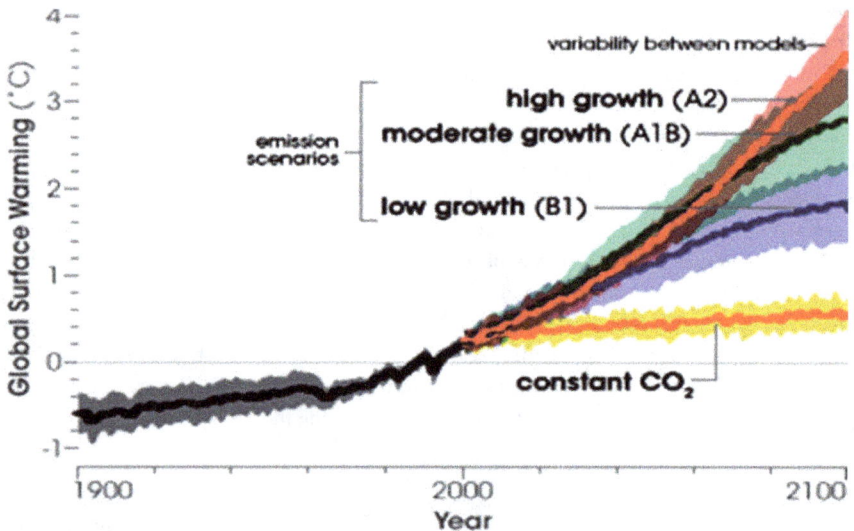

**Figure 1.** Temperature projections to the year 2100, based on a range of emission scenarios and global climate models. The orange line ("constant $CO_2$") projects global temperatures with greenhouse gas concentrations stabilized at year 2000 levels. Source: NASA Earth Observatory, based on IPCC Fourth Assessment Report (2007)

| Sources | 1990 | 1995 | 2000 | 2005 | Avg. 2001-2005 |
|---|---|---|---|---|---|
| | Million metric tons $CO_2$ equivalent (MMTCO$_2$- Eq) | | | | |
| U. S. Agricultural Activities | | | | | |
| GHG Emissions (CH$_4$ and N$_2$O) | | | | | |
| Agriculture soil management[a] | 366.9 | 353.4 | 376.8 | 365.1 | 370.9 |
| Enteric fermentation[b] | 115.7 | 120.6 | 113.5 | 112.1 | 115.0 |
| Manure management | 39.5 | 44.1 | 48.3 | 50.8 | 45.6 |
| Rice cultivation | 7.1 | 7.6 | 7.5 | 6.9 | 7.4 |
| Agricultural residue burning | 1.1 | 1.1 | 1.3 | 1.4 | 1.2 |
| Subtotal | 530.3 | 526.8 | 547.4 | 536.3 | 540.1 |
| Carbon sinks | | | | | |
| Agricultural soils | (33.9) | (30.1) | (29.3) | (32.4) | (31.7) |
| Other | NA | NA | NA | NA | NA |
| Subtotal | (33.9) | (30.1) | (29.3) | (32.4) | (31.7) |
| | | | | | |
| Net emissions, Agriculture | 496.4 | 496.7 | 518.1 | 503.9 | 508.4 |
| | | | | | |
| Attributable CO$_2$ emissions[c] Fossil fuel/mobile combustion | 46.8 | 57.3 | 50.9 | 45.5 | 52.6 |
| | | | | | |
| % All emissions, Agriculture[d] | 8.5% | 8.0% | 7.7% | 7.4% | 8.0% |
| % Total sinks, Agriculture | 4.8% | 3.6% | 3.9% | 3.9% | 4.0% |
| | | | | | |
| % Total emissions, forestry | 0.2% | 0.2% | 0.2% | 0.3% | 0.3% |
| % Total sinks, forestry[e] | 94.3% | 92.0% | 94.8% | 94.7% | 95.0% |
| | | | | | |
| Total GHG emissions, All sectors | 6,242.0 | 6,571.0 | 7,147.2 | 7,260.4 | 6,787.1 |
| Total carbon sinks, All sectors | (712.8) | (828.8) | (756.7) | (828.5) | (801.0) |
| Net emissions, All sectors | 5,529.2 | 5,742.2 | 6,390.5 | 6,431.9 | 5,986.1 |

Source: EPA, Inventory of U.S. Grenhouse Gas Emissions and Sinks: 1990-2005, April 2007,
[http://epa.gov/climatechange/emissions/usinventoryreport.html].
a. N$_2$O emissions from soil management and nutrient/chemical applications on croplands.
b. CH$_4$ emissions from ruminant livestock.
c. Emissions from fossil fuel/mobile combustion associated with energy use in the U.S. agriculture sector (excluded from EPA's reported GHG emissions for agricultural activities).
d. Does not include attributable CO$_2$ emissions from fossil fuel/mobile combustion.
e. Change in forest stocks and carbon uptake from urban trees and land filled yard trimmings.

**Table 1.** Greenhouse gas emissions and carbon sinks in agricultural activities, 1990-2005 (CO$_2$ equivalent).

## 2. Soil as a source of carbon storage

Soils are the fundamental foundation of our food security, global economy and environmental quality, the degradation of soil conditions can affect the on-farm environment. The soil environment is a principal component of the global carbon (C) cycle where key interactions between biotic and abiotic components take place to regulate the flow of materials to and from the pedosphere, atmosphere and hydrosphere. There is general agreement that although soil is part of the climate change problem, it is also an integral part of the solution. Soils altogether contain an estimated 1,700 Gt (billion metric tons) to a depth of 1 m and as much as 2,400 Gt to a depth of 2 m (Fig.2). An estimated additional 560 Gt is contained in terrestrial biota (plants and animals). The carbon in the atmosphere is estimated to total 750 Gt. The amount of organic carbon in soils is more than four times the amount of carbon in terrestrial biota and three times that in the atmosphere. Lal et al., (1999) estimated historic loss of ecosystem C due to desertification at 9–14 Pg of SOC pool, with losses from the biotic/vegetation pool at 10–15 Pg. Ojima et al., (1993) estimated that grasslands and drylands of the world have lost 13–24 Pg C due to desertification.

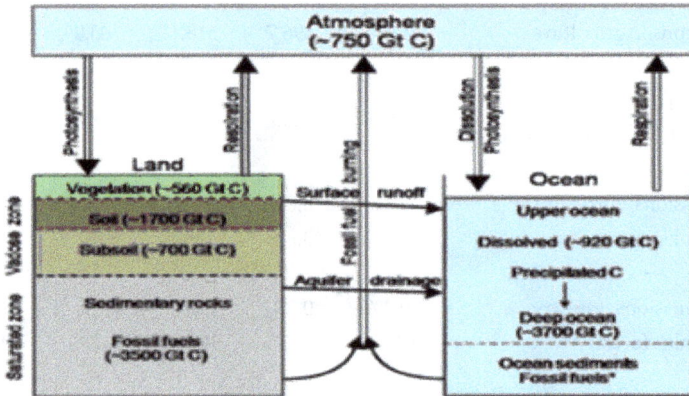

**Figure 2.** Carbon reserve and exchange in the land- ocean- atmosphere continuum (Quantitative estimates regarding fossil fuels in ocean sediments vary widely)

Carbon dioxide is removed from the atmosphere and converted to organic carbon through the process of photosynthesis. As organic carbon decomposes, it is converted back to carbon dioxide through the process of respiration. The quantity of organic carbon in soils is spatially and temporally variable, depending on the balance of inputs versus outputs. The inputs are due to the absorption of carbon dioxide from the atmosphere in the process of photosynthesis and its incorporation into the soil by the residues of plants and animals. Some of the dead plant matter is incorporated into the soil in humus, thereby enhancing the soil organic carbon pool. Decomposition of soil organic matter, releases carbon dioxide under aerobic conditions and methane under anaerobic conditions. In certain conditions, decomposition of organic matter may also cause the release of nitrous oxide, which is another powerful greenhouse gas. The content of organic carbon in soils in most cases constitutes less than 5% of the mass of soil material and is generally concentrated mainly in the upper 20 to 40 cm (the so-called topsoil). However, that content varies greatly, from less than 1% by mass in some arid-zone soils (Aridisols) to 50% or more in waterlogged organic soils such as Histosols (Table 2). Changes in agricultural activities and land use system during the past centuries have made soils act as net sources of atmospheric $CO_2$. Evidence from long-term experiments suggests that carbon losses due to oxidation and erosion can be reversed with soil management practices that minimize soil disturbance and optimize plant yield through fertilization. Appropriate land management practices can result in a significant increase in the rate of carbon into the soil. Because of the relatively long turnover time of some soil carbon fractions, this could result in storage of a sizable amount of carbon in the soil for several decades. Maintaining soil quality can reduce problems of land degradation, decreasing soil fertility and rapidly declining production levels that occur in large parts of the world which lack the basic principles of good farming practices. The loss of rain water that cannot infiltrate in the soils to replenish the ground water reserves might be the more serious long-term result of excessive tillage. Thus, the way soil is cultivated must be drastically changed.

| Soil Order | Area $10^3$ km$^2$ | Organic C Gt |
|---|---|---|
| Alfisols | 13,159 | 90.8 |
| Andisols | 975 | 29.8 |
| Aridisols | 15,464 | 54.1 |
| Entisols | 23,432 | 232.0 |
| Gelisols | 11,869 | 237.5 |
| Histosols | 1,526 | 312.1 |
| Inceptisols | 19,854 | 323.6 |
| Mollisols | 9,161 | 120.0 |
| Oxisols | 9,811 | 99.1 |
| Spodosols | 4,596 | 67.1 |
| Ultisols | 10,550 | 98.1 |
| Vertisols | 3,160 | 18.3 |
| Other orders | 7,110 | 17.1 |
| Total | 130,667 | 1,699.6 |

Source USDA

**Table 2.** Estimated mass of carbon in the worlds soils resources

## 3. Degradation of dryland

Degradation of soil is especially important in drylands of the world where desertification is a serious problem (UNEP, 1992). The world's drylands, 6.31 billion hectares (Bha) or 47% of the earth's land area, are found in a wide range of climates spanning from hot to cold.

According to FAO (1993), drylands comprise four ecoregions covering land area of 0.98 Bha in hyper-arid, 1.57 Bha in arid, 2.31 Bha in semi-arid and 1.29 Bha in dry sub-humid climates (Table 3). Soils of the drylands also vary widely, but are mostly Aridisols (2.12 Bha) and Entisols (2.33 Bha). Dryland soils also include Alfisols (0.38 Bha), Mollisols (0.80 Bha), Vertisols (0.21 Bha) and others (0.47 Bha) (Dregne, 1976; Noin and Clark, 1997). The arid zones cover about 15 percent of the land surface. The annual rainfall in these areas is up to 200 mm in winter-rainfall areas and 300 mm in summer rainfall areas. Interannual variability is 50–100 percent. Africa and Asia have the largest extension of arid zones (Table 4).

| Region | Hyper-arid(<0.05)[a] | Arid (0.05-0.20) | Semi-arid (0.20-0.50) | Dry sub-humid (0.50-0.65) | Total | % of Earth's land area |
|---|---|---|---|---|---|---|
| Africa | 67 | 0.5 | o.51 | 0.27 | 1.96 | 15.0 |
| Asia | 0.28 | 0.63 | 0.69 | 0.35 | 1.95 | 14.9 |
| Australia | 0 | 0.30 | 0.31 | 0.05 | 0.66 | 5.1 |
| Europe | 0 | 0.01 | 0.11 | 0.18 | 0.30 | 2.3 |
| N. America | 0.003 | 0.08 | 0.42 | 0.23 | 0.74 | 5.6 |
| S. America | 0.03 | 0.05 | 0.27 | 0.21 | 0.54 | 4.2 |
| Total | 0.98 | 1.57 | 2.31 | 1.29 | 6.15 | |
| % of Earth's land area | 7.5 | 12.1 | 17.7 | 9.9 | 47.2 | |

* Aridity Index = P/PET

**Table 3.** Global distribution of drylands of the world (modified from Middleton and Thomas 1992, Noin and Clarke 1997, Reynolds and Smith 2002)

| Continent | Extension | | | Percentage | | |
|---|---|---|---|---|---|---|
| | Arid | Semi-arid | Dry subhumid | Arid | Semi-arid | Dry subhumid |
| | | | Million ha | | | |
| Africa | 467.60 | 611.35 | 219.16 | 16.21 | 21.20 | 7.60 |
| Asia | 704.30 | 727.97 | 225.51 | 25.48 | 26.34 | 8.16 |
| Oceanta | 459.50 | 211.02 | 38.24 | 59.72 | 27.42 | 4.97 |
| Europe | 0.30 | 94.26 | 123.47 | 0.01 | 1.74 | 2.27 |
| North/Central America | 4.27 | 130.71 | 382.09 | 6.09 | 17.82 | 4.27 |
| South America | 5.97 | 122.43 | 250.21 | 7.11 | 14.54 | 5.97 |
| Total | 1641.95 | 1897.74 | 1238.68 | | | |

Source FAO (2002a)

**Table 4.** The global dryland areas by continent

Desertification is defined as destruction of the biological potential of land which can lead ultimately to desert-like conditions' (UNEP, 1977). In this context, the term 'land' includes whole ecosystems comprising soil, water, vegetation, crops and animals. The term 'degradation' implies reduction of resource potential by one or a combination of degradative processes including erosion by water and wind and the attendant sedimentation. The process of desertification is not confined to the drylands of the tropics it also occurs in developed countries (U.S.A.), high latitude humid ecoregions (Iceland) and even humid regions (tropical rainforest). Traditionally desertification has been defined as land degradation in arid, semi-arid, and dry sub-humid areas resulting from climatic variations and human activities (Le Hou'erou, 1975, Warren, 1996, UNEP, 1992), but it has also been observed in cool, humid climates such as iceland (Arnalds, 2000). The land area prone to desertification has been estimated at 3.5–4.0 Bha or 57%–65% of the total land area of dryland ecosystems (UNEP, 1991). Of this, the land area affected by soil degradation alone (excluding vegetation degradation) ranges from 1.02 (UNEP, 1991) to 1.14 Bha (Oldeman and Van Lynden, 1998). The estimates of current rate of desertification also vary widely. Mainguet (1991) estimated the annual rate of desertification at about 5.8 million hectares (Mha), with 55% occurring in rangeland and 45% on rainfed cropland. Desertification in humid areas results mainly from land misuse and soil mismanagement. Estimates of the extent of desertification range widely and are highly subjective. UNEP estimated 3.97 Bha in 1977, 3.48 Bha in 1984 and 3.59 Bha in 1992 (UNEP, 1992). Land area affected by desertification was estimated at 3.25 Bha by Dregne (1983) and 2.0 Bha by Mabbutt (1984). According to the GLASOD methodology (Oldeman and Van Lynden, 1998), land area affected by desertification due to soil degradation is estimated at 1.14 Bha (Table 5). As with the area affected, estimates of the current rates of desertification also vary widely. The annual rate of desertification is estimated at 5.8 million hectares (Mha) or 0.13% of the dryland in mid latitudes. Also desertification is considered as a biophysical process driven by socio-economic and political factors (Mortimore, 1994; Mainguet and Da Silva, 1998). Two principal biophysical processes leading to desertification are erosion and salinization. Accelerated soil erosion by wind and water are severe in semi-arid and arid regions (Balba 1995; Baird, 1997), especially those in the Mediterranean climates (Brandt and Thornes, 1996; Conacher and Sala, 1998a,b). Salinization is a major problem on irrigated lands. The irrigated land area in the world has increased 50 fold during the last three centuries which was 5 Mha in 1700, 8 Mha in 1800, 48 Mha in 1900, and 255 Mha in 2000. Risks of secondary salinization are exacerbated by use of poor quality water, poor drainage and excessive irrigation, leakage of water due to a defective delivery system, impeded or slow soil drainage and other causes. Salinization is a severe problem in China, India, Pakistan, and in countries of Central Asia (Babaev, 1999). The extent of land area salinized is 89% in Turkmenistan, 51% in Uzbekistan, 15% in Tadjikstan, 12% in Kyrgyzstan and 49% of the entire region (Pankova and Solovjev, 1995; Esenov and Redjepbaev, 1999). Salinization is also a problem in southwestern U.S.A., northern Mexico and dry regions of Canada (Balba, 1995). Lal et al., (1999) estimated that soil erosion in drylands leads to emission of 0.21–0.26 Pg C/y, with an additional 0.02–0.03 Pg C/y due to exposure of carbonaecous material to climatic elements caused by surface soil erosion. Therefore, total annual emission of C due to erosion- induced land degradation in dryland ecosystems may be 0.23–0.29 Pg C/y.

| Land type | Area (Bha) | Type of soil degradation | Area (Bha) |
|---|---|---|---|
| Degraded irrigated lands | 0.043 | Water erosion | 0.478 |
| Degraded rainfed croplands | 0.216 | Wind erosion | 0.513 |
| Degraded rangelands (Soil and vegetation) | 0.757 | Chemical degradation | 0.111 |
| Sub-total | 1.016 | Physical degradation | 0.035 |
| Degraded rangelands (Soil and vegetation) | 2.576 | Total | 1.137 |
| Total | 3.592 | Light | 0.489 |
| Total land area | 5.172 | Moderate | 0.509 |
| % degraded | 69.5 | Severe and extreme | 0.139 |
|  |  | Total | 1.137 |

UNEP (1991); Oldeman and Van Lynden (1998). The estimates by Oldeman and Van Lynden does not include the vegetation degradation on rangeland. (Bha= $10^9$ ha)

**Table 5.** GLASOD estimates of desertification (e.g. land degradation in dry areas excluding hyber-arid areas)

## 4. Soil organic carbon storage

The soil organic C storage decrease with increase in temperature and increases with increase in soil water content. Studies show that a $3^0C$ increase in temperature is projected to decrease soil organic C concentration by about 11% in the upper 30 cm soil depth and increase $CO_2$ emission by 8 %. This may to some extent be counteracted by higher uptake of carbon dioxide by plants as they grow faster in warmer conditions and store carbon as biomass both in the soil and the plant. The world's dryland soils contain 241 Pg of soil organic carbon (SOC) (Eswaran et al., 2000), which is about 40 times more than what was added into the atmosphere through anthropogenic activities, estimated at 6.3 Pg C/y during the 1990s (Schimel et al., 2001 IPCC, 2001). In addition, dryland soils contain at least as much as or more soil inorganic carbon (SIC) than SOC pool (Batjes, 1998; Eswaran et al., 2000). Total dryland soil organic carbon reserves comprise 27% of the global soil organic carbon reserves (MA, 2005). The soil properties, such as the chemical composition of soil organic matter and the matrix in which it is held, determine the different capacities of the land to act as a store for carbon that has direct implications for capturing greenhouse gases (FAO, 2004). Management of both SOC and SIC pools in dryland ecosystems can play a major role in reducing the rate of enrichment of atmospheric $CO_2$ (Lal, 2002).

Most soils may lose one-half to two-thirds of their SOC pool within 5 years in the tropics and 50 years in temperate regions. The new equilibrium may be attained after losing 20–50 Mg C/ha. Several studies has estimated the global loss at 40 Pg by Houghton (1995), 55 Pg by IPCC (1996) and Schimel (1998), 66–90 Pg by Lal (1999) and 150 Pg by Bohn (1978). Rozanov et al., (1990) observed that world soils have lost humus (58% C) at a rate of 25.3 Tg/year ever since agriculture began 10,000 years ago, 300 million tons/year in the past 300 years and 760 million tons per year in the last 50 years. The SOC is easily transported by runoff water or wind because it is of relatively low density (< 1.8 Mg/m$^3$) and is concentrated in the vicinity of the soil surface. A study on wind erosion in southwest Niger showed that wind-borne material trapped at 2-m height contained 32 times more SOC relative to the antecedent topsoil (Sterk et al., 1996).

In dryland soils, SOC declines with cultivation and even more so with desertification. In East Africa, Swift et al., (1994) reported that continuous cultivation for 14 years without recommended inputs of fertilizers and manures decreased SOC content by half from 2% to 1%. Pieri (1991) reported that continuous cropping without application of fertilizers and/or manure leads to rapid decline in SOC content. The experimental results revealed a marked decline in soil C, reaching some 13 tonnes/ha (Fig. 3) when aboveground material is harvested and removed and some FYM is applied at different times, equivalent to 3.9 tonnes/ha/year. The rate of depletion of SOC content is accentuated by soil erosion, because of the preferential removal of the finer soil fractions comprised of clay and organic matter. The SOC is often bound with the clay fraction (Quiroga et al., 1996, 1998) which is preferentially removed by erosion. Adoption of inappropriate land use and farming practices can deplete SOC content (Table 6). These trends, if unchecked, accentuate the process of desertification. Swift et al., (1994) indicated that land degradation around the world has led to an SOC loss of 8 to 12 Mg C ha$^{-1}$on land area of 1.02 Bha (UNEP, 1991), the total historic C loss would be 8 to 12 Pg C.

## 5. C sequestration to combat land degradation in drylands

Drylands are considered to be areas where average rainfall is less than the potential moisture losses through evaporation and transpiration. About 47 percent of the surface of the earth can be classified as dryland (UNEP, 1992). Droughts are characteristic of drylands and can be defined as periods (1–2 years) where the rainfall is below the average. The main characteristic of drylands is lack of water. This constrains plant productivity severely and therefore affects the accumulation of C in soils. The problem is aggravated because rainfall is not only low but also generally erratic. Therefore, good management of the little available water is essential. In addition, the SOC pool tends to decrease exponentially with temperature (Lal, 2002a). Consequently, soils of drylands contain small amounts of C (between 1 percent and less than 0.5 percent) (Lal, 2002b). The SOC pool of soils generally increases with the addition of biomass to soils when the pool has been depleted as a consequence of land uses (Rasmussen and Collins, 1991; Paustian et al., 1997; Powlson et al., 1998, Lal, 2001a). Soils in drylands are prone to degradation and desertification, which lead

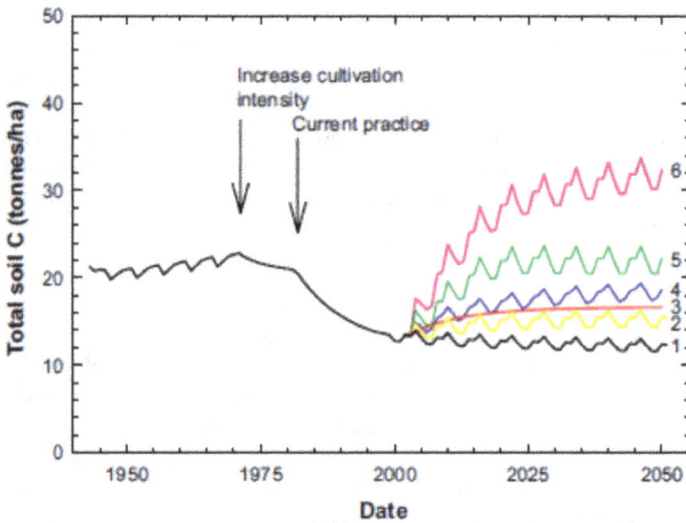

**Figure 3.** Change in total soil carbon for a rainfed farm

| Traditional practices | Recommended |
| --- | --- |
| Plough till | Conservation till/ no till |
| Residue removal/ burning | Residue return as mulch |
| Summer fallow | Growing cover crop |
| Low off-farm input | Judicious use of fertilizers and integrated nutrient management |
| Regular fertilizer use | Soil- site specific management |
| No water controlFence-to fence cultivation | Water management/ conservation, irrigation, water table management |
| Fence-to fence cultivation | Conservation of marginal lands to nature conservation |
| Monoculture | Improved farming systems with several crop rotations |
| Land use along poverty lines and political boundaries | Integrated watershed management |
| Draining wetland | Restoring wetlands |

**Table 6.** Agricultural practices for enhancing productivity and increasing the amount of carbon in soils

to dramatic reductions in the SOC pool. Soil-quality improvement as a consequence of increased soil C will have an important social and economic impact on the livelihood of people living in these areas. The ability of agriculture lands to store or sequester carbon depends on several factors, including climate, soil type, type of crop or vegetation cover and management practices. Carbon sequestration in the agriculture sector refers to the capacity of agriculture lands and forests to remove carbon dioxide from the atmosphere. Carbon dioxide is absorbed by trees, plants and crops through photosynthesis and stored as carbon in biomass in tree trunks, branches, foliage and roots and soils (EPA, 2008b). Forests and stable grasslands are referred to as carbon sinks because they can store large amounts of carbon in their vegetation and root systems for long periods of time. Soils are the largest terrestrial sink for carbon on the planet. The amount of carbon stored in soil organic matter is influenced by the addition of carbon from dead plant material and carbon losses from respiration, the decomposition process and both natural and human disturbance of the soil. By employing farming practices that involve minimal disturbance of the soil and encourage carbon sequestration, farmers may be able to slow or even reverse the loss of carbon from their fields. In the United States, forest and croplands currently sequester the equivalent of 12 percent of U.S. carbon dioxide emissions from the energy, transportation and industrial sectors (EPA, 2008b).

The sequestration of atmospheric C in the soil and biomass not only reduces greenhouse effect but also helps maintain or restore the capacity of the soil to perform its production and environmental functions on sustainable basis. Dry soils are less likely to lose C than wet soils (Glenn et al., 1992) as a lack of water limits soil mineralization and therefore the flux of C to the atmosphere. Consequenlty, the residence time of C in dryland soils is long, sometimes even longer than in forest soils.

## 6. Carbon sequestration potential

Several studies have attempted to assess the potential for carbon sequestration in drylands (Table 7). Lal (2001) estimated that they had the potential to sequester up to 0.4–0.6 Gt of carbon a year if eroded and degraded dryland soils were restored and their further degradation were stopped. Glenday (2008) measured forest carbon densities of 58 to 94 tonnes C/ha in the dry Arabuko-Sokoke Forest in Kenya and concluded that improved management of wood harvesting and rehabilitation forest could substantially increase terrestrial carbon sequestration. Farage et al., (2007) in dryland farming systems in Nigeria, Sudan and Argentina showed that it would be possible to change current farming systems to convert these soils from carbon sources to net sinks without increasing farmers' energy demand. Hülsbegen and Küstermann et al., (2008) compared 18 organic and 10 conventional farms in Bavaria, Germany and calculated the organic farms annual sequestration at 402 kg carbon, while the conventional farms had losses of 202 kg. Hepperly et al., (2008) estimated that compost application and cover crops in the rotation were particularly adept at increasing soil organic matter, also compared to no tillage techniques (Table 8).

| Technological options | Sequestration potential (Tonnes C/ha/year ) |
|---|---|
| Croplands | 0.10 – 0.20 |
| Conservation tillage | 0.05 – 0.10 |
| Mulch farming (4-6 Mg/ha/year) | 0.10 – 0.20 |
| Compost (20 Mg/ha/year) | 0.05 – 0.10 |
| Elimination of bare fallow | 0.10 – 0.20 |
| Integrated nutrient management | 0.10 – 0.20 |
| Restoration of eroded soils | 0.05 – 0.10 |
| Restoration of salt effected soils | 0.10 – 0.20 |
| Agricultural intensification | 0.10 – 0.30 |
| Water conservation and management | 0.05 – 0.10 |
| AfforestationGrassland and pastures | 0.05 – 0.10 |

Lal et al. (1998)

**Table 7.** Effects from land management practices or land use on carbon sequestration potential in drylands

| Practices | Soil Carbon sequestration (kg/ha) |
|---|---|
| Compost | 1000 to 2000 |
| Cover crop | 800 to 1200 |
| No-till | 100 to 500 |
| Rotation | 0 to 200 |
| Manure | 0 to 200 |
| Cover+rotation | 900 to 1400 |
| Compost + Cover + Rotation + No till | 2000 to 4000 |

**Table 8.** Soil carbon sequestration estimates for different agricultural practices. Data projected from Rodale long-term trials

## 7. Management options to control sequester carbons

Adoption of recommended management practices (RMPs) on favorable soils with good soil moisture regime and the possibility of supplemental irrigation can increase SOC concentration. Enhancing water use efficiency (WUE), by reducing losses due to surface runoff, evaporation and decreasing soil temperature by residue mulching, is important. Application of fertilizers, irrigation and manuring are all common practices that consume C. Innovative farming practices such as conservation tillage, organic production, improved cropping systems, land restoration, land use change, irrigation and water management are the strategies to increase the C storage. Organic systems of production increase soil organic matter levels through the use of composted animal manures and cover crops

(Rodale Institute, 2008). Organic cropping systems also eliminate the emissions from the production and transportation of synthetic fertilizers. Land restoration and land use changes that encourage the conservation and improvement of soil, water and air quality typically reduce greenhouse gas emissions. Soil quality is largely governed by soil organic matter (SOM) content, which is a dynamic pool and responds effectively to changes in soil management, primarily tillage and carbon inputs resulting from biomass production.

## 8. Conservation agriculture

Conservation tillage as an integral part of conservation agriculture includes a minimum 30% soil cover after planting to reduce soil erosion implies conformity with all three of its pillars: (i) minimum soil disturbance (ii) diverse crop rotations and/or cover crops and (iii) continuous plant residue cover. Reducing tillage reduces soil disturbance and helps mitigate the release of soil carbon into the atmosphere. Conservation tillage also improves the carbon sequestration capacity of the soil. The amount of carbon released from soils depends directly on the volume of soil disturbed during tillage operations. Therefore, lesser the soil is disturbed, better the conservation of soil carbon. Additional benefits of conservation tillage include improved water conservation, reduced soil erosion, reduced fuel consumption, reduced compaction, increased planting and harvesting flexibility, reduced labour requirements and improved soil tilth. Stewart and Robinson (2000) indicated one of the gratifying consequences of the no-till system is increase in SOC concentration in soil, which may range from 60 to over 600 kg C/ha/y. In northern Colorado, Potter et al., (1997) observed 560 kg C/ha/y accumulation during 10 years of no-till continuous cropping wheat system. Kihani et al., (1984) conducted soil analyses on a 45-year old tillage experiment and reported that incorporation of biosolids improved SOC concentration. Murillo et al., (1998) reported that SOC concentration in 0 to 5 cm depth was 0.84% in traditional tillage and 1.1% in conservation tillage after 2 years, and 0.89% in traditional tillage compared with 1.34% in conservation tillage after 4 years. Holland (2004) gives the interactive processes as a consequence; conservation agriculture generate the soil's structural stability and have a substantial impact on the environment (Fig 4).

## 9. Organic input and manuring

Mineral nitrogen in soils may contribute to the emission of nitrous oxides and is one of the main drivers of agricultural emissions. The efficiency of fertilizer use decreases with increasing fertilization, when a great part of it is not taken up by the plant but emitted into the water bodies and the atmosphere. In summary, the emission of GHG in $CO_2$ equivalents from the production and the application of nitrogen fertilizers from fossil fuel amounts at approximately 480 million tonnes (1 percent of total global GHG emissions) in 2007. In 1960, 47 years earlier, it was less than 100 million tones. In dryland areas, several studies demonstrated the importance of judicious use of fertilizer, compost and nutrient management (Fuller, 1991; Traore and Harris, 1995; Singh and Goma, 1995; Pieri, 1995; Miglierina et al., 1996; Laryea et al., 1995). Application of nitrogen fertilizer is important to obtaining high yields, but may have little impact on SOC concentration unless used in conjunction with no-till and residue management (Russell 1981; Dalal 1992; Skjemstad et al.,

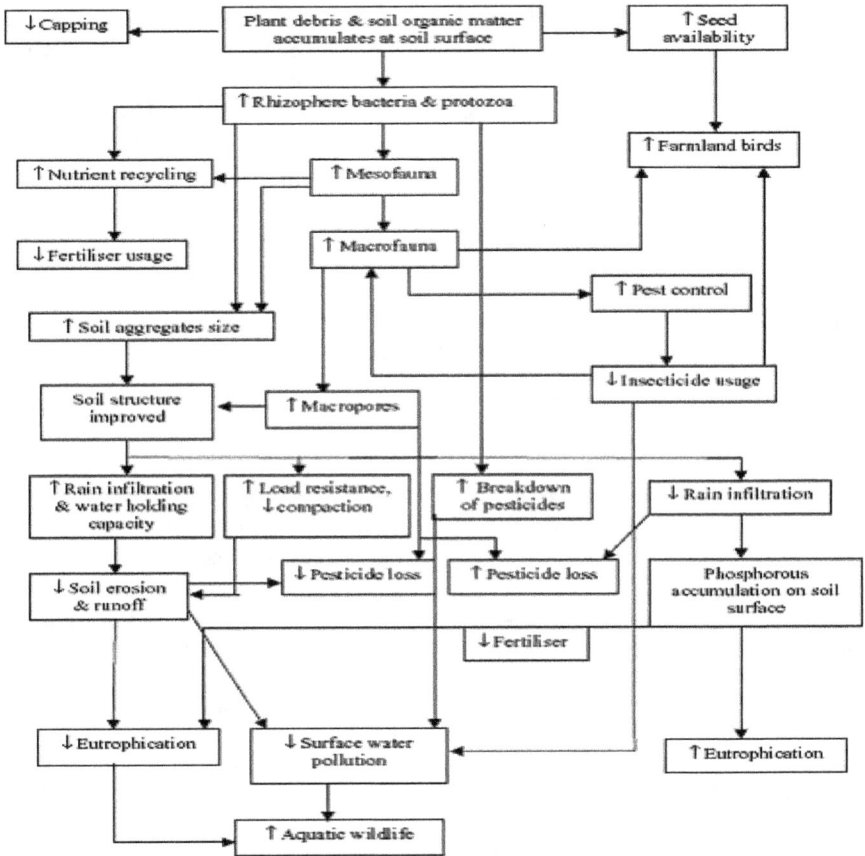

**Figure 4.** Interactive process through which conservation tillage can generate environmental benefits (Holland, 2004)

1994; Dalal et al., 1995). Recycling nitrogen on the farm by using manure and nitrogen fixing plants (the predominant technique of organic and low external input agriculture) enhances soil quality and provides nutrients. However, timing and management of its use are essential. Nambiar (1995) reported increase in SOC content with manuring from 0.20% to 0.25% in 1997-1989 for a sandy soil. Ryan (1998) observed a significant increase in SOC concentration by application of recommended rates of fertilizers. Mäder et al., (2002) compare the relative input and output of three farming systems: organic agriculture; integrated production with farmyard Manure and stockless integrated production in a 28 years experiment. Input of nutrients, organic matter, pesticides and energy as well as yields were calculated. Crop sequence was potatoes, winter wheat followed by fodder intercrop, vegetables (soybean), winter wheat (maize), winter barley (grass-clover for fodder production, winter wheat), grass-clover for fodder production, grass-clover for fodder production. Crops in brackets are alterations in 1 of the 4 crop rotations. The results indicated an increased efficiency of organic agriculture for most arable crops, with grain crops showing a yield reduction of only 20 percent while fertilizer inputs were lower by 50-60 percent (Fig. 5). Mishra et al. (1974) reported that application of manure at the rate of 9–30 Mg ha$^{-1}$ y$^{-1}$ caused significant increase in SOC content. Dalal (1989) observed a positive effect on SOC concentration after 13 years of no-till, residue retained and N application (34.5 Mg C/ha vs. 35.8 Mg C/ha). In semi-arid conditions, the SOC sequestration is limited by the input of biomass carbon. Although, crop yields are sufficiently increased by N application, the residue input is not sufficient enough to balance the mineralization rate. Mathieu et al., (2006) pointed out that higher soil carbon levels may lead to $N_2$ emission rather than $N_2O$. Petersen et al., (2005) found lower emission rates for organic farming compared to conventional farming in five European countries. In a long-term study in southern Germany, Flessa, et al. (2002) also found reduced $N_2O$ emission rates in organic agriculture, although yield-related emissions were not reduced. A reduction of the Global Warming Potential (GWP, 64 %) has also been found at Michigan State University for organic crops as compared to the conventional (Robertson et al., 2000). In India, Gupta and Venkateswarlu (1994) observed that application of manure at 10 Mg/ha increased SOC concentration. For Vertisols in the Ethiopian Highlands, Wakeel and Astartke (1996) recommended adoption of improved agricultural practices (nutrient management, water conservation, new varieties and crop rotation) to minimize risks of soil degradation. Use of high-lignin amendments, recalcitrant to decomposition, increases SOC concentration.

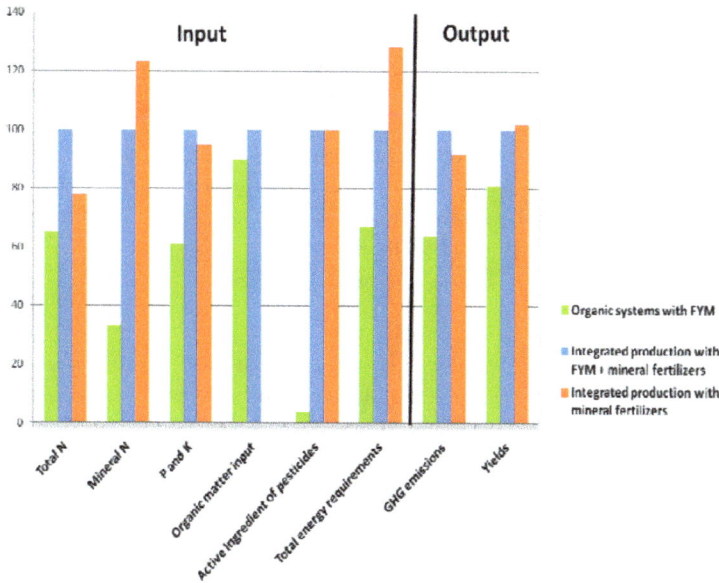

**Figure 5.** Comparison of GHG and crop productivity in different farming systems in long term field experiments

## 10. Crop rotation

Numerous case studies show that in comparison to traditional subsistence farming, organic yields were 112 percent higher due to crop rotation, legumes and closed circuits. Miglierina et al., (1993, 1996) observed that SOC content was high in wheat-grassland and wheat-alfalfa (*Medicago sativa*) rotations, especially with a conservation tillage system. In another study, Ryan et al., (1997) reported the beneficial effects of using reduced tillage for enhancing SOC concentration are accentuated when used in conjunction with rotations based on appropriate cover crops or pastures. Skjemstad et al., (1994) reported an increase of 550 kg C/ha/y in a Vertisol under Rhodes grass. In Saudi Arabia, Shahin et al., (1998) observed that introducing alfalfa in rotation with wheat grown on a sandy soil increased SOC concentration threefold as compared with continuous wheat. In Syria, Ryan (1998) reported that incorporation of *Medicago* in rotation increased SOC concentration to 1-m depth. Jenkinson et al., (1999) assessed the SOC pool under different rotations on a calcareous soil in Syria. The SOC pool in wheat-meadow rotation increased by 1.6 Mg/ha at a mean rate of 0.17 Mg C/ha/y in comparison with wheat-wheat rotation and by 3.8 Mg/ha at the mean rate of 0.38 Mg C/ha/y in comparison with wheat-fallow rotation. In Australia, Whitehouse and Littler (1984) observed an increase in SOC concentration from 1.18% to 1.37% in 0 to 15 cm depth after 2–4 years of lucerne-prairie grass pasture. In a Vertisol in central India, Mathan et al., (1978) reported that continuous cropping and manuring increased SOC concentration by 20%–40% over 3 years. In northern India, Singh et al., (1996) observed that incorporation of legumes in a rice-wheat rotation increased SOC concentration. Growing crops with a deep and prolific root system generally has a favorable impact on SOC concentration in the sub-soil. Barber (1994) observed that sub-soiling and incorporation of cover crops in rotation enhanced soil quality. Lomte et al., (1993) reported that intercropping sorghum (*Sorghum bicolor*) with legumes and application of manure increased SOC content and aggregation. Some examples of soil management practices that may lead to SOC sequestration are listed in Table 9 and 10. Activity of soil fauna, especially termites, improves soil structure and enhances the SOC pool in the long run. An appropriate use of stone cover and gravel mulch can also improve soil moisture regime and enhance the SOC pool.

## 11. Grazing management

Excessive and uncontrolled grazing are a major cause of the acceleration of the desertification process. Grazing is the predominant land use in dryland ecosystems, and adoption of improved grazing practices can improve C sequestration through conservation and better management of surface residue. In the Sahel, deposition of droppings ranges from 1 tonne/ha to 50 tonnes/ha depending on the time that animals are kept on the same field (Sagna-Cabral, 1989; Hoffmann and Gerling, 2001). However, direct exposure to the elements can reduce the nutrient value of dung and droppings considerably. Although stubble grazing has a long tradition in drylands, increasing land scarcity, limited purchasing power among many smallholders and increased risks of animal theft in many areas have contributed to a general decline in herd sizes and in some cases, led to the abandonment of

| Strategy/technique | Practice | Location/region | Reference |
|---|---|---|---|
| Erosion control/water conservation | a) No-Till farming | Bushland, TX, USA Northern CO, USA Queensland, Australia West Africa Sahel Southern Spain | Jones and others 1997 Potter and others 1997 Dalal and others 1997 Bationo and others 2000 Murillo and others 1998 |
| | b) Mulching • stone cover • residue mulch • mulch | Negev Desert Chihuahuan Desert Suriname | Lahav and Steinberg 2001 Rostagno and Sosebal 2001 Breeman and Protz 1988 |
| Crop Diversification | a) Rotations | Saudi Arabia, West Asia, Alegria, North Africa | Shahin and others 1998 Arabi and Roose 1989 |
| | b) Legumes | Syria, West Asia Australia Northern India Argentina | Jenkinson and others 1999 Whitehouse and Littler 1984 Singh and others 1996 Galantini and Rosell 1997 |
| Integrated nutrient management and recycling | a) Manuring | Maiduguri, Nigeria | Aweto and Ayub 1993 |
| | b) Organic by-products | Spain | Pascual and others 1998 |
| | c) Soil fauna | Chihuahuan Desert | Nash and Whitford 1995 |
| | d) Sewage sludge | Spain | Pedreno and others 1996 |
| Water management | a) Irrigation and conservation tillage | Mexico | Folleu and others 2003 |
| | b) Irrigation with sewage | Israel | Hillel 1998 |
| | c) Irrigation with silt-laden water | China | Fullen and others 1995 |
| | d) Saline aquaculture | Drylands | Glen and others 1993 |

**Table 9.** Strategies of soil management in dryland ecosystems for carbon sequestration

stubble grazing altogether. Pluhar et al., (1987) observed that grazing caused a significant decline in infiltration capacity by reducing the protective vegetal cover and increasing the surface area of the bare ground. Thurow et al., (1988) also observed that infiltration capacity decreased and inter-rill erosion increased in the heavily stocked pastures. In Alice, Texas, Weltz and Blackburn (1995) observed that the saturated hydraulic conductivity was the least for the bare soil. Biomass burning also affects soil hydrological properties. Hester et al., (1997) showed that fire reduced water infiltration capacity in case of the oak and juniper vegetation types. Therefore, controlled grazing, fire management and planting improved species are important considerations of enhancing biomass production and improving soil quality. Some examples of improved practices with positive impact on the SOC pool are listed in Table 11. Important among these are grazing management through controlled stocking and rotational grazing, fire management, and agroforestry practices involving legume species (Conarc et al., 2001).

| Strategy/technique | Location/region | Reference |
|---|---|---|
| Surface application of biosolids | Chihuahuan Desert | Rostagno and Sosebal 2001 |
| Stone cover | Negev Desert | Lahav and Steinberg 2001 |
| Enhancing termites activity | Chihuahuan Desert | Nash and Whitford 1995 |
| Manuring | Maiduguri, Nigeria | Aweto and Ayub 1993 |
| Desert soil macrofauna (termites/ants) | Chihuahuan Desert | Whitford 1996 |
| Sewage sludge | Spain | Pedreno and others 1996 |
| Organic by-products | Spain | Pascual and others 1998 |

**Table 10.** Soil management options for C sequestration in soils of dryland ecosystems

| Strategy/technique | Practice | Location/region | Reference |
|---|---|---|---|
| Improved species | Sowing legumes | Vertisols, Australia | Chan and others 1997 |
| | | Northern Colorado | Havlin and others 1990 |
| | | Sadore, Niger | Hiernaux and others |
| | Agroforestry | West African Sahel | 1999 |
| | | | Breeman and Kessler 1997 |
| Fire management | Prescribed burning | Wyoming, USA | Schuman and others |
| | Stocking rate | Negev, Israel | 2002 |
| | | | Zaady and others 2001 |
| Grazing management | Controlled grazing | Kawas, USA | Rice and Owensby 2001 |
| Improving grasslands | Integrated management | World's drylands | Conant and others 2001 |
| Erosion management | Integrated management | World's drylands | Lal 2001 |

**Table 11.** Strategies of pasture and range land management for soil carbon sequestration.

## 12. Erosion control

Soil C losses can occur both as a result of mineralization as well as through erosion often making it a complex relationship. Where water erosion dominates, a high proportion of soil C may be washed into alluvial deposits close to the erosion site and stored there in forms which decay more slowly than in the parent soils. Therefore, this kind of erosion may have a positive effect on soil CS. In Western Nigeria Gabriels and Michiels (1991) observed C losses from bare fallow Alfisol plots with slopes of 1, 5 and 10 % , varied from 54 to 3080 kgha$^{-1}$. Erosion does not always decrease productivity, but if it could be shown to do so, it would be perverse to favour decreased productivity for a medium term and perhaps one-off gain in sequestered C. The same arguments probably do not apply where wind erosion is the main erosional process, for organic matter is usually blown great distances and dispersed to places where it may decay rapidly and release its C. Management options that increase the amount of live and dead biomass left in agricultural areas decrease erosion in general while simultaneously increasing the C input to the soil (Tiessen and Cuevas, 1994). Assuming that 20% of the C displaced is emitted to the atmosphere ( Lal et al., 1998), erosion (e.g., light, moderate, severe and extreme forms) leads to emission of 0.206 to 0.262 Pg C y$^{-1}$. Erosion also leads to exposure of the sub-soil rich in calciferous materials. These areas, severely affected by strong and extreme wind erosion, are estimated at about 103.6 Mha. If 10% of these areas have calciferous horizons exposed at the soil surface, about 10 Mha are subject to the impact of anthropogenic perturbations and environmental factors (e.g., ploughing, application of fertilizers, root exudates, acid rain, etc.). These factors may lead to dissolution of carbonates and emission of $CO_2$. If this exposed layer containing high amounts of carbonates and bicarbonates leads to emissions of C at the rate of 0.2 to 0.4 Kg C ha$^{-1}$ yr$^{-1}$, the annual rate of emissions of C from SIC is 2 to 4 × 10$^6$ Kg C y$^{-1}$. Therefore, total C emission due to soil erosion and exposure of calciferous horizon is 0.21 to 0.26 Pg C y$^{-1}$. Three main type of erosion preventive techniques are (Lal, 1990) i) those that increase the soils' resistance against agents of erosion; ii) soil surface management techniques that help establish quick ground cover and; iii) techniques that provide a buffer against rainfall and runoff erosivity.

## 13. Summary

Many of the factors affecting the flow of C into and out of the soil are affected by land-management practices. The soils of drylands have lost a significant amount of C and, therefore, offer a great potential for rehabilitating these areas. There are vast areas of dryland ecosystems in developing countries where improvements in farming systems could add C to soils. Tillage-based agriculture damages the soil, conservation agriculture builds soil quality, protects water quality, increases biodiversity and sequesters carbon. Considering the growing concern of elevated atmospheric greenhouse gases, the complex economics and availability of fossil fuels, the deterioration of the environment and health conditions, a shift away from intense reliance on heavy chemical inputs to an intense biologically based agriculture and food system is possible today. Sustainable and conservation agriculture offer multiple opportunities to reduce greenhouse gas emissions and counteract global warming. Improving energy efficiency by managing agricultural and food inputs can make a positive contribution to reducing agricultural greenhouse gas emissions. This environmentally beneficial and economically viable method of production agriculture should be supported and endorsed through policy mechanisms so that worldwide adoption is increased and global benefits are realized. Mitigation of atmospheric $CO_2$ by increase CS in the soil, particularly make sense in the scope of other global challenges such as combating land degradation, improving soil quality and preserving biodiversity. Effective mitigation policies will likely be based on a combination of modest and economically sound reduction which confer added benefits to society.

## Author details

Peeyush Sharma
*Division of Soil Science & Ag-Chemistry, SKUAST-J, India*

Vikas Abrol
*Dryland Research Substation, RakhDhiansar, SKUAST-J, India*

Shrdha Abrol
*GGM Science College, Jammu, J&K, India*

Ravinder Kumar
*KVK, Rampur, Sardar Vallabhbhai Patel University of Agriculture and Technology, Meerut, India*

## 14. References

Arnalds, O. 2000. Desertification: An appeal for a broader perspective. Pages 5–15 *in* O. Arnalds, and S. Archer. Eds, Rangeland desertification. Kluver Academic Publishers, Dordrecht, The Netherlands.

Babaev, A. G. (ed.): 1999, Desert Problems and Desertification in Central Asia, Springer, Berlin,p. 293.

Baird, A. J.: 1997. 'Overland Flow Generation and Sediment Mobilization by Water', in Arid Zone Geomorphology: Process, Form and Change in Drylands, J.Wiley and Sons, U.K., pp. 165–184.

Balba, A. M.: 1995. Management of Problem Soils in Arid Ecosystems, CRC/Lewis Publishers, Boca Raton, FL, p. 250.

Barber, R. G.: 1994, 'Soil Degradation in the Tropical Lowlands of Santa Cruz, Eastern Bolivia', Land Degrad. Rehab., 6: 95–108.

Batjes, N. H. 1998. Mitigation of atmospheric carbon dioxide concentrations by increased carbon sequestration in the soil. Biology & Fertility of Soils. 27:230–235.

Blum, C. Valentine, and B. A. Stewart. Eds, Methods for assessment of soil degradation. CRC Press, Boca Raton, Florida.

Brandt, C. J. and Thornes, J. B. (eds.): 1996, Mediterranean Desertification and Land Use, J. Wiley & Sons, Chichester, U.K., p. 554.

Change 2007: Mitigation', Contribution of Working Group III to the Fourth Assessment Report of the IPCC [B. Metz, O. R. Davidson, P. R. Bosch, R. Dave, L. A. Meyer (eds)],

Cambridge and NY, USA: Cambridge University Press, Cambridge, United Kingdom and New York, NY, USA.

Conacher, A. J. and Sala, M. (eds.): 1998., 'Land Degradation in Mediterranean Environments of the World: Nature and Extent, Causes and Solutions', J. Wiley & Sons, Chichester, U.K., p. 491.

Conacher, A. J. and Sala, M.: 1998b. 'The Causes of Land Degradation 2: Vegetation Clearing and Agricultural Practices', in Conacher, A. J. and Sala, M. (eds.), *Land Degradation in Mediterranean Environments of the World: Nature and Extent, Causes and Solutions*, J. Wiley & Sons,Chichester, U.K., pp. 285–307.

Conant, R. T., K. Paustian, and E. T. Elliott. 2001. Grassland management and conversion into grassland: effects on soil carbon. Ecological Applications, 11:343–355.

Dalal, R. C. 1989. Long-term effects of no-tillage, crop residue and nitrogen application on properties of a Vertisol. Soil Science Society of America J., 53:1511–1515.

Dalal, R. C. 1992. Long-term trends in total nitrogen of a Vertisol subjected to zero tillage, nitrogen application and stubble retention. Australian Journal of Soil Res., 30:223–231.

Dalal, R. C., W. M. Strong, E. J. Weston, J. E. Cooper, K. J. Lehane, A. J. King, and C. J. Chicken. 1995. Sustaining productivity of a Vertisol at Warra, Queensland, with fertilizers, no-till or legumes. I. Organic matter status. Australian Journal of Experimental Ag., 37:903–913.

Dregne, H. E.: 1983. *Desertification of Arid Lands,* Harwood Academic Publishers, New York.

Dregne, H.E. 1976. Soils of Arid Regions. Elsevier, Amsterdam. p. 237.

EPA. 2008b. Carbon Sequestration in Agriculture and Forestry. www.epa.gov/sequestration/index.html

Esenov, P. E. and Redjepbaev, K. R.: 1999. 'The Reclamation of Saline Soils', in Babaev, A. G. (ed.), *Desert Problems and Desertification in Central Asia,* Springer, Berlin, pp. 167–177.

Eswaran, H., P. F. Reich, J. M. Kimble, F. H. Beinroth, E. Padamnabhan, P. Moncharoen, and J. M. Kimble. 2000.Global carbon stocks. Pages 15–250 *in* R. Lal, H. Eswaran,and B. A. Stewart. Eds, Global climate change and pedogeniccarbonates. Lewis Publishers, Boca Raton, Florida.540.

FAO 2004. Carbon Sequestration in Dryland Soils, World Soils Resources Reports 102. Food and Agriculture Organization Of The United Nations, Rome.

FAO: 1993, *Key Aspects of Strategies for Sustainable Development of Drylands,* FAO, Rome, Italy

Flessa, H., Ruser, R., Dörsch, P., Kamp, T., Jimenez, M.A., Munch, J.C., Beese, F. 2002. Integrated evaluation of greenhouse gas emissions ($CO_2$, $CH_4$, $N_2O$) from two farming systems in southern Germany. Agriculture, Ecosystems and Environment, 91: 175-189.

Fuller,W. H.: 1991, 'Organic Matter Applications', in Skujins, J. (ed.), *Semi-Arid Lands and Deserts:Soil Resource and Reclamation,* Marcel Dekker Inc., New York, pp. 507–542.

Glenn, E.P., Pitelka, L.F. & Olsen, M.W. 1992. The use of halophytes to sequester carbon. Wat. Air Soil Poll., 64: 251–263.

Gupta, J. P., and J. Venkateswarlu. 1994. Soil management and rainwater conservation and use. V. Aridisols*in* Soil management for sustainable agriculture in dryland areas. Bulletin on Indian Society of Soil Science ,16:Pages 66–77.

Hepperly, P., Moyer, J., Pimentel, D., Douds Jr, D., Nichols, K. And Seidel, R. (2008) Organic Maize/Soybean Cropping Systems Significantly Sequester Carbon and Reduce Energy Use. In: Neuhoff, D. *et al.*:(Eds): Cultivating the Future Based on Science. Volume 2 -

Livestock, Socio-economy and Cross disciplinary Research in Organic Agriculture. Proceedings of the Second Scientific Conference of the International Society of Organic Agriculture Research (ISOFAR), held at the 16th IFOAM Organic World Congress in Cooperation with the International Federation of Organic Agriculture Movements (IFOAM) and the Consorzio ModenaBio, 18 – 20 June 2008 in Modena, Italy.

Hester, J. W., Thurow, T. L., and Taylor, C. A., Jr.: 1997, 'Hydrologic Characteristic of Vegetation Types as Affected by Prescribed Burning', J. Range Manage., 50: 199–204.

Hoffmann, I. & Gerling, D. 2001. Farmers' management strategies to maintain soil fertility in a remote area in northwest Nigeria. Ag. Ecosys. Env., 86: 263–275.

Holland, J.M. 2004.The environmental consequences of adopting conservation tillage in Europe: reviewing the evidence Agriculture, Ecosystems and Environment 103 :1–25

Houghton, R.A. 1995. Changes in the storage of terrestrial carbon since 1850. In R. Lal, J. Kimble, E. Levin & B.A. Stewart, eds. Soils and global change, pp. 45–65. CRC Press.

Hülsbergen, K.-J., Küstermann, B. 2008. Optimierung der Kohlenstoffkreisläufe in Ökobetrieben. Ökologie und Landbau, 145: 20-22.

IPCC. 2001. Climate Change 2001: The scientific basis. Executive summary. Cambridge Univ. Press, Cambridge, U.K.

IPCC. 1996. Climate change 1995. Impacts, adaptations and mitigation of climate change: scientific, technical analyses. Working Group II. Cambridge, UK, Cambridge University Press.

IPCC. 2001. Climate change: the scientific basis. Cambridge, UK, Cambridge University Press.

IPCC. 2007a. Climate Change 2007: The Physical Science Basis. Contribution of Working Group I to the Fourth Assessment Report of the Intergovernmental Panel on Climate Change [Solomon, S., D. Qin, M. Manning (eds)].http://ipcc-wg1.ucar.edu/wg1/wg1-report.html

Jenkinson, D. S., H. C. Harris, J. Ryan, A. M. McNeil, C. J. Pilbeam, and K. Coleman. 1999. Organic matter turnover in calcareous soil from Syria under a two-course cereal rotation. Soil Biology & Biochemistry, 31:643–649.

Kihani, K. N., and D. A. More. 1984. Long-term effect of tillage operation and farmyard manure application on soil properties and crop yield in a Vertisol. Journal of Indian Society of Soil Science ,32:392–393.

Lal, R. 2001a. World cropland soils as a source or sink for atmospheric C. Adv. Agron., 71: 145–174.

Lal, R. 2001a. World cropland soils as a source or sink for atmospheric C. Adv. Agron.,71: 145–174.

Lal, R. 2002. Carbon sequestration in dryland ecosystems of West Asia and North Africa. Land Degradation & Development 13:45–59.

Lal, R. 2002a. Soil carbon dynamics in crop land and rangeland. Env. Poll., 116: 353–362.

Lal, R. 2002b. Carbon sequestration in dryland ecosystems of west Asia and north Africa. Land Deg. Dev., 13: 45–59.

Lal, R., Hassan, H. M. and Dumanski, J.: 1999, 'Desertification Control to Sequester Carbon and Mitigate the Greenhouse Effect', in Rosenberg, N., Izaurralde, R. C., and Malone, E. (eds.), Carbon Sequestration in Soils: Science, Monitoring and Beyond, Proc. St. Michael Workshop, December 1998, Battelle Press, Columbus, OH, pp. 83–151.

Lal, R., Kimble, J.M., Follet, R.F. & Cole, C.V. 1998. The potential of U.S. cropland to sequester carbon and mitigate the greenhouse effect. Chelsea, USA, Ann. Arbor. Press.

Laryea, K. B., Andres, M. M., and Pathak, P.: 1995, 'Long-Term Experiments on Alfisols and Vertisols in the Semi-Arid Tropics', in Lal, R. and Stewart, B. A. (eds.), *Soil management:Experimental Basis for Sustainability and Environmental Quality*, CRC/Lewis Publishers, Boca Raton, FL, pp. 267–292.

Le Hou'erou, H. N. 1975. Science, power and desertification. Meeting on desertization. Dept. Geogr. Univ. Cambridge,22–28 Sept, 1975, 26pp.

Lomte, M. H., Ateeque, M., Bharambe, P. R., and Kawarkhe, P. K.: 1993, 'Influence of Sorghum Legume Association on Physico-Chemical Properties of Soil', J. Maharashtra Agric,. 18: 388–390.

M.A. 2005. Millennium Ecosystem Assessment. Ecosystems and Human Well-being:Desertification Synthesis. World Resources Institute, Washington, DC.

Mabbutt, J. A.: 1984. 'A New Global Assessment of the Status and Trends of Desertification', Env.Cons., 11: 100–113.

Mäder, P., Fließbach, A., Dubois, D., Gunst, L., Fried, P., Niggli, U. (2002): Soil Fertility and Biodiversity in Organic Farming. Science, 296: 1694-1697.

Mainguet, M. 1991. Desertification: Natural background and human mismanagement. Springer Verlag, Berlin.

Mainguet, M. and Da Silva, G. G.: 1998. 'Desertification and Dryland Development: What Can Be Done?', Land Degrad. Develop., 9: 375–382.

Mathan, K. K., K. Sankaran, K. Kanakabushan, and K. K. Krishnamoorthy. 1978. Effect of continuous rotational cropping on the organic carbon and total nitrogen contents in a black soil. Journal of Indian Society of Soil Science, 26:283–285.

Middleton, N. J., and D. S. G. Thomas. 1992. World atlas of desertification. Edward Arnold/UNEP, Seven Oaks.

Miglierina, A. M., Galantini, J. A., Iglesias, J. O., Rosell, R. A., and Glave, A. 1996, 'Crop Rotation and Fertilization in Production Systems of the Semi-Arid Region of Buenos Aires', Revista de la Faculatd de Agronomia, Universidad de Buenos Aires, 15: 9–14.

Miglierina, A.M., Landriscini,M. R., Galantini, J. A., Iglesias, J. O.,Glave, A., and Gallardo Lancho, J. F. 1993, 'Fifteen Years of Crop Rotation and Fertilization in the Semi-Arid Pampas Region. I. Effect on the Chemical Properties of the Soil', *El estudio del suelo y de su degradacion en relacion con la desertification. Actas del 12 Congreso Latinamericano de la Ciencia del Suelo*,Sevilla (Espana), 19–26 Sept. 1993, pp. 751–758.

Mishra, M. M., Neelakantan, S., and Khandelwal, K. C. 1974, 'Buildup of Soil Organic Matter Continuous Manuring and Cropping Practices', *J. Res. Hayana Agric. Univ.* 4, 224–230.

Mortimore, M.: 1994. *Roots in the African Dust: Sustaining the Sub-Saharan Drylands*, Cambridge University Press, Cambridge, U.K., p. 219.

Murillo, J. M., F. Moreno, F. Pelegrin, and J. E. Fernandez. 1998. Response of sunflower to traditional and conservation tillage in southern Spain. Soil & Tillage Research, 49:223–241.

Nambiar, K. K. M.: 1995, 'Major Cropping Systems in India', in Barnett, V., Payne, R., and Steiner, R. (eds.), *agricultural Sustainability, Economic, Environmental and Statistical Considerations*, J. Wiley & Sons, Chichester, U.K., pp. 133–170.

Noin, D. and Clarke, J. I.: 1997. 'Population and Environment in Arid Regions of the World', in Clarke, J. and Noin, D. (eds.), *Population and Environment in Arid Regions*, MAB/UNESCO,Vol. 19, The Parthenon Publishing Group, New York, pp. 1–18.

Ojima, D. S., Parton, W. J., Schimel, D. S., and Scurlock, J. M. O.: 1993, 'Modeling the Effects of Climatic and CO2 Changes on Grassland Storage of Soil C', Water Air Soil Pollut,. 70: 643–657.

Oldeman, L. R. and Van Lynden, G. W. J.: 1998. 'Revisiting the GLASOD Methodology', in Lal, R., Blum, W. H., Valentine, C., and Stewart, B. A. (eds.), *Methods for Assessment of Soil Degradation*, CRC Press, Boca Raton, FL, pp. 423–440.

Pankova, E. I. and Solovjev. D. S.: 1995. 'Remote Sensing Methods of Dynamics of Irrigated Land Salinity for Reconstruction of Geotechnical Systems', *Prubliemy Osvoieniia Pustin*, 1995, No. 3.

Paustian, K., Collins, H.P. & Paul, E.A. 1997. Management control on soil carbon. *In*:E.A. Paul, K. Paustian, E.T. Elliot & C.V. Cole, eds. *Soil organic matter in temperate agroecosystems. Long term experiments in north America*, pp. 15–49. Boca Raton, USA,CRC Press.

Petersen, S.O., Regina, K., Pöllinger, A., Rigler, E., Valli, L., Yamulki, S., Esala, M., Fabbri, C., Syväsalo, E., Vinther, F.P. 2005. Nitrous oxide emissions from organic and conventional crop rotations in five European countries. Agriculture, Ecosystems and Environment, 112: 200-206.

Pieri, C.: 1991. *Fertility of Soils: A Future for Farming in the West African Savannah*, Springer Verlag, Berlin, p. 348.

Pieri, C.: 1995, 'Long-Term Soil Management Experiments in Semi-Arid Francophone Africa', in Lal, R. and Stewart, B. A. (eds.), *Soil Management: Experimental Basis for Sustainability and Environmental Quality*, CRC/Lewis Publishers, Boca Raton, FL, pp. 225–266.

Pluhar, J. J., Knight, R.W., and Heitschmidt, R. K.: 1987. 'Infiltration Rate and Sediment Production as Influenced by Grazing Systems in the Texas Rolling Plains', J. Range Manage., 40: 240–243.

Polwson, S.D., Smith, P. & Coleman, K. 1998. European network of long-term sites for studies on soil organic matter. Soil Till. Res. 47: 309–321.

Potter, K. N., O. R. Jones, H. A. Torbert, and P. W. Unger. 1997. Crop rotation and tillage effects on organic carbon sequestration in the semi-arid southern Great Plains. Soil Science ,162:140–147.

Quiroga, A. R., Buschiazzo, D. E., and Peinemann, N.: 1996. 'Soil Organic Matter Particle Size Fractions in Soils of the Semi-Arid Argentinian Pampas', Soil Sci., 161, 104–108.

Quiroga, A. R., Buschiazzo, D. E., and Peinemann, N.: 1998. 'Management Discriminate Properties in Semi-Arid Soils', Soil Sci. 163, 591–597

Reynolds, J. F., and D. M. S. Smith. 2002. Do humans cause deserts?. Pages 1–21 *in* J. F. Reynolds, and D. M. Stafford Smith. Eds, Global desertification: do humans cause deserts?. Dahlem Univ. Press, Berlin.

Robertson, G.P., Paul, E.A., Harwood, R.R. 2000. Greenhouse gases in intensive agriculture: contributions of individual gases to the radioactive forcing of the atmosphere. Science, 289: 1922-1925.

Rodale. 2003. Farm Systems Trial, The Rodale Institute 611 Siegfriedale Road Kutztown, PA 19530-9320 USA

Rozanov, B. G., Targulian, V., and Orlov, D. S.: 1990. in Turner, B. L., Clark, W. C., Kates, R. W.,Richards, J. F., Mathews, J. T., and Meyer, W. B. (eds.), *The Earth as Transformed by Human Action*, Cambridge University Press, Cambridge, U.K., pp. 203–214.

Russell, J. S. 1981. Models of long-term soil organic nitrogen change. Pages 222–232 *in* M. J. Frissel, and J. A. Van Veen. Eds, Simulation of nitrogen behavior of soil-plant systems. Center for Agricultural Publishing and Documentation, Wageningen, Holland.

Ryan, J. 1998. Changes in organic carbon in long-term rotation and tillage trials in northern Syria. Pages 285–296 *in* R. Lal, J. M. Kimble, R. F. Follett, and B. A. Stewart. Eds, Management of carbon sequestration in soils. CRC/Lewis Publishers, Boca Raton, Florida.

Ryan, J., S. Masri, S. Garabet, and H. Harris. 1997. Changes in organic matter and nitrogen with a cereal legume rotation trial. Pages 79–87 *in* J. Ryan Eds, Accomplishment and future challenges in dryland soil fertility research in the Mediterranean area. ICARDA, Aleppo.

Sagna-Cabral, M.A. 1989. *Utilisation et gestion de la matière organique d'origine animale dans un terroir du centre nord du Sénégal. Cas du village de Ndiamsil.* Montpellier,France, Centre National d'Etudes gronomiques des Régions Chaudes.

Schimel, D. S., J. I. House, K. A. Hubbard, and P. and others Bousquest. 2001. Recent patterns and mechanisms of carbon exchange by terrestrial ecosystems. Nature

Schimel, D.S. 1998. The carbon equation. Nature, 393: 208 - 209.

Shahin, R. R., M. A. El-Meleigi, A. A. Al-Rokiba, and A. M. Eisa. 1998. Effect of wheat stubble management pattern on organic carbon content and fertility of sandy soils under pivot irrigation system. Bulletin of the Faculty of Agriculture,Univ. of Cairo. *Cairo* 42:283–296.

Singh, G. and Goma, H. C. 1995. 'Long-Term Soil Fertility Management Experiments in East Africa', in Lal, R. and Stewart, B. A. (eds.), *Soil Management: Experimental Basis for Sustainability and Environmental Quality*, CRC/Lewis Publishers, Boca Raton, FL, pp. 347–384.

Singh, Y., D. C. Chaudhary, S. P. Singh, A. K. Bhardwaj, and D. Singh. 1996. Sustainability of rice-wheat sequential cropping through introduction of legume crops and green manure crops in the system. Indian Journal of Agronomy,41:510–514.

Skjemstad, J. O., V. R. Catchpoole, and R. P. LeFeuvre. 1994. Carbon dynamics in Vertisols under several crops as assessed by natural abundance 13C. Australian Journal of Soil Research, 32:311–321.

Skjemstad, J. O., V. R. Catchpoole, and R. P. LeFeuvre. 1994. Carbon dynamics in Vertisols under several crops as assessed by natural abundance 13C. Australian Journal of Soil Research, 32:311–321.

Sterk, G., Herrmann, L., and Bationo, A.: 1996. 'Wind-Blown Nutrient Transport and Soil Productivity Changes in Southwest Niger', Land Degrad. Develop., 7: 325–336.

Stern, N. 2006. The Stern Review on the Economics of Climate Change, Cambridge: Cambridge University Press.

Stewart, B. A., and C. A. Robinson. 2000. Land use impact on carbon dynamics in soils of the arid and semi-arid tropics. Pages 251–257 *in* R. Lal, J. M. Kimble, and B. A. Stewart. Eds, Global climate change and tropical ecosystems. CRC/Lewis Publishers, Boca Raton, Florida.

Swift, M. J., Seward, P. D., Frost, P. G. H., Qureshi, J. N., and Muchena, F. N.: 1994, 'Long-Term Experiments in Africa: Developing a Database for Sustainable Land Use under Global Change',in Leigh, R. A. and Johnston, A. E. (eds.), *Long-Term Experiments in Agricultural and Ecologica Sciences*, CAB International, Wallingford, U.K., pp. 229–251.

Swift, M.J., Seward, P.D., Frost, P.G.H., Qureshi, J.N. & Muchena, F.N. 1994. Longterm experiments in Africa: developing a database for sustainable land use under global change. *In* R.A. Leigh & Johnston A.E., eds. *Long-term experiments in agricultural and ecological sciences*, pp. 229–251. Wallingford, UK, CAB International.

Thurow, T. L., Blackburn, W. H., and Taylor, C. A., Jr. 1988. 'Infiltration and Inter-Rill Erosion Responses to Selected Livestock Grazing Strategies, Edwards Plateau Texas', J. Range Manage., 41: 296–302.

Tiessen, H. & Cuevas, E. 1994. The role of organic matter in sustaining soil fertility. Nature, 371: 783–785.

Traoré, S. and Harris, P. J. 1995, 'Long-Term Fertilizer and Crop Residue Effects on Soil and Crop Yields in the Savanna Region of Côte d'Ivoire', in Lal, R. and Stewart, B. A. (eds.), *Soil Management: Experimental Basis for Sustainability and Environmental Quality*, CRC/Lewis Publishers, Boca Raton, FL, pp. 141–180. UN: 1977, *Status of Desertification in the Hot, Arid Regions,Climatic Aridity Index Map and Experimental World Scheme of Aridity and Drought Probability, at a Scale of 1:25,000,000. Explanatory note*, UN Conference on Desertification A/Conf. 74/31, New York.

UNCED: 1992. Earth Summit Agenda 21: Programme of Action for Sustainable Development, UNEP, New York.

UNEP 1991. Status of desertification and implementation of the plan of action to combat desertification. UNEP, Nairobi, Kenya.

UNEP. 1992. World atlas of desertification. Nairobi.

UNEP. 1992. World atlas of desertification. Nairobi.

UNEP. 1997. World atlas of desertification. 2nd Edition. Nairobi.

UNEP: 1990. Desertification Revisited, UNEP/DC/PAC, Nairobi, Kenya.

US-EPA 2006 Global Anthropogenic Non-CO2 Greenhouse Gas Emissions: 1990-2020. United States Environmental Protection Agency, EPA 430-R-06-003, June 2006.

Verstraete, M. M. 1986, 'Defining Desertification: A Review', Clim. Change, 9: 5–18.

Wakeel, A. El and Astartke, A. 1996, 'Intensification of Agriculture on Vertisols to Minimize Land Degradation in Parts of the Ethiopian Highlands', Land Degrad. Develop., 7: 57–68.

Warren, A. 1996. Desertification. Pages XX–XX *in* . Eds, The physical geography of africa. Oxford Univ. Press, New York.

Weltz, M. A. and Blackburn, W. H. 1995, 'Water Budget for South Texas Rangelands', J. Range Manage,. 48: 45–52.

Whitehouse, M. J., and J. W. Littler. 1984. Effect of pasture on subsequent wheat crops on a black earth soil of the Darling Downs. II. Organic carbon, nitrogen and pH changes. Queensland Journal of Agriculture & Animal Science, 41:13–20.

# Soil and Nutrient Management
# for Sustainable Agriculture in Drylands

# Chapter 7

# Soil Quality and Productivity Improvement Under Rainfed Conditions – Indian Perspectives

K.L. Sharma, Biswapati Mandal and B. Venkateswarlu

Additional information is available at the end of the chapter

## 1. Introduction

India, predominantly has agrarian economy with an about 83 m ha area without irrigation and totally dependent on rainfall. This rainfed area constitutes about 58 % of net cultivated area of 142 m ha. The rainfed area supports about 44% of the total food production in the country. Most of the essential commodities such as coarse cereals (90%), pulses (87%), and oil seeds (74%) are produced from the rainfed lands. These statistics emphasise the role that rainfed regions play in ensuring food for the ever-increasing population. Owing to diversity in rainfall pattern, temperature, parent material, vegetation and relief or topography, this country is bestowed with different soil types predominantly alluvial soils, black soil, red soils, laterites, desert soils, mountainous soils etc. Taxonomically, soils in India represent Entisols (80.1 m ha), Inceptisols (95.8 m ha), Vertisols (26.3 m ha), Aridisols (14.6), Mollisols (8.0 m ha), Ultisols (0.8 m ha), Alfisols (79.7 m ha), Oxisols (0.3 m ha) and non-classified soil (23.1 m ha). Based on the rainfall pattern, 15 m ha area falls in a rainfall zone of <500mm, 15 m ha under 500 to 750 mm, 42 m ha under 750 to 1150 mm and 25 m ha under > 1150 mm rainfall. Predominant soil orders which represent semi-arid tropical region are Alfisols, Entisols, Vertisols and associated soils. Other soil orders such as Oxisols, Inceptisols and Aridisols also form a considerable part of rainfed agriculture. Most of the soils in rainfed regions are at the verge of degradation having low cropping intensity, relatively low organic matter status, poor soil physical health, low fertility, etc.

Moisture stress accompanied by other soil related constraints result in low productivity of majority of the crops (Sharma et al 1999). Besides natural causes, agricultural use of land is causing serious soil losses in many places across the world including India. It is probable that human race will not be able to feed the growing population, if this loss of fertile soils continues at the existing rate. In many developing countries, hunger is compelling the community to cultivate land that is unsuitable for agriculture and which can only be converted to agricultural use through enormous efforts and costs, such as those involved in the construction of terraces and other surface treatments. India represents wide spectrum of climate ranging from arid to semi arid, sub humid and humid with wider variation in rainfall amount and pattern. Seasonal temperature fluctuations are also enormous.

## 2. Constraints in improving the productivity in rainfed agriculture

The major constraints in improving the productivity and returns from rainfed farming in India are as follows: (i) erratic and uncertain rainfall, leading to moisture scarcity, droughts and failure of crops, especially annual crops, (ii) soil degradation and poor soil quality (iii) fragmented and low holding size, leading to constraints in mechanization, (iv) poverty among growers and constraints in availability and purchase of essential inputs, such as seeds and fertilizers, bullock-drawn small seed-cum-fertilizer drills, etc., (v) lack of assured credit and financial support and marketing, (vi) inadequate infrastructure for post-harvest value-addition and storage of produce, (vi) low procurement prices of agricultural commodities, in general, and (vii) inadequate earnings for livelihood from the farming profession because of low volume of business due to small holding size, low productivity and low produce prices, etc. The consequences of these constraints are likely to lead the marginal and small-farming communities towards distraction from agriculture, migration to cities to look for alternate assured wages, suicides, etc. To mitigate these constraints and transform the rainfed farming to an attractive option, there is a strong need for strategic planning and policy changes in a phased manner.

### 2.1. Specific causes of land degradation and soil quality deterioration

Out of the 329 m ha of total geographical area in the country, the total degraded area accounts for120.7 m ha, of which 73.3 m ha area is affected by water erosion, 12.4 m ha by wind erosion, 6.73 m ha by salinity and alkanity and 25 m ha by soil acidity. The predominant reasons which degrade land and deteriorate soil quality could be enumerated as : i) washing away of topsoil and organic matter associated with clay size fractions due to water erosion resulting in a 'big robbery in soil fertility', ii) intensive deep tillage and inversion tillage with moldboard and disc plough resulting in a) fast decomposition of remnants of crop residues which is catalyzed by high temperature, b) breaking of stable soil aggregates and aggravating the process of oxidation of entrapped organic C and, c) disturbance to the habitat of soil micro flora and fauna and loss in microbial diversity, iii) dismally low levels of fertilizer application and widening of removal-use gap in plant nutrients, iv) mining and other commercial activities such as use of top soil for other than agricultural purpose, v) mono cropping without following any suitable rotation, vi) nutrient imbalance caused due to disproportionate use of primary, secondary and micronutrients, vii) no or low use of organic manures such as FYM, compost, vermi-compost and poor recycling of farm based crop residues because of competing demand for animal fodder and domestic fuel, viii) no or low green manuring as it competes with the regular crop for date of sowing and other resources, ix) poor nutrient use efficiency attributing to nutrient losses due to leaching, volatilization and denitrification, x) indiscriminate use of other agricultural inputs such as herbicides, pesticides, fungicides, etc., resulting in poor soil and water quality, xi) water logging, salinity and alkalinity and acid soils. Among the various causes of degradation mentioned here, the first predominant cause of soil degradation in rainfed regions undoubtedly is water erosion. In fact, the process of water erosion sweeps away the topsoil along with organic matter and exposes the subsurface horizons. The second major indirect cause of degradation is loss of organic matter by virtue of temperature mediated fast decomposition owing to high temperature prevailing in these regions. Above all, the several other farming practices such as reckless tillage methods, harvest of every small component of biological produce and virtually no return of any plant residue back to the soil, burning of the existing residue in the field itself for preparation of clean seed bed, open grazing etc aggravate the process of soil degradation.

As a result of several above-mentioned reasons, soils encounter diversity of constraints broadly on account of physical, chemical and biological soil quality and ultimately end up with poor functional capacity (Sharma et al., 2007). In order to restore the quality of degraded soils and to prevent them from further degradation, it is of paramount importance to focus on restorative practices and conservation agricultural practices on long-term basis.

There is no doubt that, agricultural management practices such as crop rotations, inclusion of legumes in cropping systems, addition of animal based manures, adoption of soil water conservation practices, various permutations and combinations of deep and shallow tillage, mulching of soils with leafy materials grown in-situ grown and brought externally always remained the part and parcel of agriculture in India. Despite all these efforts, the concept of conservation farming could not be followed in an integrated manner to expect greater impact in terms of protecting the soil resource from degradative processes. In the context of likely changes in climate in the years to come , threats to agriculture in general and land and soil resources in particular, will be more and more , hence concrete strategies to protect the land resource and mange the soil effectively are must.

## 2.2. Climatic threat to Agriculture – Indian perspective

According to Rao et al (2010), the major weather related risks in Agriculture could be as follows: Monsoons in India exhibits substantial inter-seasonal variations, associated with a variety of phenomena such as passage of monsoon disturbances related with active phase and break monsoon periods whose periodicities vary from 3-5 and 10-15 days respectively. It is well noticed that summer monsoon rainfall in India varied from 604 to 1020 mm. The inter-seasonal variations in rainfall cause floods and droughts, which are the major climate risk factors in Indian Agriculture. The main unprecedented floods in India are mainly due to movement of cyclonic disturbances from Bay of Bengal and Arabian Sea on to the land masses during monsoon and post-monsoon seasons – and during break monsoon conditions in some parts of Uttar Pradesh and Bihar States. The thunderstorms due to local weather conditions also damages agricultural crops in the form of flash floods. Beside floods, drought is a normal, repetitive feature of climate associated with deficiency of rainfall over extended period of time to different dryness levels describing its severity. Rao et al (2010), have reported that during the period 1871 to 2009, there were 24 major drought years, defined as years with less than one standard deviation below the mean. Another important adverse effect of climate change could be unprecedented heat waves. Heat waves generally occur during summer season where the cropped land is mostly fallow, and therefore, their impact on agricultural crops is limited. However, these heat waves adversely affect orchards, livestock, poultry and rice nursery beds. The heat wave conditions during 2003 May in Andhra Pradesh and 2006 in Orissa are recent examples that have affected the economy to a greater extent. Also occurrence of heat waves in the northern parts during summer is common every year resulting in quite a good number of human deaths. Further, the water requirements of summer crops grown under irrigated conditions increase to a greater extent. Another adverse effect of climate change is cold waves which mostly occur in northern states. The Northern states of Punjab, Haryana, U.P., Bihar and Rajasthan experience cold wave and ground frost like conditions during winter months of December and January almost every year. The occurrence of these waves has significantly increased in the recent past due to reported climatic changes at local, regional and global scales. Site-specific short-term fluctuations in lower temperatures and the associated phenomena of chilling, frost, fogginess and impaired sunshine may sometimes play havoc in an otherwise fairly stable cropping/farming system of a region. All these apprehensions, however, are based on the data base generated in India. The reports of the Non-Governmental International Panel on Climate Change (NIPCC, 2009), has dispelled many fears about the global warming and its consequent adverse effects on climate change an in-turn influence on agriculture.

## 2.3. Likely climate change effects on soil

It is anticipated that climate change is likely to have a variety of impacts on soil quality. Soils vary depending on the climate and show a strong geographical correlation with climate.

The key components of climate in soil formation are moisture and temperature. Temperature and moisture amounts cause different patterns of weathering and leaching. Wind redistributes sand and other particles especially in arid regions. The amount, intensity, timing, and kind of precipitation influence soil formation. Seasonal and daily changes in temperature affect moisture effectiveness, biological activity, rates of chemical reactions, and kinds of vegetation. Soils and climate are intimately linked.

Climate change scenarios indicate increased rainfall intensity in winter and hotter, drier summers. Changing climate with prolonged periods of dry weather followed by intense rainfall could be a severe threat to soil resource.

Climate has a direct influence on soil formation and cool, wet conditions and acidic parent material have resulted in the accumulation of organic matter.

A changing climate could also impact the workability of mineral soils and susceptibility to poaching, erosion, compaction and water holding capacity. In areas where winter rainfall becomes heavier, some soils may become more susceptible to erosion. Other changes include the washing away of organic matter and leaching of nutrients and in some areas, particularly those facing an increase in drought conditions, saltier soils, etc. Not only does climate influence soil properties, but also regulates climate via the uptake and release of greenhouse gases such as carbon dioxide, methane and nitrous oxide. Soil can act as a source and sink for carbon, depending on land use and climatic conditions. Land use change can trigger organic matter decomposition, primarily via land drainage and cultivation. Restoration and recreation of peat lands can result in increased methane emissions initially as soils become anaerobic, whereas in the longer term they become a sink for carbon as organic mater accumulates. Climatic factors have an important role in peat formation and it is thus highly likely that a changing climate will have significant impacts on this resource.

No comprehensive study has yet been made of the impact of possible climatic changes on soils.

Higher temperatures could increase the rate of microbial decomposition of organic matter, adversely affecting soil fertility in the long run. But increases in root biomass resulting from higher rates of photosynthesis could offset these effects.

Higher temperatures could accelerate the cycling of nutrients in the soil, and more rapid root formation could promote more nitrogen fixation. But these benefits could be minor compared to the deleterious effects of changes in rainfall.

For example, increased rainfall in regions that are already moist could lead to increased leaching of minerals, especially nitrates. In the Leningrad region of the USSR a one-third increase in rainfall (which is consistent with the GISS 2 x CO2 scenario) is estimated to lead to falls in soil productivity of more than 20 per cent. Large increases in fertilizer applications would be necessary to restore productivity levels.

Decreases in rainfall, particularly during summer, could have a more dramatic effect, through the increased frequency of dry spells leading to increased proneness to wind erosion. Susceptibility to wind erosion depends in part on cohesiveness of the soil (which is affected by precipitation effectiveness) and wind velocity.

Nitrogen availability is important to soil fertility and N cycling is altered by human activity. Increasing atmospheric $CO_2$ concentrations, global warming and changes in precipitation patterns are likely to affect N processes and N pools in forest ecosystems. Temperature, precipitation, and inherent soil properties such as parent material may have caused differences in N pool size through interaction with biota. Keller et al., 2004 reported that climate change will directly affect carbon and nitrogen mineralization through changes in temperature and soil moisture, but it may also indirectly affect mineralization rates through changes in soil quality.

Climate change is having a major impact on biodiversity and in turn biodiversity loss (in the form of carbon sequestration trees and plants) is a major driver of climate change. Land degradation such as soil erosion, deteriorating soil quality and desertification are driven by climate variability such as changes in rainfall, drought and floods. Degraded land releases more carbon and greenhouse gases back into the atmosphere and slowly kills off forests and other biodiversity that can sequester carbon, creating a feed back loop that intensifies climate change.

Soil is our most fundamental terrestrial asset and natural resource. Along with sunlight and water, it provides the basis for all terrestrial life viz., the biodiversity around us, the field crops that we harvest to meet our food and fiber demands, animal products, etc. Healthy soils provide us with a range of 'ecosystem services' - they support healthy plant growth, resist erosion, receive and store water, retain nutrients and act as an environmental buffer in the landscape. Soils supply nutrients, water and oxygen to plants, and are inhabited by soil biota which are essential for decomposition and recycling processes. According to Arshad and Martin (2002), like air and water, the soil is an integral component of our environment and constitutes the most important natural resource together with water. The intellectual and efficient use of this vital resource is essential for sustainable development and feeding the growing world population. In the recent decade, soil is perceived as an important environmental component and the need to maintain or improve its ability to perform the multitude of functions has been recognized. At the same time, it has also been recognized that the soil is not an inexhaustible resource, and if used inappropriately or mismanaged it may be deteriorated in a relatively short period of time, with very limited opportunity for regeneration or replacement.

## 3. Sustainability of agriculture – General concepts and scenario

Agricultural sustainability is defined as the ability of agricultural systems to remain productive, efficiently and indefinitely. Quantitatively, it implies trends in agricultural production over time. A non-negative trend in production of a system over time implies that the system is sustainable (Lal, 1998). According to Lockeretz (1988) 'sustainable agriculture' is a loosely defined term that encompasses a range of strategies for addressing many of the problems that afflict agriculture worldwide. These problems include loss of soil productivity from excessive erosion and associated plant nutrient losses, surface and groundwater pollution from pesticides fertilizers and sediment, impending shortages of nonrenewable resources and low farm income because of low market price and high production costs. Herdt and Steiner (1995) have given three dimensions of assessing the sustainability and it is essential to assess the sustainability in relation to all three dimensions. These dimensions include biophysical, economic and social. Among these, the biophysical dimension is related to the quantity of output per unit area (may be Tonnes or Mg of yield ha[-1]), the economic dimension to the gross or net value of the output, and the social dimension to the capacity of the system to support the farming community. The biophysical output (biomass or grain yield per ha) may change due to change in soil properties over time (erosion, compaction, salinization, waterlogging, etc), introduction of new cultivars, and change in the input used. The economic output may change over time independent of the biophysical output, and the social carrying capacity may change due to change in food habits, preferences and standard of living. Lal (1994) emphasized more specific indicators for measuring sustainability. First is the productivity which indicates productivity per unit of resources used. This indicator is influenced by several factors. To cite an example, the unsustainability in rice –wheat system could be attributed to decline in organic matter owing to (i) removal of wheat straw for feeding the animals, (ii) burning or removal of rice straw for ensuring clean fields and trouble free cultivation, (iii) puddling process for transplanting the rice seedlings which breaks the soil aggregates and subjects the entrapped organic matter fractions to further loss and ultimately lead to poor soil structure. Irrespective of straw removal and burning, wheat –rice system can be compared with the analogy of 'making' the house in the morning and 'demolishing' it in the evening where wheat season during which soil gets time for aggregation can be said as 'making of the house' and rice season where puddling of soil assumes the shape of colloidal solution is just like demolishing of the house. Another important indicator emphasized is the total factor productivity which considers the total output in relation to the cost of all the inputs used to get that output. The third indicator suggested for assessing agricultural sustainability is total natural resource productivity. This indicator of sustainability takes care of the indirect cost incurred on account of quantitative depreciation or wear and tear (e.g. decrease in soil depth due to loss of top soil owing to erosion, build up of salinity, fall in groundwater table by irrigation, increase of nutrient load in water bodies etc.,) of the natural resources for achieving the specific output.

Campbell *et al.* (1995), emphasized that a sustainable agricultural system is that which is economically viable, provides safe, nutritious food, and conserves or enhances the environment. The ultimate goal or the ends of sustainable agriculture is to develop farming systems that are productive and profitable, conserve the natural resource base, protect the environment, and enhance health and safety, and to do so over the long-term. The means of achieving this is low-input methods and skilled management, which seek to optimize the management and use of internal production inputs (i.e., on-farm resources) in ways that provide acceptable levels of sustainable crop yields and livestock production and result in economically profitable returns. This approach emphasizes such cultural and management practices as crop rotations, recycling of animal manures, and conservation tillage to control soil erosion and nutrient losses and to maintain or enhance soil productivity. Low-input farming systems seek to minimize the use of external product inputs (i.e., off-farm resources), such as purchased fertilizers and pesticides, wherever and whenever feasible and practicable; to lower production costs; to avoid pollution of surface and groundwater: to reduce pesticide residues in food; to reduce farmer's overall risk; and to increase both short- and long-term farm profitability (Parr *et al.*, 1989, Parr and Hornick, 1990). According to Parr *et al.*, (1990) , another reason for the focus on low-input farming systems is that most high- input systems, sooner or later, would probably fail because they are not either economically or environmentally sustainable over the long-term. How we achieve "Sustainable agriculture" depends on creative and innovative conservation and production practices that provide farmers with economically viable and environmentally sound alternatives or options in their farming systems. Stewart *et al.* (1990) emphasized that climate and soils are the two most critical factors that will determine the ultimate sustainability of agricultural systems. Jodha (1994) opined that despite significant growth and refinements in the definitions of the term "sustainability", its operationalization continued to be a major problem that reduces its practical utility. One practical way to handle this problem could be 'to approach sustainability through unsustainability' This implies identification and analysis of indicators of unsustainability and their underlying processes and focused efforts to reverse them to restore sustainability to a system. Based on the definitions of the term by ecologists, economists, environmentalists, development experts, etc, (Conway 1985, Markande and Pearce, 1988, Lynam and Herdt, 1988, Graham-Tomasi, 1991) sustainability would mean the ability of a system (say dryland agriculture) to maintain or enhance its performance, output, services, (even though linkage with other systems), without damaging own long term production potential. Hence, to halt any further deterioration of the natural resource base, that is, agricultural land, and the associated loss of soil productivity, the key to improving the sustainability of rainfed/dryland farming systems could be implementing sound soil and water management practices. In many cases, improvements can be achieved by the application of established principles of soil and water management to crop and livestock production. In other situations, new concepts and methodologies appropriate to the unique aspects of dryland areas will be required (Steiner *et al.*, 1988; Parr *et al.*, 1990).

India has been working hard since 1950 to produce adequate food to feed its increasing population and to become self dependent. Unfortunately, the growth in the food production is getting neutralized by the growth in the population. If the food production history of the country is traced back, country increased it food production from 53.87 Mt during 1950-51 and jumped to 78.61 Mt during 1960-61 which was followed by 100.64 during 1970-71, 123.7 Mt during 1980-81 and 172.39 in 1990-91, 206 Mt in 2001 and finally to 230.7 Mt in 2007-08. This made the country self sufficient in food production for the time being. But at the same time, the population growth demands more growth in food production with shrinking availability of land resource and degrading land and water resources. Lal (2008), based on a critical perusal of food production data brought out that agronomic production in India between 1960 and 2002 increased by a factor of about 2.5 for rice, 6.4 for wheat, and 2.5 for all food grains. He cautioned that country must not be satisfied with this increase in food production as the country is going to face a big mandate of feeding an expected 1.59 billion population by 2050. This rise in population from the existing 1.1 billion during 2007 to 1.59 billion by 2050 will be approximately 45%. According to some other estimates, there would be a need to increase food grain production from 206 million tonnes (Mt) in 2001 to 301 Mt with low food demand, 338 Mt with medium food demand and 423 Mt with high food demand by 2025 (Sekhon, 1997; USDA, 2004), which is a matter of growing concern for all those involved in agricultural research, planning and policy making. Country has to gear up to meet these difficult targets among all odds such as (i) stagnating yields levels even under irrigated systems, (ii) degrading land and deteriorating soil quality, (iii) increasing cost on energy, (iv) extreme climatic variations and uncertainties in rainfall pattern and (iv) frequent droughts, non-stretchable irrigation potential and major dependence on rainfed lands, (v) distraction of the farmers from agriculture because of marginal size of holdings and non remunerative nature of agriculture due to low minimum support prices etc. Lessons learnt from the success of green revolution clearly indicate that four pillars of achieving higher production were: high yielding input responsive varieties, assured irrigation, adequate amount of fertilization and appropriate plant protection measures. Under irrigated agriculture, the response to these inputs has slowed down and yield levels have stagnated in some of the crops may be due to land degradation and deterioration of soil quality. On the other hand, rainfed regions which constitute 83 m ha comprising of about 58% of net cropped area (142.2 m. ha) still do not get adequate inputs like water, fertilizer and good seed and continue to depend on 'Rain-God' even for seeding the crops. Above all, rainfed regions encounter several other productivity related constraints which may or not be common to those of irrigated agriculture, and ultimately limit the yield to a miserably low levels leading to poverty, migration of the communities to the urban areas in search of livelihoods and even suicides under most distressful condition. The situation has become so grim that out of the 110 million farm holders in the country, about 40% wish to quit the farming, if they get any alternative source of livelihood earnings (Katyal, 2008).

If we look in the world perspective, crop yields in India are not only low in comparison with developed countries viz., U.S.A., Canada, Europe, Australia, Japan, but also with those in China, South East Asia, and South America. Similar to food grains, the yields of vegetables in India are about 50% lower than those of the world average, and 60-100% lower than those of China (Pain, 2007; Lal, 2008). Besides several other factors, low yields of crops and cropping systems are also attributed to poor soil fertility and inadequate replenishment of the nutrients. The situation is grimmer in rainfed areas where the crops are poorly nourished or fertilized owing to low soil fertility, low fertilizer use due to poor economic condition of the farmers, monsoon uncertainties, etc., which has resulted in multi-nutrient deficiencies in soils. On the other hand, the changing price policies on fertilizers have made the resource poor farmers of rainfed areas to feed their crops with only certain type of fertilizer, which has resulted in low nutrient use efficiency and profitability due to deficiency/antagonistic relationships of certain essential nutrients. It has been estimated that only 9% of the districts in India use more than 200 kg of $N + P_2O_5 + K_2O$ per hectare (Tiwari et al., 2006). On the other hand, only 32% districts used < 50 kg nutrients ($N + P_2O_5 + K_2O$) per hectare. Most of the rainfed regions fall in this category. The average fertilizer use in the country as a whole in the recent years is about 117 kg NPK $ha^{-1}$ $yr^{-1}$ (Tiwari, 2008) which is very low when compared with the neighboring countries, like China (277.7 kg $ha^{-1}$), Japan (290.6 kg $ha^{-1}$) and Korea Republic (409.7 kg $ha^{-1}$) as stated earlier. At the same time, the yield levels of some of the crops such as paddy, wheat, maize, etc., in India are significantly lower compared to these countries. Hence, apart from many other reasons, low fertilizer use in India is definitely one of the important causes of low yields. According to Tiwari (2008), the major challenge ahead to the country is that it needs to have about 30-35 Mt of NPK from different sources to produce 300 Mt food grains to feed its expected population of about 1.4 billion by the end of 2025. Further, he added that, if the nutrient removal by other crops like horticulture, vegetables, plantation crops, sugarcane, cotton, oilseeds and potato is considered, nutrient demand curve will touch 40-45 Mt. Further, Katyal (2008) has emphasized that, because of existence of wider gap between nutrient addition and mining by the crops, almost 50% of the Indian soils have reached below the critical limit of plant available zinc in soil. The corresponding deficiency in case of iron is about 25%. The severity of the deficiency has increased because of negligible or no application of organic sources of nutrients. Therefore, to ensure sustainability in production, all possibilities need to be explored to narrow down the nutrient removal use gap in future.

## 3.1. Relationship between soil quality and agricultural sustainability

Beside several other factors influencing crop production, better soil quality is definitely one of the key players influencing sustainability. At the same time, sustainable management practices are those which do not deteriorate soil quality on long term basis. Soil quality and sustainability evaluation is a fundamental concept bridging between the utilization and protection aspects of soil. In terms of agricultural production, soil quality refers to its ability to sustain productivity. There exists a strong link between soil quality and agricultural sustainability. If an agricultural system is unsustainable, it may partly due to the fact that soil quality is declining over time. Understanding soil quality means, assessing and managing soil so that it functions optimally now and is not degraded for future use. Therefore, understanding the whole concept of soil quality including methods of assessment, delineation of key indicators and their related soil functions, transformation of indicators in to a single value soil quality index etc., assumes importance.

Soil quality, in short, has been defined as the "capacity of the soil to function" (Doran and Parkin, 1994; Karlen et al., 1997). But broadly, soil quality has been defined as 'the capacity of a living soil to function within natural or managed ecosystem boundaries, to sustain plant and animal productivity, maintain or enhance water and air quality, and promote plant and animal health' (Doran et al., 1996, 1998). A slightly modified definition of soil quality was given by Seybold et al. (1999), in which soil quality was defined as 'the capacity of a specific kind of soil to function, within natural or managed ecosystem boundaries, to sustain plant and animal productivity, maintain or enhance water and air quality and support human health and habitation'. Soil quality acts as a major linkage between the strategies of conservation management practices and achievement of the major goals of sustainable agriculture (Acton and Gregorich, 1995). The terms 'soil health' and 'soil quality' are interchangeable. 'Soil quality' is generally used more by soil scientists and 'soil health' by others, but they do have different emphasis (Doran et al., 1996). Some prefer the term 'soil health' as it portrays soil as a living, dynamic system whose functions are mediated by a diversity of living organisms that require management and conservation. 'Soil quality' is the capacity of soils within landscapes to sustain biological productivity, maintain environmental quality, and promote plant and animal health. 'Soil health' is the fitness (or condition) of soil to support specific uses (e.g. crop growth) in relation to its potential - as dictated by the inherent soil quality and is more sensitive to anthropogenic disturbance and is severely limited in extreme environments (Freckman and Virginia, 1997). So, soil health and soil quality are functional concepts that describe how fit the soil is to support the multitude of roles that can be defined for it. Therefore, soil quality can be regarded as soil health (Doran et al., 1996).

Quality with respect to soil can be viewed in two ways: (1) as inherent properties of a soil; and (2) as the dynamic nature of soils as influenced by climate, and human use and management. Inherent soil quality is a soil's natural ability to function and the inherent soil characteristics are those directly linked with the basic soil forming factors and these characteristics determine why any two soils will always be different. These generally focus on the entire soil profile (~ 2 m deep), and is the reason why there can be no single value describing soil quality for all soil resources and land uses. Such soils can be compared with regard to inherent differences in productivity and with regard to their capacity for a specific land use in the absence of human interventions. (Karlen *et al*, 2001). Attributes of inherent soil quality usually show little change over time. Generally, dynamic soil quality changes in response to soil use and management (Larson and Pierce, 1994). Management choices affect the amount of soil organic matter, soil structure, soil depth, water and nutrient holding capacity. Soils respond differently to management depending on the inherent properties of the soil and the surrounding landscape. According to Carter (1996), attributes of dynamic soil quality are subjected to change over a period of years to decades, while pH and labile organic matter fractions may change over a period of months to years. In comparison, microbial biomass and populations, soil respiration, nutrient mineralization rates, and macroporosity can change over a period of hours to days. Thus, maintenance and/or improvement of dynamic soil quality deal primarily with those attributes or indicators that are most subject to change, loss, depletion, and strongly influenced by agronomic practices. The distinction between inherent and dynamic soil quality can be characterized by the genetic (or static) pedological processes versus the kinetic (or dynamic) processes in soil as proposed by Richter (1987).

## 4. Soil quality indicators

Brejda and Moorman (2001) stated that soil quality can not be measured directly but can be measured through some sensitive indicators. Further, they emphasized that the changes in these indicators are used to determine whether soil quality is improving, stable, or declining with changes in management, land-use, or conservation practices. Indicators of soil quality can be defined loosely as those soil properties and processes that have greatest sensitivity to changes in soil functions (Andrews *et al.*, 2004). Indicators are a composite set of measurable attributes which are derived from functional relationships and can be monitored via field observation, field sampling, remote sensing, survey or compilation of existing information (Walker and Reuter, 1996). Indicators signal desirable or undesirable changes in land and vegetation management that have occurred or may occur in the future. These indicators may directly monitor the soil, or monitor the outcomes that are affected by the soil, such as increases in biomass, improved water use efficiency, and aeration. Soil quality indicators can also be used to evaluate sustainability of land-use and soil management practices in agroecosystems (Shukla *et al.* 2006). The predominant soil quality indicators at micro and macro farm scale as suggested by Singer and Ewing (2000) have been listed in Table 1.

Several researchers have observed different set of key indicators for assessing soil quality depending upon the soil types and other variations. Mairura *et al.* (2007) reported the integration of scientific and farmer's evaluation of soil quality indicators and emphasized that the indicators for distinguishing productive and non-productive soils include crop yields and performance, soil colour and its texture. Parr *et al.* (1992) suggested that increased infiltration, aeration, macropores, aggregate distribution and their stability and soil organic matter and decreased rate of bulk density, soil resistance, erosion and nutrient runoff are some of the important indicators for improved soil quality. Further, Chaudhury *et al.* (2005) identified total soil N, available P, dehydrogenase activity and mean weight diameter of the aggregates as the key indicators for alluvial soils. While working in rainfed Alfisols in semiarid tropical India under sorghum - mungbean system, Sharma et al. (2008) identified easily oxidizable N ($KMnO_4$ oxidizable -N) DTPA extractable zinc (Zn) and copper (Cu), microbial biomass carbon (MBC), mean weight diameter (MWD) of soil aggregates and hydraulic conductivity (HC) as the key indicators of soil quality. In another study in Alfisols under sorghum–castor system, the key soil quality indicators identified were available N, K, S, microbial biomass carbon (MBC) and hydraulic conductivity (HC) (Sharma *et al.*, 2005). Karlen *et al.* (1992) suggested biological measurements viz., microbial biomass, respiration, and ergosterol concentrations as very effective indicators for assessing long-term soil and crop management effects on soil quality. Assessment of soil-test properties from time to time has also been emphasized for evaluating the chemical aspects of soil quality (Karlen *et al.* 1992; Arshad and Coen 1992). The indicators used or selected by different researchers in different regions may not be the same because soil quality assessment is purpose and site specific (Wang and Gong 1998; Shukla *et al.* 2006). However, while selecting the indicators, it is important to ensure that the indicators should i) correlate well with natural processes in the ecosystem (this also increases their utility in process-oriented modelling, ii) integrate soil physical, chemical, and biological properties and processes, and serve as basic inputs needed for estimation of soil properties or functions which are more difficult to measure directly, iii) be relatively easy to use under field conditions, so that both specialists and producers can use them to assess soil quality, iv) be sensitive to variations in management and climate and v) be the components of existing soil databases wherever possible (Doran *et al.* 1996; Doran and Parkin 1996; Chen 1998). Interpreting soil quality by merely monitoring changes in individual soil quality indicators may not give complete information about soil

| Physical indicators | Chemical indicators | Biological indicators |
|---|---|---|
| Passage of air | BSP | Organic carbon |
| Structural stability | Cation exchange capacity | Microbial biomass carbon |
| Bulk density | Contaminant availability | C and N/Oxidizable carbon |
| Clay mineralogy | Contaminant concentration | Total biomass |
| Colour | Contaminant mobility | Bacterial |
| Consistence (dry, moist, wet) | Contaminant presence | Fungal |
| Depth of root limiting layer | Electrical conductivity | Potentially mineralizable N |
| Hydraulic conductivity | Exchangeable sodium | Soil respiration |
| Oxygen diffusion rate | percentage | Enzymes |
| Particle size distribution | Nutrient cycling rates | Dehydrogenase |
| Penetration resistance | pH | Phosphatase |
| Pore conductivity | Plant nutrient availability | Arlysulfatase |
| Pore size distribution | Plant nutrient content | Biomass C/total organic |
| Soil strength | Sodium adsorption ratio | carbon/ |
| Soil tilth | | Respiration /biomass |
| Structure type | | Microbial community |
| Temperature | | fingerprinting |
| Total porosity | | Substrate utilization |
| Water holding capacity | | Fatty acid analysis |
| | | Nucleic acid analysis |

Source: Singer and Ewing (2000)

**Table 1.** Predominant soil quality indicators at micro and macro farm scale

quality. Therefore, combining them in a meaningful way to a single index may assess soil quality more precisely (Jaenicke and Lengnick, 1999; Bucher, 2002) which is used to gauge the level of an improving or declining soil condition (Wienhold, 2004).

## 4.1. Soil quality indicators influences soil functions and sustainability

Every soil attribute or soil quality indicator has an important role to play in influencing various soil processes and functions. Hence, to understand the changes in processes and functions, quantitative measurement of attributes or indicators is inevitable. The predominant soil physical, chemical and biological attributes or indicators and corresponding processes influenced by them as suggested by Lal (1994) are given in Table 2.

| Attributes / Indicators | Processes and soil functions |
| --- | --- |
| **Physical attributes** | |
| **A. Mechanical** | |
| Texture | Crusting, gaseous diffusion, infiltration |
| Bulk density | Compaction, root growth, infiltration |
| Aggregation | Erosion, crusting, infiltration, gaseous diffusion |
| Pore size distribution and continuity | Water retention, and transmission, root growth and gaseous exchange |
| **B. Hydrological** | |
| Available water capacity | Drought stress, biomass production, soil organic matter content |
| Non-limiting water range | Drought, water imbalance, soil structure |
| Infiltration rate | Runoff, erosion leaching |
| **C. Rooting zone** | |
| Effective rooting depth | Root growth, nutrient and water use efficiencies |
| Soil temperature | Heat flux, soil warming activity and species diversity of soil fauna |
| **Chemical Attributes** | |
| pH | Acidification and soil reaction, nutrient availability |
| Base saturation | Absorption and desorption, solublization |
| Cation exchange capacity | Ion exchange, leaching |
| Total and plant available nutrients | Soil fertility, nutrient reserves |
| **Biological Attributes** | |
| Soil organic matter | Structural formation, mineralization, biomass carbon, nutrient retention |
| Earthworm population and other soil, macro fauna and activity | Nutrient cycling, organic matter decomposition, formation of soil structure |
| Soil biomass carbon | Microbial transformations and respiration, formation of soil structure and organo-mineral complexes |
| Total soil organic carbon | Soil nutrient source and sink, bio-mass carbon, soil respiration and gaseous fluxes |

Source (Lal, 1994)

**Table 2.** Predominant soil physical, chemical and biological indicators and associated functions

## 4.2. Chemical indicators and their soil functions

Of the various indicators, pH is one of the important indicator, which influence some of the soil functions. It can provide trends in change in soil health in terms of soil acidification (surface and sub surface) (Moody and Aitken, 1997), soil salinization, electrical conductivity, exchangeable sodium (soil structural stability) (Rengasamy and Olsson, 1991), limitations to root growth, increased incidence of root disease, biological activity, and nutrient availability (e.g. P availability at either high pH > 8.5 or low pH < 5; Zn availability at high pH > 8.5) (Doran and Parkin, 1996). Soil pH trends also provide changed capacity of the soil for pesticide retention and breakdown as well as the mobility of certain pesticides through soil. These processes affect soil health on-farm and have effects beyond farm gate (Karlen *et al.* 1997). Electrical conductivity is a measure of salt concentration and therefore, its measure can provide trends in salinity for both soil and water, limitations to crop growth and water infiltration, and along with pH (indicating soil sodicity), it can be a surrogate measure of soil structural decline (eg. high pH > 8.5 and low electrical conductivity, < 0.1 dSm$^{-1}$) (Rengasamy and Olsson, 1991).

It is a well known fact that, the organic matter is fundamental to the maintenance of soil health because it is essential to the optimal functioning of a number of processes important to sustainable ecosystems. Soil organic matter is a source and sink of carbon and nitrogen and partly of phosphorus and sulphur. It affects micronutrient availability through complexation, chelation and production of organic acids, thus altering soil pH. Conversely, it ties up metals present in toxic amounts (e.g. Cu, As, Hg) (Doran and Parkin, 1996). Organic matter is essential for good soil structure especially in low clay content soils, as it contributes towards both formation and stabilization of soil aggregates (Dalal and Mayer, 1986). Other functions include: contribution to low cation exchange capacity, especially in low clay content soil, pesticide retention (Kookana *et al.*, 1998), microbial biodiversity, water retention in sandy and sandy-loam soils, and provision of carbon sink and source for greenhouse gases. Trends in soil organic matter content provide an integrated measure of sustainable ecosystem (Karlen *et al.*, 1997). Status of plant available nutrients, for example, N, P, S and K indicate the systems sustainable land use, especially, if the nutrient concentration and availability are approaching but remain above the critical or threshold values. In the long-term, nutrient balance of the system (e.g. Input efficiency =output) is essential to sustainability. Thus, available nutrients are indicators of the capacity to support crop growth, potential crop yield, grain protein content (Dalal and Mayer, 1986), and conversely, excessive amounts may be a potential environmental hazard (e.g. algal biomass).

## 4.3. Physical indicators and their soil functions

The physical indicators of soil health reflect the capacity to accept, store, transmit and supply water, oxygen and nutrients within ecosystem. This includes monitoring of soil structure through pore size distribution, aggregate stability, saturated hydraulic conductivity, infiltration, bulk density, and surface crust. Rooting depth provides a good indicator of buffering against water, air and nutrient stress. Soil surface cover can be used as an indicator of soil surface protection against raindrop impact, and hence enhanced infiltration, reduced surface crust, and reduced soil erosion and runoff. Soil water infiltration measures the rate at which water enters soil surface, and transmitted through the immediate soil depth (Arshad *et al.* 1996). Rainfall is rapidly absorbed by soil with high infiltration rate, but as the soil structure deteriorates, usually with the loss of organic matter, increase in exchangeable sodium and low electrolyte concentration, infiltration rate of a soil becomes low (Rengasamy and Olsson, 1991). This increases the tendency for soil erosion and runoff in sloping soils and water logging in flat soils. Unfortunately, current procedures for measuring infiltration rates are cumbersome, and subject to large errors. A modified disc permeameter could make infiltration rate and hydraulic conductivity a routine procedure (Bridge 1997). Soil aggregate stability is a measure of structural stability and refers to the resistance of soil aggregates to breakdown by water and mechanical force. Aggregate stability is affected by health and quantity of organic matter, types of clays, wetting and drying, freezing and thawing, types and amounts of electrolyte, biological activity, cropping systems and tillage practices (Arshad *et al.* 1996). For monitoring trends in soil health, sampling procedures for aggregate stability need to be standardized. Bulk density varies with the structural condition of the soil. It is altered by cultivation, loss of organic matter (Dalal and Mayer, 1986), and compression by animals and agricultural machinery, resulting in compact plough layer. It generally increases with depth in the soil profile. In cracking clay soils such as Vertisol, it varies with water content (Bridge and Ross, 1984). In Vertisols, bulk density should be corrected for soil water content at the time of sampling, and bulk density values adjusted at field capacity moisture content assuming three dimensional matrix shrinkage.

Effective soil depth is a good indicator of plant available water capacity, subsoil salinity and other root growth constraints in the soil profile. It is not known whether trends can be discerned over relatively long periods (Walker and Reuter, 1996; Doran and Parkin, 1996). Surface crust retards seed germination and reduces aeration and water entry. It provides an indication of soil structure decline (Aggarwal *et al.* 1994, Bridge, 1997). However, it needs to be quantitatively measured or alternatively photographed over time and the extent of area quantified. Surface cover by either crop residues or vegetation protects soil surface from raindrop impact, enhances infiltration, reduces soil erosion and may decrease runoff (Freebairn and Wockner, 1986). The extent of surface cover therefore provides an integrated indicator of soil physical management, organic matter input and the effects beyond farm gate. It can be measured by satellite imagery (currently expensive), and by combining with the terrain and digital elevation mapping, may provide an indicator of erosion hazard. However, correct timing of monitoring in relation to cropping and vegetation cycle and erosive rainfall periods is essential.

## 4.4. Biological indicators and their soil functions

In the set of biological soil quality indicators, soil microbial biomass and/or respiration, potentially mineralizable N, enzyme activity, fatty acid profile or microbial biodiversity, nematode communities and earthworm populations are quite predominant. Soil microbial biomass is a labile source and sink of nutrients. It affects nutrient availability as well as nutrient cycling and is a good indicator of potential microbial activity (Dalal and Mayer, 1987) and capacity to degrade pesticides (Perucci and Scarponi, 1994). Although useful as a research tool, its cumbersome measurement and variability with short-term environmental conditions makes it difficult as a routine soil quality indicator (Sparling, 1997; Dalal, 1998). Respiration measurements are also similarly affected. However, respiration rates can be measured in the field using portable $CO_2$ analysers. Easily oxidizable N and potentially mineralizable N are measured by alkaline-$KMnO_4$ method and aerobic or anaerobic incubation respectively. Anaerobic method is considered to be more effective and is recommended as routine procedure. Potentially mineralizable N measures soil N supplying capacity and is also a surrogate measure of microbial biomass and a labile fraction of soil organic matter (Rice et al. 1996). Soil enzyme activity is often closely related to soil organic matter, microbial activity and microbial biomass. It is sensitive to change in management practice and can readily be measured. Of numerous soil enzymes, dehydrogenase is a potential indicator of active soil microbial biomass. However, it is very sensitive to seasonal variability. Potentially useful indicators of soil quality could be beta-glucosidase, urease, amidase, phosphatase, and aryl-sulphatase and fluorescein diacetate hydrolyzing enzymes. Since enzyme activity is operationally defined, it requires strict protocol (Dick et al. 1996). Soil fauna (soil meso and macro fauna), including nematode communities, affect soil structure, alter patterns of microbial activity and influence soil organic matter dynamics and nutrient cycling (Heal et al., 1996), and are sensitive to soil disturbance and contamination. Of the soil invertebrates, earthworms and nematodes are the potential indicators of soil quality (Pankhurst, 1994; Blair et al. 1996). It has been understood that some of the soil indicators do not change immediately and take some time for getting influenced through management practices. Hence, for to be more objective in the approach, these indicators need to be monitored after a specific intervals only.

## 5. Assessment of soil quality- Recent approaches

Assessment of soil quality is a sensitive and dynamic way to document soils condition, its response to management, or its resistance to stress imposed by natural forces or human uses (Larson and Pierce, 1991). It is needed to identify problem production areas, make realistic estimates of food production, monitor changes in sustainability and environmental quality as related to agricultural management, and to assist government agencies in formulating and evaluating sustainable agricultural and land-use policies (Granatstein and Bezdicek, 1992). As stated earlier, soil quality can be assessed by measuring soil attributes or properties that serve as soil quality indicators. The changes in these indicators signal the changes in soil quality (Brejda and Moorman, 2001). The first step is selecting the appropriate soil quality indicators to efficiently and effectively monitor critical soil functions as determined by the specific management goals for which an evaluation is being made. These indicators together form a minimum data set (MDS) that can be used to determine the performance of the critical soil functions associated with each management goal. In order to combine the various chemical, physical and biological measurements with totally different units, each indicator is then scored using ranges established by the soil's inherent capability to set the boundaries and shape of the scoring function. Indicator scoring can be accomplished in a variety of ways (e.g. linear or nonlinear, optimum, more is better, more is worse) depending upon the function. These unitless values are combined into an overall index of soil quality and can be used to compare effects of different practices on similar soils or temporal trends on the same soil. Andrews and Carroll (2001) suggested that dynamic soil quality assessment could be viewed as one of the components needed to quantify agro ecosystem sustainability.

In order to quantify the effects of the three tillage systems on soil quality and to test the sensitivity of various indexing procedures, Hussain et al. (1999) has adopted the soil quality framework developed by Harris et al. (1996). The overall soil quality index was computed using the equation, index = f (y nutrient + y water + y rooting) where y = weighting factor for each function. To complete the evaluation, they regressed the six overall soil quality indices and the individual function ratings against the dependent variable (erosion, yield, and plant populations). While, Andrews and Caroll (2001) and Andrews et al. (2002a and b) have described comprehensively another approach of soil quality assessment 'a comparative assessment technique' The three predominant steps adopted under this technique were i) selection of a minimum data set (MDS) of indicators that best represent soil function, ii) scoring of the MDS indicators based on their performance of soil functions, and iii) corroboration of the MDS indicators with functional goals set by the land manager or grower and iv) integration of the indicator score into a comparative index of soil quality. This method is being used widely in recent soil quality assessment studies (Hazra et al., 2004; Mandal, 2005; Sharma et al., 2005, 2008, Chaudhury et al., 2005, Masto et al., 2007). Sharma et al. (2004) has reviewed and given a brief account of various methods of assessment of soil quality such as: i) simple assessment of soil properties using quick soil test kits and to observe the changes occurred as a result of management practices, ii) issuing of soil health cards to the farmers and to advise them to observe the changes in the visible soil and crop indicators and go on recording them periodically, iii) deviation from the normal: computation of percent deviations in soil attributes with reference to control situation and to assign the score using score functions, iv) key indicator approach: identification of key indicators using functional goals and computation of soil quality index, and v) use of critical levels of indicators: identification of critical levels of indicators and assigning the rank and computation of Cumulative Rating Index (CRI).

According to Masto et al. (2007), the success and usefulness of a soil quality index mainly depends on setting the appropriate critical limits for individual soil properties. They stated that the optimum values of soil quality could be obtained from the soils of undisturbed ecosystems (Warkentin, 1996; Arshad and Martin, 2002), where soil functioning is at its maximum potential to provide critical values. They fixed the thresholds for each soil quality indicator based on the range of values measured in natural ecosystems or in best–managed systems and on critical values available in the literature. After finalizing the thresholds, they transformed the soil property values recorded into unitless scores (between 0 and 1), using the equation:

$$\text{Non-linear score (Y)} = 1/1+e^{-b(x-A)}$$

Where x is the soil property value, A the baseline or value of the soil property where the score equals 0.5 and b is the slope. Using the equation, they generated three types of standardized scoring functions as i) 'More is better', ii) 'Less is better' and iii) 'Optimum' as defined in earlier studies (Karlen and Stott, 1994; Hussian et al., 1999; Glover et al., 2000). For positive slopes, the equation defined a 'More is better' scoring curve; for negative slopes, a 'Less is better' curve; and for the combination of both, an 'Optimum' curve has been defined. They converted the numerical values for each soil quality indicator into unitless scores ranging from 0 to 1. The score for each indicator was calculated after establishing lower threshold limits, baseline values and upper threshold limits. Threshold values are soil property values where the score equals one (upper threshold) when the measured soil property is at most favorable level; or equals zero (lower threshold) when the soil property is at an unacceptable level. Baseline values are soil property values where the scoring function equals 0.5 and equal the midpoints between threshold soil property values. Baselines are generally regarded as minimum target values. In their study, to determine soil quality, they used the model primarily described by Karlen et al. (1994) with some modification, which is given as follows:

$$\text{Soil quality index (SQI)} = q_{we}(wt) + q_{wms}(wt) + q_{rsd}(wt) + q_{rbd}(wt) + q_{pns}(wt) + q_{scp}(wt)$$

Where $q_{we}$ is the rating for the soil's ability to accommodate water entry, $q_{wms}$ to facilitate water movement and storage, $q_{rsd}$ to resist surface degradation, $q_{rbd}$ to resist biochemical degradation, $q_{pns}$ to supply plant nutrients, $q_{scp}$ to sustain crop productivity and wt is a numerical weighting for each soil function. These were set according to the function's importance in fulfilling the overall goal of maintaining soil quality.

With the progressive development in the methodology of soil quality assessment, many new tools of soil quality assessment viz., Soil Conditioning Index (SCI), Soil Management Assessment Framework (SMAF), the Agroecosystem Performance Assessment Tool (AEPAT) and the New Cornell "Soil Health Assessment" have been recently reported. Out of these, SMAF and AEPAT were developed as malleable tools for assessing soil response to management. Weinhold et al. (2008) brought out that some of these tools can be highly useful for assessing soil quality at watershed scale. Hence, these approaches could be of importance for assessing soil quality under watershed development programme in India.

## 6. Effects of management practices on soil quality, productivity and sustainability – Recent reports

During the past, most of the research studies pertaining to soil quality and sustainability in India was centering around soil testing for only essential plant nutrients, crop response to fertilizer, manures, conjunctive use of fertilizers and organic sources of nutrient on medium and long term basis , computation of optimum levels of fertilizers and manures for recommendation, use of soil amendments to correct acidity, alkalinity, water logging and drainage, protection of top soil through effective soil and water conservation measures etc . Progressively, the research focus shifted with All India Coordinated Projects, where researchers started observing the long term influence of soil and nutrient management treatments on soil parameters through systematic analysis. Many useful results emerged from these studies. Somehow, the prime research focus remained on soil fertility or chemical soil quality indicators except in case of program such as soil structure improvement programs, organic waste recycling, biological nitrogen fixation and few others. In the process of soil quality monitoring, all the three pillars of soil quality (Physical, Chemical and Biological) did not get holistic deal. With advancement in the concept of research on soil quality and sustainability across the world, a paradigm shift in the thinking processes and research programs has come in India also. There are several reports describing the influence of soil and nutrient management treatments such as tillage, residue recycling, application of organic manures, green manuring and integrated use of organic and inorganic sources of nutrients on soil quality. The salient findings of some of the recently conducted studies on soil quality in India and abroad are presented in this section.

Manna et al. (2005) studied the potential impact of continuous cultivation of crops in rotation, and fertilizer and manure application on yield trends and predominant soil quality parameters in rice–wheat–jute, soybean–wheat and sorghum–wheat system at Barrackpore (Typic Eutrochrept), Ranchi (Typic Haplustalf) and Akola (Typic Haplustert), respectively In this study, the negative yield trend was observed in unbalanced use of inorganic N and NP application at all the three sites. The positive yield trend was observed in the NPK and NPK + FYM treatments at Ranchi and Akola. Results showed that the SOC in the unfertilized plot (control) decreased by 41.5, 24.5, and 15.5% compared to initial values at Barrackpore, Ranchi and Akola, respectively, wherein the treatment receiving NPK and NPK + FYM either maintained or improved it over initial SOC content in these sites reported. In a critical study, Mandal *et al.* (2007) observed that crop species and cropping systems that are cultivated may also play an important role in maintaining SOC stock because both quantity and quality of their residues that are returned to the soils vary greatly affecting their turnover or residence time in soil and thus its quality. Further, they reported that conjunctive use of organic and inorganic source of nutrients make significant contribution of carbon inputs to soil.

The impact of land configuration in combination with nutrient management treatments was studied by Selvaraju *et al.* (1999) in rainfed Alfisols. From this study, it was observed that tied ridging and application of FYM in combination with inorganic N and P fertilizer can increase the soil water storage and yield of crops compared to traditional flat bed cultivation. Mohanty *et al.* (2007) recorded that soil quality in rice-wheat cropping system is governed primarily by the tillage practices used to fulfill the contrasting soil physical and hydrological requirements of the two crops. and observed that Soil Quality Index (SQI) values of 0.84 to 0.92, 0.88 to 0.93 and 0.86 to 0.92 were found optimum for rice, wheat and the combined system (rice + wheat), respectively. Kusuma (2008) has established the quantitative relationship between Relative Soil Quality Indices (RSQI) and functional goal such as long term average yields and Sustainability Yield Indices (SYI) of sorghum and mungbean system (Fig 1). The simultaneous contribution of the key indicators towards functional goals has also been studied under sorghum - castor system in rainfed Alfisol using multiple regression functions (Table 3). These relationships help in predicting the crop yield from a given value of RSQI and quantitative contribution of indicators towards long-term crop yields and SYI. While working with biological soil health, Ghoshal (2004) proved that different biological indicators contributed differently towards explaining biological soil health for different cropping systems.

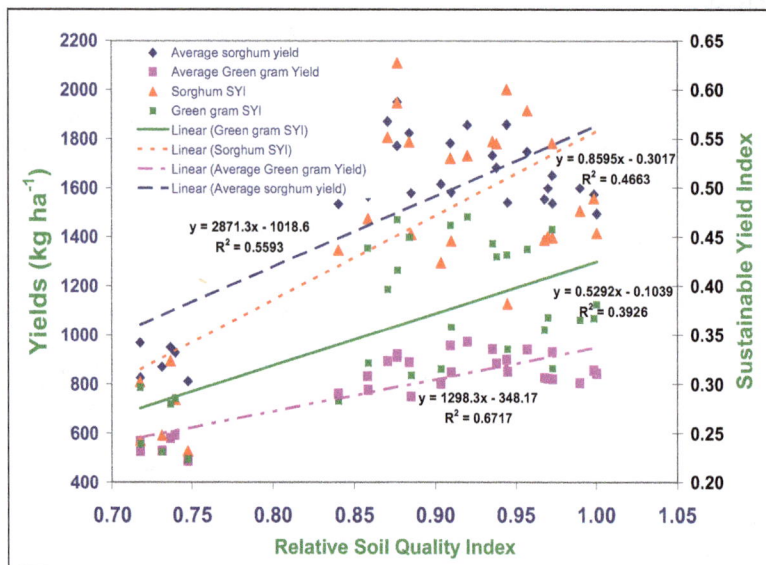

**Figure 1.** Relationships between functional goals and relative soil quality indices

| SNo | Parameter | Equation | R | $R^2$ | Level of significance |
|---|---|---|---|---|---|
| 1 | Castor average yield | $-1527.7 + 5.69$ (N)** $+ 16.25$ (S) $+ 1.83$ (K)** $-2.26$ (MBC)** $+ 252.0$ (HC)** | 0.712 | 0.507 | P= 0.01 |
| 2 | Sorghum average yield | $-1466.62 + 6.69$ (N)** $+1.20$ (K) $+22.96$ (S) $+ - 0.330$ (MBC) $+ 209.1$ (HC)** | 0.648 | 0.420 | P= 0.01 |
| 3 | Sustainability Yield Index | $-0.598 +0.003$ (N)** $+ 0.0006$ (K) $+ 0.0137$ (S)** $- 0.0006$ (MBC) $+ 0.096$ (HC)** | 0.641 | 0.411 | P= 0.01 |
| 4 | Organic matter | $- 0.324 -0.0002$ (N) $+ 0.001$(K)** $+ 0.0052$ (S) $+0.00596$ (MBC)** $+ 0.0077$ (HC) | 0.812 | 0.659 | P= 0.01 |

Source: Kusuma (2008)

**Table 3.** Relationship of key soil quality indicators with functional goal in sorghum –castor system in rainfed Alfisol

Through a collaborative study with a number of centres and involving a large number of long-term experiments under various agro-climatic zones, Mandal (2005) identified a few master variables and their relative contributions towards soil quality index calculated for different cropping systems and soil types (Table 4). Sharma *et al.,* (2008), in a long term study conducted in rainfed Alfisol, on integrated nutrient management under reduced and conventional tillage in sorghum – mungbean system reported that irrespective of conventional and reduced tillage, the sole organic treatments out-performed in aggrading the soil quality to the extent of 31.8 % over control whereas, the conjunctive nutrient use treatments aggraded the soil quality by 24.2 to 27.2 %, and the sole inorganic treatment could aggrade only to the extent of 18.2 % over unamended control. The extent of percent contribution of the key indicators towards soil quality index (SQI) as presented in Fig 2 was: microbial biomass carbon (MBC) (28.5%), available nitrogen (28.6%), DTPA- Zn (25.3%), DTPA- Cu (8.6%), HC (6.1%) and MWD (2.9%). Conjunctive use of organic and inorganic sources of nutrients also proved effective in realizing significantly higher grain yields and sustainability of both sorghum and mungbean crops (Sharma *et al.*, 2009). The predominant soil quality indicators identified for Vertisols under cotton + green gram system were: pH, electrical conductivity (EC), organic carbon (OC), available K, exchangeable magnesium (Mg), dehydrogenase assay (DHA), and microbial biomass carbon (MBC). The soil quality indices as influenced by different long-term soil and nutrient-management treatments in this study varied from 1.46 to 2.10. Among the treatments, the conjunctive use of 25 kg P2O5 ha$^{-1}$ + 50 kg N ha$^{-1}$ through leuceana green biomass maintained significantly higher soil quality index with a value of 2.10 followed by use of 25 kg N +25 kg P2O5 + 25 kg N ha$^{-1}$ through FYM (T5) (2.01) (Sharma et al 2011). In a review on effects of tillage, Ishaq *et al.* (2002) reported that tillage affects soils physical, chemical and biological properties. Tillage-induced changes in these properties depend on antecedent soil properties, type of tillage, and climate. A proper tillage can alleviate soil related constraints while improper tillage leads to a range of degradative processes, e.g., decline in soil structure, accelerated erosion, depletion of soil organic matter (SOM) and fertility and disruption in cycles of water, organic carbon and plant nutrients (Lal, 1993).

| Location | Soil classification /textural type | Cropping system | Indicators identified | Indicator's contributions towards SQI | | |
|---|---|---|---|---|---|---|
| | | | | > 25% | 15-25% | < 15% |
| Varanasi | Typic Haplustept/ Sandy loam | Rice – Lentil | Available P and organic C | - | Available P | Organic C |
| Barrack-pore | Eutrochrept/ Sandy loam | Jute – Rice – Wheat | Mean weight diameter Available P, Microbial biomass C, and organic C | Mean weight diameter, Available P | Microbial biomass C | Organic C |
| Mohanpur | Aeric Haplaquept/ Sandy loam | Rice – Wheat | Alkaline phosphatase, organic C, mineralizable N | Alkaline phosphatase | Organic C | Minerali-zable N |

Source: Mandal, 2005

**Table 4.** Soil quality indicators identified for different soil types and cropping systems and their contributions towards soil quality index (SQI)

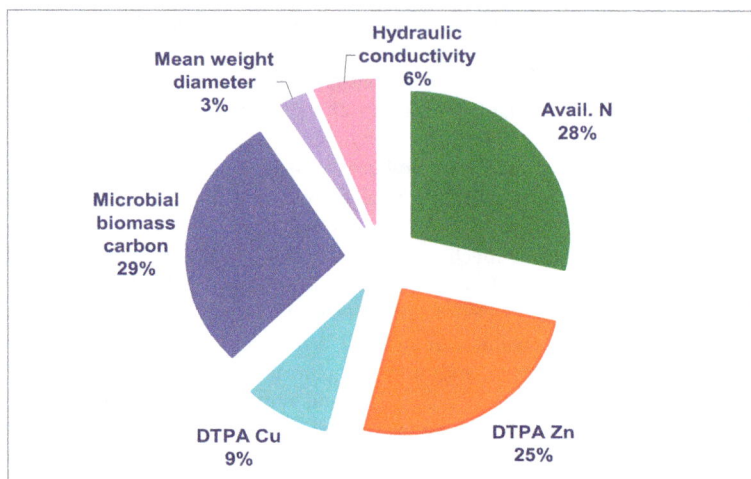

**Figure 2.** Per cent contribution of different soil quality indicators towards relative soil quality index (RSQI) in Alfisol

According to Ghuman and Sur (2001), reduced tillage in conjunction with crop residue improves soil properties and subsequent wheat yield on sandy-loam soils in the subtropical climate of northwestern Punjab. Roldan *et al.* (2005) observed that conservation tillage systems, in particular no-tillage, increased crop residue accumulation on the soil surface. Soil electrical conductivity and pH were not affected by the tillage practices. The no-tilled soil had higher values of water soluble C, dehydrogenase, urease, protease, phosphatase and β-glucosidase activities and aggregate stability than tilled soils, but had lower values than the soil under native vegetation. The enzyme activity and aggregate stability showed higher sensitivity to soil management practices than did physical–chemical properties. They finally concluded that no tillage system was the most effective for improving soil physical and biochemical qualities.

Beside conservation tillage, positive effect of other soil and nutrient management practices such as green manuring, integrated nutrient management practices, manure application, crop residue recycling, legume based crop rotations, balanced fertilization etc have been observed on predominant soil quality indicator, overall soil quality indices and Sustainability Yield Indices of crops. Sharma *et al.*, (2005) reported that organic carbon in the soil was significantly influenced by long term application of crop residues such as sorghum stover and gliricidia @ 2 tons ha$^{-1}$ under minimum and conventional tillages in sorghum-castor rotation in rainfed Alfisols. Further, they reported that increase in nitrogen levels from 0 to 90 kg N ha$^{-1}$ also helped in significantly improving the organic carbon status in these soils over a period of 8 years. From these studies, they concluded that continuous application of organic residues is inevitable to see the significant effect on organic carbon status in soils. Green manuring, which is considered as one of the important practice for improving soil fertility and soil health, is the process of turning a crop into the soil, whether originally intended or not, irrespective of its state of maturity, for the purpose of affecting some agronomic improvement (Mac Rae and Mehuys, 1985). This practice has been found to increase soil N and P availability for the following crop and at the same time, contribute to the conservation of soil organic matter and soil biological, physical and chemical properties (Astier *et al.*, 2006). While studying the management of residues in cropping systems, Smith and Elliot (1990) and Rasmussen and Collins (1991) emphasized the importance of residue application in conserving soil and water and thus ensuring sustainable production, especially in the semi-arid regions where soil and water conservation are of utmost importance. Prasad and Power (1991) emphasized that no one-residue management practice is superior under all conditions. It is important, therefore, to establish under local conditions the beneficial and detrimental consequences associated with a residue management practice before it is propagated among farmers for implementation. It was reported that the application of manure as amendment proved quite effective in improving the soil nutrient status and increasing soil organic C (SOC) levels (Rochette and Gregorich, 1998). The conjunctive use of urea and organics such as loppings of leuceana and gliricidia (1:1 ratios on N equivalent basis) had considerable effects on raising the sorghum grain yield to the levels of 16.9 and 17.2 q ha$^{-1}$ respectively and thus revealed that a minimum of 50 % N requirement of sorghum can be easily met from farm based organic sources of nutrients (Sharma *et al.*, (2002). Based on a long term study conducted in rainfed Alfisol soils prone to hardsetting and crusting, it was observed that integrated nutrient management treatments. viz. , 2t gliricidia loppings + 20 kg N and 4 t compost + 20 kg N were found to be most effective in increasing the sorghum grain yield by 84.62 and 77.7 percent over control. However, the highest amount of organic carbon content (0.74%) was recorded in 100 % organic treatment (4 t compost + 2 t gliricidia loppings). Some of these options of managing nutrients by using farm based organics can save expenditure on fertilizer and help in improving organic C in soil (Sharma *et al.*, 2004).

Chaudhury *et al.* (2005) in a study of identifying several biological, chemical, and physical indicators of soil quality concluded that the highest SQI was found in 100% NPK+FYM treatment followed by 100% NPK, 100% NP, 100% N, and control treatment, respectively. A collaborative study coordinated by Mandal (2005) indicated that cultivation without any fertilization (control) or only with N caused a net degradation of soil quality. Cultivation even with application of balanced NPK could hardly maintain such quality at the level where no cultivation was practiced. Only integrated use of organic and inorganic sources of nutrients could aggrade the system (Table 5). Manna *et al.* (2007) compared the fertilizer treatments in a long-term study for 30 Years in Alfisol (Typic Haplustalf) Ranchi, India. They reported that yield increased with time for NPK +FYM and NPK + lime treatments in wheat. Biological soil health indicators such as Soil Microbial Biomass Carbon (SMBC), nitrogen (SMBN) and acid hydrolysable carbohydrates (HCH) were greater in NPK + FYM and NPK + lime as compared to other treatments. Findings of this study suggested that continuous use of NPK + FYM or NPK + lime would sustain yield in a soybean – wheat system without deteriorating soil quality. Soil degradation occurs due to nutrient depletion, soil structure degradation, acidification and sub–optimal addition of organic and inorganic fertilizer to soil. Masto *et al.* (2007) quantified the effects of 10 fertilizer and farm yard manure (FYM) treatments applied for 31 years to a rotation that included maize, pearl millet, wheat and cowpea on an Inceptisol in India. A soil quality index (SQI) based on six soil functions was derived for each treatment using bulk density, water retention, pH, electrical conductivity (EC), plant-available nutrients, soil organic matter (SOM), microbial biomass, soil enzymes and crop yield. SQI ratings ranged from 0.552 (unfertilized control) to 0.838 for the combined NPK fertilizer plus manure treatment.

| Treatment / Cropping system | Rice - wheat | Rice - Lentil | Jute-Rice-Wheat |
|---|---|---|---|
| Control | - 56.0 | - 8.0 | - 49.0 |
| N only | - | - 11.7 | - 35.0 |
| NPK only | -10.8 | -9.7 | 19.0 |
| NPK+FYM | 18.7 | 8.6 | 45.1 |

Source: Mandal, 2005

**Table 5.** Soil quality change (as % over fallow) under different nutrient management practices and cropping systems

Soil quality indices have been used to compare tillage practices, organic and conventional vegetable production systems, litter management practices, and spatially large regions Plains, Hills, and several other practices (Andrews *et al.*, 2003). These varied uses suggest that SQ indices may be applicable not only to different soil types but also to multiple regions and management systems. The impact of long term soil and nutrient management treatments on soil quality using 19 soil chemical , physical and biological indicators has been assessed by Sharma (2009 a, b) at All India Coordinated Research Project Centers spread across the country and reported that conjunctive nutrient use as well as sole organic nutrient treatments found superior to 100 % inorganic nutrient application. Further, he has also suggested the set of key soil quality indicators for each location depending upon the soil type and cropping system. Mandal *et al.* (2001) worked out a crop specific land quality index (LQI) for sorghum [Sorghum bicolor (L.) Moench] under semiarid tropics of India. The method developed as LQI is a function of climatic quality index (CQI) and soil quality index (SQI). The LQI was correlated with the actual sorghum yield obtained from benchmarks soils and it was found that LQI bears good agreement with the yield. Doran and Parkin (1994) described a performance based index of soil quality that could be used to provide an evaluation of soil function with regard to the major issues of (i) sustainable production, (ii) environmental quality, and (iii) human and animal health. They proposed a soil quality index consisting of six elements: SQ = f (SQE1, SQE2, SQE3, SQE4, SQE5, SQE6); Where SQET is the food and fibre production, SQE2 the erosivity, SQE3 the ground water quality, SQE4 the surface water quality, SQE5 the air quality, and SQE6 is the food quality. Awasthi *et al.* (2005) computed integrated soil quality indices in four dominant land uses [forest, upland maize and millet (Bari), irrigated rice (Khet), and grazed systems). Integrated soil quality index (SQI) values varied from 0.17 to 0.69 for different land uses, being highest for undisturbed forest and lowest for irrigated rice. The SQI demonstrated the degradation status of land uses in the following ascending order: irrigated rice > grazed system > forest with free grazing > upland maize and millet > managed forest > grass land > undisturbed forest. The irrigated rice, grazed system, upland maize and millet, and freely grazed forestlands need immediate attention to minimize further deterioration of soil quality in these land uses. From the information presented in this section, it is evident that (i) effective soil and nutrient management practices can help in a long to improve soil quality indicators and overall indices of soil quality and crop yield sustainability. Hence soil management practices assume great importance in improving soil quality and sustainability.

# 7. Effective steps for improving soil quality, productivity and sustainability with emphasizes in rainfed areas

The following steps are suggested for effective land care and soil quality improvement for higher productivity and sustainability in rainfed areas.

## 7.1. Controlling soil erosion through effective soil and water conservation (SWC) measures

It is well accepted connotation that 'Prevention is better than cure'. In order to protect the top soil, organic mater content contained in it and associated essential nutrients, it is of prime importance that there should be no migration of soil and water out of a given field. If this is controlled, the biggest robbery of clay-organic matter -nutrients is checked. This can be easily achieved, if the existing technology on soil and water conservation is appropriately applied on an extensive scale. The cost for in-situ and ex-situ practices of SWC has been the biggest concern in the past. There is a need to launch 'Land and Soil Resource Awareness Program' (LSRAP) at national level to educate the farming community using all possible communication techniques. It is desirable to introduce the importance of soil resource and its care in the text books at school and college levels. The subject at present is dealt apparently along with geography. Farming communities too need to be made aware about soil, its erosion, degradation, benefits and losses occurred due to poor soil quality. This can be done through various action learning tools which explain the processes of soil degradation in a simple and understandable manner

## 7.2. Rejuvenation and reorientation of soil testing program in the country

About more than 600 Soil testing labs situated in the country need to be reoriented, restructured and need to be given fresh mandate of assessing the soil quality in its totality including chemical, physical, biological soil quality indicators and water quality. The testing needs to be on intensive scale and recommendations are required to be made on individual farm history basis. Special focus is required on site specific nutrient management (SSNM). Soil Health Card (SHC) system needs to be introduced. Soil fertility maps of intensive scale need to be prepared. District soil testing labs need to be renamed as 'District Soil Care Labs' and required to be well equipped with good equipments and qualified manpower for assessing important soil quality indicators including micronutrients. Fertilizer application needs to be based on soil tests and nutrient removal pattern of the cropping system in a site specific manner. This will help in correcting the deficiency of limiting nutrients. Keeping in mind the sluggish and inefficient activities of regional soil testing labs of the states, private sector can also be encouraged to take up Soil Care Programs with a reasonable costs using a analogy of 'Soil Clinics for Diagnosis and Recommendation' (SCDR).

### 7.3. Promotion of agricultural management practices which enhance soil organic matter

Enhancing organic matter in soils in semi-arid tropics and tropics is indomitable task. However regular additions of organics without hastening their decomposition process can provide some relief. Management practices such as application of organic manures (composts, FYM, vermi-composts), legume crop based green manuring, tree-leaf based green manuring, crop residue recycling, sheep-goat penning, organic farming, conservation tillage, inclusion of legumes in crop rotation need to be encouraged (Sharma et al., 2002, 2004). Similar to inorganic fertilizer, provision for incentives for organic manures including green manuring can also be made so that growers should be motivated to take up these practices as inbuilt components of integrated nutrient management (INM) system.

### 7.4. Development and promotion of other bio-resources for enhancing microbial diversity and ensuring their availability

In addition to organic manures, there is a huge potential to develop and promote bio-fertilizers and bio-pesticides in large scale. These can play an important role in enhancement of soil fertility and soil biological health. Use of toxic plant protection chemical can also be reduced. In addition to this, there is a need to focus on advance research for enhancing microbial diversity by identifying suitable gene pools.

### 7.5. Ensuring availability of balanced multi-nutrient fertilizers

Fertilizer companies need to produce multi-nutrient fertilizers containing nutrients in a balanced proportion so that illiterate farmers can use these fertilizers without much hassle.

### 7.6. Enhancing the input use efficiency through precision farming

The present level of use efficiency of fertilizer nutrients, chemicals, water and other inputs is not very satisfactory. Hence, costly inputs go waste to a greater extent and result in monetary loss and environmental (soil and water) pollution. More focus is required to improve input use efficiency. The components required to be focused could be suitable machinery and other precision tools for placement of fertilizers, seeds and other chemicals in appropriate soil moisture zone so that losses could be minimized and efficiency could be increased. This aspect has a great scope in rainfed agriculture. This will also help in increasing water use efficiency (WUE) too.

## 7.7. Amelioration of problematic soils using suitable amendments and improving their quality to a desired level

History has a record that poor soil quality or degraded soils have taken toll of even great civilizations. No country can afford to let its soils be remaining degraded by virtue of water logging, salinization, alkalinity, erosion etc. Lots of efforts have already gone into the research process in relation to soil amendments. There is a need to ameliorate the soils at extensive scale on regular basis. No matter, how much it costs. Soil amelioration programs should be national programs linked with 'state agricultural departments'

## 7.8. Land cover management

The concept of land cover management is still ridiculed at some quarters in India may be because of lack of understanding. The lessons of United States Department of Agriculture (USDA) regarding soil erosion due to wind and water during dusty storms and torrential rains are adequate to understand the concept of land cover management. Covering the land with cover crops such as legumes , natural and pasture grasses, mulches with separable crop residues will help in protecting the land from the direct hits of high energy raindrops, ill effects of extreme temperatures during summer and winter, reduction in evaporation, enhanced biological activity due to congenial soil habitat conditions, higher C sequestration etc. Hence, this concept needs to be propagated extensively among the farming community.

## 7.9. Need for organized functional statutory bodies at Centre and in the States on Land Care and Soil Resource Health

State Soil and Water Conservation departments restrict their activities only up to construction of small check dams, plugging of gullies etc in common lands. State Soil testing labs are almost sluggish in action, poorly equipped and are with under-qualified manpower. Mostly, no tests are done except for Organic C, P and K. State agricultural universities (SAU) only adopt few villages, and consequently, no extensive testing of soil health is done. ICAR institutions also take up few watersheds covering few villages. Then, there will be no one to work for Land Care and Soil Health program at large scale. Hence, **organized functional statutory bodies at Centre and in the States on Land Care and Soil Resource Health** are necessary to effectively coordinate the Land Care and Soil Health Restoration and maintenance programs . It is beyond the capacity of research organizations to take up such giant and extensive task in addition to their regular research mandates. Some of the activities of land care can be linked with National Rural Employment Guarantee Program.

## 7.10. More intensive research on soil quality

There is a need for developing critical levels of some of the soil quality indicators for which this information is not available for Indian condition. Research experiments should be planned keeping in view three aspects viz. soil quality restoration, improvement and maintenance. The subject of soil resilience is still not explored much world over. Systematic research is needed to study soil resilience for diversity of edaphic, climatic and management conditions. Conservation agricultural practices such as conservation tillage, residue recycling, land cover management, appropriate crop rotations have shown the proven benefit to improve soil quality across the world. The quantum of impact may vary depending upon the variations in soil, climate, duration of the practice and level of overall management of the farms. It would be relevant to study soil quality, resilience and sustainability quantitatively under long term restorative management practices in different crop growing environments. There is a need to develop Soil health Cards covering important visible and easily understandable indicators so that even illiterate farmers should be able to use them periodically

## Author details

K.L. Sharma and B. Venkateswarlu
*Central Research Institute for Dryland Agriculture, Hyderabad (Andhra Pradesh), India*

Biswapati Mandal
*Bidhan Chandra Krishi Vishwavidalaya, Mohanpur, Nadia (West Bengal), India*

## 8. References

Acton, D.F. and Gregorich, L.J. (eds) (1995) The health of our soils. Towards sustainable agriculture in Canada. Agriculture and Agri-Food Canada, Ottawa, Canada. Soil Science Society of America. Statement on soil quality. *Agronomy News*, p 7.

Aggarwal, R. P., Phogat, V.K., Chand, T. and Grewal, M.S. (1994) *Improvements of Soil Physical Conditions in Haryana. Dept. of Soil Sci.*, CCS Haryana Agric. Univ., Hisar, India.

Andrews, S.S., and Carroll, C.R. (2001) Designing a soil quality assessment tool for sustainable agro-ecosystem management. *Ecological Applications* 11, 1573-1585.

Andrews, S.S., Flora, C.B., Mitchell, J.P., and Karlen, D. L. (2003) Grower's perceptions and acceptance of soil quality indices. *Geoderma* 114, 187-213.

Andrews, S.S., Karlen, D.L., and Cambardella, C.A. (2004) The soil management assessment framework: A quantitative evaluation using case studies. *Soil Science Society of America Journal* 68, 1945–1962.

Andrews, S.S., Karlen, D.L., and Mitchell, J.P., (2002a) A comparison of soil quality indexing methods for vegetable production systems in northern California. *Agriculture, Ecosystems and Environment* 90, 25–45.

Andrews, S.S., Mitchell, J.P., Mancineelli, R., Karlen, D.L., Hartz, T.K., Horwath, W.R., Pettygrove, G.S., Scow, K.M. and Munk, D.S. (2002 b) On-farm assessment of soil quality in California's central valley. *Agronomy Journal* 94, 12-23.

Arshad, M.A., and Coen, G.M. (1992) Characterization of soil quality: physical and chemical criteria. *American Journal of Alternative Agriculture* 7, 25-31.

Arshad, M.A., and Martin, S. (2002) Identifying critical limits for soil quality indicators in agro-ecosystems. *Agriculture, Ecosystems and Environment* 88, 153-160.

Arshad, M.A., Lowery, B., and Grossman, B. (1996) Physical tests for monitoring soil quality. In: *Methods for Assessing Soil Quality* (J.W. Doran and A.J. Jones, Eds.) Special Publ. 49. *SSSA, Madison*, WI. p. 123–141.

Astier, M. Maass, J.M., Etchevers-Barra, J.D., Pena, J.J., and de Leon Gonzalez. F. (2006) Short-term green manure and tillage management effects on maize yield and soil quality in an Andisol. *Soil and Tillage Research* 88, 153-159.

Awasthi, K.D., Singh, B.R., and Sitaula, B. K. (2005) Profile carbon and nutrient levels and management effect on soil quality indicators in the Mardi watershed of Nepal.: Acta-Agriculturae-Scandinavica-Section-B, *Soil and Plant Science* 55(3), 192-204.

Blair, J.M., Bohlen, P.J., and Freckman, D.W. (1996) Soil invertebratesas indicators of soil quality. In: *Methods for Assessing Soil Quality* (J.W. Doran and A.J. Jones, Eds.) Soil Science Society of America, Special Publication 49, Madison, WI, pp 273–291.

Brejda, J. J., and Moorman, T.B. (2001) Identification and interpretation of regional soil quality factors for the Central High Plains of the Midwestern USA. In: *Sustaining the global farm—Selected papers from the 10th International Soil Conservation Organization Meeting (ISCO99)* (D. E. Stott, R. H. Mohtar, and G. C. Steinhardt, Eds), West Lafayette, In: International Soil Conservation Organization in cooperation with the USDA and Purdue University. Pp. 535-540.

Bridge, B.J. (1997) Soil physical deterioration. In: *Sustainable Crop Production in Subtropics: An Australian Perspective* (A.L. Clarke, and P. Wylie, Eds.), Queensland Department of Primary Industries: Brisbane. pp 64-78.

Bridge, B.J. and Ross, P.J. (1984) Relations among physical properties of cracking clay soils. In: *Properties and Utilization of Cracking Clay Soils* (J.W. McGarity, E.Hoult and H.B. So, Eds.), Reviews in Rural Science, University of New England, Armidale. Pp 97-104.

Bucher, E. (2002) Soil quality characterization and remediation in relation to soil management. PhD Thesis, Department of Crop and Soil Sciences, Pennsylvania State University, USA.

Campbell, C.A., and Zentner, R.P. (1995) Soil organic matter as influenced by crop rotations and fertilization. *Soil Science Society of America Journal* 57, 1034- 1040.

Carter, M.R. (1996) Concepts of soil quality. In: *Soil Quality is in the Hands of the Land Manager* (R.J. MacEwan, M.R. Carter, Eds), Symposium Proceedings, University of Ballarat, Ballarat.

Chaudhury, J., Mandal, U.K., Sharma, K.L., Ghosh, H., and Biswapati Mandal (2005) Assessing soil quality under long-term rice based cropping system. *Communications in Soil Science and Plant Analysis* 36(9/10), 1141-1161.

Chen, Z. S. (1998) Selecting indicators to evaluate the soil quality of Taiwan soils and approaching the national level of sustainable soil management pp. 131-171. In FFTC ASPAC (Editor). Proceedings of International Seminar on soil management fro sustainable agriculture in the tropics. Dec. 14-19, 1998. Taichung, Taiwan, ROC. (Invited Speaker).

Conway, G. (1985) Agro ecosystem analysis. *Agricultural Administration* 20 (1), 31-55.

Dalal, R. C. (1998) Soil microbial biomass-what do the numbers really mean? *Australian Journal of Experimental Agriculture* 38, 649-665.

Dalal, R.C. (1989) Long term effects of no-tillage, crop residues and nitrogen application on properties of a Vertisol, *Soil Science Society of America Journal* 53, 1511–1515.

Dalal, R.C., and Mayer, R.J. (1986) Long term trends in fertility of soils under continuous cultivation in southern Queensland. 11. Total organic carbon and its rate of loss from the soil profile. *Australian Journal Soil Research* 37, 265–279.

Dalal, R.C., and Mayer, R.J. (1987) Long-term trends in fertility of soils under continuous cultivation and cereal cropping in Southern Queensland. VI. Loss of total nitrogen from different particle-size and density fractions. *Australian Journal Soil Research* 25, 83–93.

Dick, R.P., Breakwell, D.P., and Parkin, T.B. (1996) Soil enzyme activities and biodiversity measurements as integrative microbiological indicators, In: *Methods for Assessing Soil Quality* (J.W. Doran et al., (eds.), SSSA Spec. Pub. 49. Madison, WI. pp. 247-271.

Doran, J. W., and Parkin. T. B. (1996) Quantitative indicators of soil quality: A minimum data set. In: *Methods for Assessing Soil Quality*, (J. W. Doran and A. J. Jones, Eds.). Soil Sci. Am. Special Publication No. 49, Madison, Wisconsin, USA, pp. 25-37.

Doran, J. W., Sarrantonio, M., and Liebig, M. A. (1996) Soil health and sustainability. *Advances in Agronomy* 56, 1-54.

Doran, J.W., Liebig, M.A., and Santana, D.P. (1998) Soil health and global sustainability. In: *Proceedings of the 16th World Congress of Soil Science.* Montepellier, France, 20-26 August 1998.

Doran. J.W., and Parkin, T. B. (1994) Defining and assessing soil quality. In: *Defining Soil Quality for a Sustainable Environment*, (J. W. Doran, D. C. Coleman, D. F. Bezdicek, and B. A. Stewart, Eds.), Soil Sci. Soc.Am. Special Publication No.35, Madison, Wisconsin, USA, pp. 3-21.

Freckman, D.W., and Virginia, R.A. (1997) Low-diversity Antarctic soil nematode communities: distribution and response to disturbance. *Ecology* 78, 363-369.

Freebairn, D.M., and Wockner, G.H. (1986) A study on soil erosion on Vertisols of the eastern Darling Downs, Queensland. I. Effect of surface conditions on soil movement within contour bay catchment. *Australian Journal of Soil Research* 24, 135-158.

Ghoshal, S.K. (2004) Studies on biological attributes of soil quality for a few long-term fertility experiments with rice base cropping systems. PhD Thesis, Bidhan Chandra Krishi Viswavidyalaya, pp.146.

Ghuman B.S., and Sur, H.S. (2001) Tillage and residue management effects on soil properties and yields of rainfed maize and wheat in a subhumid subtropical climate, *Soil and Tillage Research* 58, 1–10.

Glanz, J.T. (1995) *Saving Our Soil: Solutions for Sustaining Earth's Vital Resource*, Johnson Books, Boulder, CO, USA.

Glover, J.D., Reganold, J.P., and Andrews, P.K. (2000) Systematic method for rating soil quality of conventional, organic, and integrated apple orchard in Washington State, *Agriculture Ecosystem and Environment* 80, 29–45.

Graham – Tomasi, T. (1991) Sustainability: concept and implications for agricultural research policy. In: *Agricultural Research Policy: International Quantitative Perspectives* (P.C., Pardey, J. Roseboom, and J.R. Anderson, Eds.). Cambridge University Press. Pp 81-102.

Granatstein , D., and Bezdicek, D. F. (1992) The need for a soil quality index: Local and regional perspectives. *American Journal of Alternative Agriculture.* 17, 12-16.

Harris, R.F., Karlen, D. L., and Mulla, D.J. (1996) A conceptual framework for assessment and management of soil quality and health. In: *Methods for Assessing Soil Quality* (J. W. Doran, and A.J. Jones, Eds.), Soil Sci. Soc. Am. Spec. Publ. # 49. SSSA, Madison, WI, USA, pp. 61-82.

Hazra, G.C., Ghoshal, S.K. and Mandal, Biswapati (2004) Soil quality evaluation for a long-term fertility experiment with organic and inorganic inputs under rice-wheat cropping system. *Proc. Intl. Symp. on Agro environ.* pp 311-321 held during October 20-24, 2004 at Udine, Italy.

Heal, O.W., S. Struwe, and Kjoller, A. (1996) Diversity of soil biota and ecosystem function. Annual Conference, Australian Agricultural and Resource Economics Society, Amidale, New South Wales. pp 1-10.

Herdt, R.W. and Steiner, R.A. (1995) Agricultural Sustainability: Concepts and Conundrums. In: *Agricultural Sustainability: Economic, Environmental and Statistical Considerations,* (V. Barnett, R. Payne, and R.Steiner, Eds.), J. Wiley & Sons, Chichester, U.K., pp.1-13.

Hussain, I., Olson, K. R., Wander, M. M., and Karlen, D.L. (1999) Adaptation of soil quality indices and application to three tillage systems in southern Illinois. *Soil and Tillage Research* 50, 237-249.

Ishaq, M., Ibrahim, M., and Lal, R. (2002) Tillage effects on soil properties at different levels of fertilizer application in Punjab, Pakistan. *Soil and Tillage Research* 68 (2), 93-99.

Jaenicke, E.C., and Lengnick, L.L. (1999) A soil –quality index and its relationship to efficiency and productivity growth measures two decompositions. *American Journal of Agricultural Economics* 81, 881-893.

Jodha, N.S. (1994) Indicators of Unsustainability. In: *Stressed Ecosystems and Sustainable Agriculture.* (S.M. Virmani, J.C. Katyal, H. Eswaran and I.P. Abrol, Eds.), Oxford & IBH Publishing Co.PVT.LTD. Pap. No 65-77.

Karlen, D.L., and Stott, D.E. (1994) A framework for evaluating physical and chemical indicators of soil quality. In: *Defining Soil Quality for a Sustainable Environment* (J. W. Doran, D. C. Coleman, D. F. Bezdicek, and B. A. Stewart, Eds.), Madison, WI. *Soil Science Society of America Journal* 35, 53-72 (special publication).

Karlen, D.L., Andrews, S.S., Doran, J.W. (2001) Soil quality: current concepts and application. *Advances in Agronomy* 74, 1-40.

Karlen, D.L., Eash, N.S., Unger, P.W. (1992) Soil and crop management effects on soil quality indicators. *American Journal of Alternative Agriculture* 7, 48–55.

Karlen, D.L., Mausbach, M.J., Doran, J. W., Cline, R. G., Harres, R. F., and Schuman. G. E. (1997) Soil quality: A concept definition, and framework for evaluation. *Soil Science Society of America Journal* 61, 4-10.

Katyal, J.C. (2008) Presidential Address. *Journal of the Indian Society of Soil Science* 56 (4), 322-324.

Keller, K., B.M. Bolker, D.F. Bradford (2004). Uncertain climate thresholds and optimal economic growth, Journal of Environmental Economics and Management, 48, pp. 723- 741.

Kookana, R.S., Baskaran, S., and Naidu, R. (1998) Pesticide fate and behaviour in Australian soils in relation to contamination and management of soil and water - a review. *Australian Journal of Soil Research* 36:715-764.

Kusuma Grace, J. (2008) Assessment of soil quality under different land use treatments in rainfed Alfisol. Ph.d Thesis, Jawaharlal Nehru Technological University, Hyderabad, Andhra Pradesh, India. p. 250.

Lal R. (1993) Tillage effects on soil degradation, soil resiliency, soil quality and sustainability. *Soil and Tillage Research* 27(1-4), 1-8.

Lal, R. (1994) Date analysis and interpretation. In: *Methods and Guidelines for Assessing Sustainable Use of Soil and Water Resources in the Tropics*, (R. Lal, Ed.). Soil Management Support Services Technical. Monograph. No. 21. SMAA/SCS/USDA, Washington D. C, pp. 59-64.

Lal, R. (2008) Soils and India's Food Security. *Journal of the Indian Society of Soil Science*, 56 (2), 129-138.

Lal, R. (Ed.). (1998) Soil quality and agricultural sustainability. In: *Soil Quality and Agricultural Sustainability*. Ann Arbor Press, Chelsea, MI, pp. 3-12

Lal, R., Hall, G.F., Miller, P. (1989) Soil degradation: I. Basic processes. *Land Degradation and Rehabilitation* 1, 51-69.

Larson, W. E., and Pierce. F. J. (1994) The dynamics of soil quality as a measure of sustainable management. In: *Defining Soil Quality for a Sustainable Environment*. (J. W. Doran, D. C. Coleman, D. F. Bezdicek, and B. A. Stewart, Eds.), Soil Science Society of America Special Publication No. 35. Madison, Wisconsin, USA, pp. 37-51.

Larson, W.E., and Pierce. F.J. (1991) Conservation and enhancement of soil quality. In: *Evaluation for Sustainable Land Management in the Developing World* (J. Dumanski, Ed.). Proceedings of the International Workshop, Chiang Rai, Thailand, 15-21 Sept. 1991. Technical papers, vol. 2 Int. Board for Soil Res.and Management, Bangkok, Thailand, pp. 175-203.

Lockeretz, W. (1988) Open questions in sustainable agriculture. *American Journal of Alternative Agriculture*. 3, 174—181.

Lynam, J.K., and Herdt, R.W. (1988) Sense and sustainability: sustainability as an objective in international agriculture research. Presented at the CIP-Rockefeller Foundation Conference on Farmer and Food Systems, 26-30 September 1988, Lima, Peru .(Limited distribution.)

Mac Rae, R.Y., and Mehuys, G.R. (1985) The effect of green manuring on the physical properties of temperate area soils. *Advances in Soil Science*. Vol. 3, Springer-Verlag, Inc., New York. pp. 71–94.

Mairura, F.S., Mugendi, D.N., Mwanje, J.I., Ramisch, J.J., Mbugua, P.K., and Chianu, J.N. (2007) Integrating scientific and farmer's evaluation of soil quality indicators in central Kenya. *Geoderma* 139, 134 -143.

Mandal B (2005) Assessment and Improvement of Soil Quality and Resilience for Rainfed Production System. Completion Report. National Agricultural Technology Project, Indian Council of Agricultural Research, New Delhi, 30 pp.

Mandal, B., Majumder, B., Bandopadhyay, P.K. (2007) The potential of cropping systems and soil amendments for carbon sequestration in soils under long-term experiments in subtropical India. *Global Change Biology* 13, 357-369.

Mandal, D.K., Mandal, C., and Velayutham, M. (2001) Development of a land quality index for sorghum in Indian semi-arid tropics (SAT). *Agric-syst.* Oxford : Elsevier Science Ltd. Oct 2001. 70 (1), 335-350.

Manna, M.C., Swarup, A., Wanjari, R.H., Mishra, B., and Shahi D.K. (2007) Long-term fertilization, manure and liming effects on soil organic matter and crop yields. *Soil and Tillage Research,* 94 (2): 397-409.

Manna, M.C., Swarup, A., Wanjari, R.H., Ravankar, H.N., Mishra, B., Saha, M.N., Singh, Y.V., Sahi, D.K., and Sarap, P.A. (2005) Long-term effect of fertilizer and manure application on soil organic carbon storage, soil quality and yield sustainability under sub-humid and semi-arid tropical India. *Field Crops Research* 93, 264–280.

Markanday, A., and Pearce, D.W. (1988) Environment considerations and the choice of the discount rate in developing countries. *Environment Department Working Paper* no.3. Washington, DC, USA: World Bank.

Masto, R.E., Chhonkar, P.K., Dhyan Singh, and Patra, A.K. (2007) Soil quality response to long-term nutrient and crop management on a semi-arid Inceptisol. *Agriculture, Ecosystems and Environment* 118, 130 –142.

Mohanty, M., Painuli, D.K., Misra, A.K., and Ghosh, P.K. (2007) Soil quality effects of tillage and residue under rice–wheat cropping on a Vertisol in India. *Soil and Tillage Research,* 92 (1-2), 243-250.

Moody, P.W., and Aitken, R.L. (1997) Soil acidification under some tropical agricultural systems. I. Rates of acidification and contributing factors. *Australian Journal of Soil Research* 35, 163-173.

Pain, A.K. (2007) Indian agriculture: The present scenario. *Agrarian Crisis in India: An Overview* (A.K. Pain, Ed.), pp. 1-11. The ICFAI Press, Hyderabad, India.

Pankhurst, C.E. (1994) Biological indicators of soil health and sustainable productivity. In: *Soil Resilience and Sustainable Land Use* (D.J. Greenland, and I. Szabolcs, Eds.), CAB International, Wallingford, UK. pp 331-351.

Parr, J.F. and Hornick, S.B. (1990) Recent developments in alternative agriculture in the United States. In: *Proc. of Intl. Conf. on Kyusei Nature Farming,* October 17−21, 1989, Khon Kaen University, Khon Kaen. Thailand.

Parr, J.F., Papendick R.L., Hornick, S.B., and Meyer, R.E. (1990) Strategies for developing low-input sustainable farming systems for rainfed agriculture. In: *Proc. Intl. Symposium for Managing Sandy Soils,* February 6 -11, 1989, Jodhpur, India. Indian Council of Agricultural Research. New Delhi.

Parr, J.F., Papendick, R.I., Hornick, S.B., and Meyer, R. E. (1992) Soil quality: attributes and relationship to alternative and sustainable agriculture. *American Journal of Alternative Agriculture* 7, 5-11.

Parr, J.F., Papendick, R.I., Youngberg, 1G. and Meyer, R.E. (1989) Sustainable agriculture in the United States. Chap. 4, In: *Proc. Intl. Symposium on Sustainable Agricultural Systems.* Soil and Water Conservation Society, Ankeny, Iowa. pp. 50 - 67.

Perucci, P. and Scarponi, L. (1994) Effects of the herbicide imazethapyr on soil microbial biomass and various soil enzymes. *Biology and Fertility of Soils* 17, 237-240.

Prasad, R. and Power, J.F. (1991) Crop residue management. *Advances in Soil Science* 15, 205–251.

Rao, G.G.S.N., Rao, V.U.M., Vijaya Kumar, P., Rao, A.V.M.S., and Ravindra Chary, G. (2010) Climate risk management and contingency crop planning. Lead papers. In: National Symposium on Climate Change and Rainfed Agriculture, 18-20 February, 2010, CRIDA, Hyderabad, India. Organized by Indian Society of Dryland Agriculture and Central Research Institute for Dryland Agriculture. Pp. 37.

Rasmussen, P.E. and Collins. H.P. (1991) Long-term impacts of tillage, fertilizer, and crop residue on soil organic matter in temperate semiarid regions. *Advances in Agronomy* 45, 93-134.

Rengasamy, P. and Olsson, K.A. (1991) Sodicity and soil structure. *Australian Journal of Soil Research* 29, 935-952.

Rice, C.W., Moorman, T.B. and Margules C.R. (1996) Role of microbial biomass carbon and nitrogen in soil quality. In: *Methods for Assessing Soil Quality*. (J.W. Doran, and A.J. Jones, Eds.), Special Publication number 49, SSSA, Madison. pp 203-215.

Richter, J. (1987) *The Soil as a Reactor*. Catena Verlag. Cremilingen, Germany.

Rochette P. and Gregorich E.G. (1998) Dynamics of soil microbial biomass C, soluble organic C and $CO_2$ evolution after three years of manure application. *Canadian Journal of Soil Science.* 78, 283-290.

Roldán, A., Salinas-García, J.R., Alguacil, M.M., Díaz, E. and Caravaca, F. (2005) Soil enzyme activities suggest advantages of conservation tillage practices in sorghum cultivation under subtropical conditions. *Geoderma*, 129 (3-4), 178-185.

Sekhon, G.S. (1997) Nutrient needs of irrigated food cropsin India and related issues. In: *Plant Nutrient Needs, Supply, Efficiency and Policy Issues 2000-2025* (J.S. Kanwar and J.C. Katyal, Eds), National Academy of Agricultural Sciences, New Delhi, India.

Selvaraju, R., Subbian, P., Balasubramanian, A. and R. Lal. (1999) Land configuration and soil nutrient management options for sustainable crop production on Alfisols and Vertisols of southern peninsular India. *Soil and Tillage Research* 52, 203-216.

Seybold, C.A., Herrick, J.E., Brejda, J.J. (1999) Soil Resilience a fundamental component of soil quality. *Soil Science* 164 (4), 224-234.

Sharma , K.L. (2009b) Restoration of soil quality through conservation agricultural management practices and it's monitoring using Integrated Soil Quality Index (ISQI) approach in rainfed production system (s). Consolidated Report (2005-09) of National Fellow Project. Submitted to ICAR, New Delhi. Central Research Institute for Dryland Agriculture pp.1-419.

Sharma, K. L. , Grace, J. Kusuma , Mishra, P. K. , Venkateswarlu, B. , Nagdeve, M. B. , Gabhane, V. V. , Sankar, G. Maruthi , Korwar, G. R. , Chary, G. Ravindra , Rao, C. Srinivasa , Gajbhiye, Pravin N. , Madhavi, M. , Mandal, U. K. , Srinivas, K. and Ramachandran, Kausalya(2011) 'Effect of Soil and Nutrient-Management Treatments on Soil Quality Indices under Cotton-Based Production System in Rainfed Semi-arid Tropical Vertisol', Communications in Soil Science and Plant Analysis, 42: 11, 1298 — 1315

Sharma, K. L., Grace, J. Kusuma, Srinivas, K., Venkateswarlu, B., Korwar, G. R., Sankar, G. Maruthi, Mandal, Uttam Kumar, Ramesh, V., Bindu, V. Hima, Madhavi, M. and Gajbhiye, Pravin N. (2009) Influence of Tillage and Nutrient Sources on Yield Sustainability and Soil Quality under Sorghum-Mung Bean System in Rainfed Semi-arid Tropics', Communications in Soil Science and Plant Analysis,40:15,2579 – 2602

Sharma, K. L., Kusuma Grace, J., Srinivas, K., Venkateswarlu, B., Korwar, G. R. Maruthi Sankar, G. Uttam Kumar Mandal ; Ramesh, V., Hima Bindu, V., Madhavi M., Pravin N. Gajbhiye. (2009a) Influence of Tillage and Nutrient Sources on Yield Sustainability and Soil Quality under Sorghum-Mung Bean System in Rainfed Semi-arid Tropics. *Communications in Soil Science and Plant Analysis,* 40 (15), 2579-2602.

Sharma, K. L., Vittal, K. P. R., Ramakrishna, Y. S., Srinivas, K., Venkateswarlu, B., and Kusuma Grace, J. (2007) Fertilizer use constraints and management in rainfed areas with special emphasis on nitrogen use efficiency. In: (Y. P. Abrol, N. Raghuram and M. S. Sachdev (Eds)), Agricultural Nitrogen Use and Its Environmental Implications. I. K. International Publishing House, Pvt., Ltd. New Delhi. Pp 121-138.

Sharma, K.L., Kusuma Grace, J., Uttam Kumar Mandal, Pravin N. Gajbhiye, Srinivas, K., Korwar, G. R., Himabindu, V., Ramesh, V., Kausalya Ramachandran, and Yadav, S.K. (2008) Evaluation of long-term soil management practices through key indicators and soil quality indices using principal component analysis and linear scoring technique in rainfed Alfisols. *Australian Journal of Soil Research* 46: 368-377.

Sharma, K.L., Mandal, U.K., Srinivas, K., Vittal, K.P.R., Biswapati Mandal, Grace, J.K., and Ramesh, V. (2005) Long-term soil management effects on crop yields and soil quality in a dryland Alfisol. *Soil and Tillage Research* 83(2), 246-259.

Sharma, K.L., Srinivas, K., Das, S.K., Vittal, K.P.R. and Kusuma Grace, J. (2002) Conjunctive Use of Inorganic and Organic Sources of Nitrogen for Higher Yield of Sorghum in Dryland Alfisol. *Indian Journal of Dryland Agricultural Research and Development* 17 (2), 79-88.

Sharma, K.L., Srinivas, K., Mandal, U.K., Vittal, K.P.R., Kusuma Grace, J., and Maruthi Sankar, G. (2004) Integrated Nutrient Management Strategies for Sorghum and Green gram in Semi arid Tropical Alfisols. *Indian Journal of Dryland Agricultural Research and Development* 19 (1), 13-23.

Sharma, K.L., Vittal, K.P.R., Srinivas, K., Venkateswarlu, B. and Neelaveni, K. (1999) Prospects of organic farming in dryland agriculture.  *In* Fifty Years of Dryland Agricultural Research in India (Eds. H.P. Singh, Y.S. Ramakrishna, K.L. Sharma and B. Venkateswarlu) Central Research Institute for Dryland Agriculture, Santoshnagar, Hyderabad, pp 369-378.

Shukla, M.K., Lal, R. and Ebinger, M. (2006) Determining soil quality indicators by factor analysis. *Soil and Tillage Research* 87, 194-204.

Singer, M.J. and Ewing, S. (2000) Soil quality. In: *Handbook of Soil Science* (Sumner, Ed.), CRC Press, Boca Raton, FL, pp. G-271-G-298.

Smith, L.J. and Elliott, F.L. (1990) Tillage and residue management effects on soil organic matter dynamics in semiarid regions. *Advance in Soil Science* 13, 69-88.

Sparling, G.P. (1997) Soil microbial biomass, activity and nutrient cycling as indicators of soil health. In: *Biological Indicators of Soil Health.* (C. E. Pankhurst, B. M. Doube, and V. V. S. R. Gupta, Eds.), CAB International, Wallingford, pp. 97–119.

Steiner. J.L.. Day, J.C., Papendick. R.I., Meyer, R.E. and Bertrand, A.R. (1988) Improving and sustaining productivity in dryland regions of developing countries. *Soil Science* 8: 79 - 122.

Stewart, B.A., Lal, R. and El-Swaify, S.A. (1990) Sustaining the resource base of an expanding world agriculture. In: *Proc. Workshop on Mechanisms for a Productive and Sustainable Resource Base.* July 29-30. 1989. Edmonton, Canada. Soil and Water Conservation Society, Ankeny, Iowa.

Szabolcs, I. (1994) The Concept of Soil Resilience. In: *Soil Resilience and Sustainable Land Use* (D. J.Greenland and I.Szabolcs, Eds). CAB International Pap.No:33-39.

Tiwari, K. N. (2008) Future of plant nutrition research in India. *Journal of the Indian Society of Soil Science* 56, 327-336.

Tiwari, K.N., Sharma, S.K. Singh, V.K., Dwivedi, B.S. and Shukla, A.K. (2006) Site Specific Nutrient Management for Increasing Crop Productivity In India. PDCSR (ICAR) – PPIC, India Programme, pp 92.

USDA (2004) Food Security Assessment. Agriculture and Trade Reports. USDA-ERS. GFA-15, Washington, D.C. 88 pp.

Walker, J. and Reuter, D.J. (1996) *Indicators of Catchments Health: A Technical Perspective,* CSIRO, Collingwood, Australia, 174 pp.

Wang, X. and Gong, Z. (1998) Assessment and analysis of soil quality changes after eleven years of reclamation in subtropical China. *Geoderma* 81, 339-355.

Warkentin, B.P. (1996) Overview of soil quality indicators. In: *Proceedings of the Soil Quality Assessment on the Prairies Workshop,* Edmonton, 22-24 January, 1996. pp. 1-13.

Wienhold, B.J., Andrews, S.S. and Karlen, D.L. (2004) Soil quality: a review of the science and experiences in the USA. *Environmental Geochemistry and Health* 26, 89–95.

Wienhold, B.J., Andrews, S.S., Kuykendall, H. and Karlen, D.L. (2008) Recent Advances in Soil Quality Assessment in the United States. *Journal of the Indian Society of Soil Science,* 56 (3), 237-246.

# Chapter 8

# Efficient Nutrient Management Practices for Sustainable Crop Productivity and Soil Fertility Maintenance Based on Permanent Manorial Experiments in Different Soil and Agro-Climatic Conditions

G.R. Maruthi Sankar, K.L. Sharma, Y. Padmalatha, K. Bhargavi,
M.V.S. Babu, P. Naga Sravani, B.K. Ramachandrappa, G. Dhanapal,
Sanjay Sharma, H.S. Thakur, A. Renuka Devi, D. Jawahar, V.V. Ghabane,
Vikas Abrol, Brinder Singh, Peeyush Sharma, N. Ashok Kumar, A. Girija,
P. Ravi, B. Venkateswarlu and A.K. Singh

Additional information is available at the end of the chapter

## 1. Introduction

Rainfed agriculture plays an important role in contributing to the food bowl of the world. Its importance varies regionally but produces food for poor communities in developing countries. In India, rainfed agriculture is in about 85 million hectare, constituting about 60 % of net cultivated areas supporting 40% of the population of the country. In Sub-Saharan Africa, more than 95 % farm land is rainfed, while the corresponding figure for Latin America is almost 90%, for South Asia about 60 %, for East Asia 65% and for the Near East and North Africa 75% (Wani, *et al.*, 2009). Besides, the climatic constraints especially erratic and uncertain pattern of rainfall, soils in the rainfed areas are under severe grip of degradation in terms of their physical, chemical and biological properties.

In AICRPDA, permanent manorial experiments (PME) are conducted on different rainfed crops viz., upland rice, sorghum, finger millet, pearl millet, cotton, maize, soybean, groundnut crops under varying soil and agro-climatic conditions at different centers. They are conducted in alfisols, vertisols, inceptisols, entisols, aridisols and other soil types. Alfisols are most abundant soils in the semi–arid tropics and cover about 16% of tropics and 33% of semi–arid tropics (SAT). These soils are mostly found in the south Asia, west and central Africa, and many parts of South America, particularly north eastern Brazil (Cocheme and Franquin 1967). Mostly these soils are shallow with a compacted sub–surface layer that limits the root development and water percolation. The loamy sand texture of top soil and abundance of 1:1 type clay minerals viz., kaolinite, make them structurally inert (Charreau 1977). These soils are constrained by crusting and hard setting tendencies under erratic rainfall distribution and occurrence of dry spells (Bansal, Awadhwal, and Mayande 1987). Owing to less contribution of root biomass due to low crop intensity, high temperature mediated fast oxidation of organic matter, poor recycling back of crop residues, washing away of top soil, reckless tillage and imbalanced fertilizer use results in low organic carbon and low fertility of these soils (Kampen and Burford 1980; El-Swaify, Singh, and Pathak 1983). Often these soils encounter a diversity of soil physical, chemical and biological constraints and provide a low productivity of crops.

Vertisols are the predominant soil groups found across the world. The majority of the acreage of Vertisols and associated soils in the world is spread in Australia (70.5 million ha), India (70 million ha), Sudan (40 million ha), Chad (16.5 million ha), and Ethiopia (10 million ha). These five countries constitute over 80% of the total area (250 million ha) of Vertisols in the world (Dudal 1965). In India, substantial Vertisol areas are found in the states of Maharashtra, Madhya Pradesh, Gujarat, Andhra Pradesh, Karnataka, and Tamil Nadu (Murthy 1981). Most of these regions receive 500 to 1300 mm of annual rainfall, concentrated in a short period of 3 to 3½ rainy months interspersed with droughts. Crop yields in these areas are miserably low and may vary from year to year. Virmani et al., (1989) have comprehensively characterized Vertisols found in India. Their texture may vary from clay to clay loam, or silty clay loam, with the clay content generally varying from 40% to 60% or more. They have high bulk density when dry, with clod density values ranging from 1.5 to 1.8 g cm$^{-3}$); high CEC (47 to 65 cmol kg soil$^{-1}$); and pH values usually above 7.5. Tropical Vertisols are low in organic matter and available plant nutrients, particularly N, P, and Zinc. The dominant clay mineral is smectite. High clay content, better effective soil depth associated with other physical properties makes these soils to store higher amount of moisture. Low organic matter status accompanied by poor soil fertility is one of the predominant constraints in these Vertisol soils. Farmers of the rainfed SAT regions, being poor, are not able to use adequate amount of chemical fertilizers. Earlier researchers have established that the productivity of these soils can be enhanced by way of supplying adequate nutrient inputs (Virmani et al., 1989; Willey et al., 1989; Burford et al., 1989). Based on numerous agronomic experiments, it has been found that supplementation of N, P and Zinc through fertilizer is inevitable to ensure satisfactory crop production in SAT soils especially in Vertisols (Kanwar 1972; Randhawa and Tandon 1982). Despite many efforts, there is a low adoption of fertilizers in rainfed crops which could probably be attributed to many reasons viz., incapability of the farmers to purchase fertilizers, erratic and uncertain rainfall leading to risk of crop failures, uncertainty and variability in crop responses (Jha and Sarin, 1984; Kanwar et al., 1973).

The productivity of any rainfed crop is significantly influenced by the distribution of seasonal rainfall during cropping season, soil fertility status and amount of fertilizer nutrient applied

(Maruthi Sankar, 2008). Research studies have shown that among different variables, the quantity of rainfall received during crop growing period would significantly influence the response of a crop to fertilizer application under rainfed conditions (Behera et al., 2007; Mohanty et al., 2008). Vikas et al., (2007) while optimizing the fertilizer requirement of rainfed maize in a dry sub-humid Inceptisol at Jammu in north India opined that if fertilizer doses are judiciously optimized considering the rainfall distribution pattern during the cropping season, higher productivity could be achieved in rainfed crops. Nema et al., (2008) examined the effects of crop seasonal rainfall and soil moisture availability at different days after sowing on yield and identified suitable tillage and fertilizer practices for attaining sustainable pearl millet yield in a semi-arid Inceptisol at Agra in north India. Further, to attain sustainable yield of crops in any soil and agro-climatic conditions and to save on fertilizers, it is important that while optimizing the fertilizer doses, changes in soil fertility also need to be periodically monitored (Maruthi Sankar 1986; Vittal et al., 2003). Long term effects of fertilizer on crop yield and soil properties have also been examined for different crops in order to suitably restore soil fertility and prescribe soil test based fertilizer recommendation for different crops (Prasad and Goswami 1992; Bhat et al., 1991; Dalal and Mayer 1986; Mathur 1997). Permanent manorial experiments are conducted at different research centers of AICRPDA with an objective to i) assess the response of rainfed crops and changes in soil fertility (with special emphasis on nitrogen, phosphorus, and potassium) due to long term application of organic and inorganic sources of nutrients under changing crop seasonal rainfall situations and ii) identify an efficient treatment for attaining sustainable yield over long-term basis under different soils and climatic conditions. The details of PMEs conducted on (i) finger millet at Bangalore; (ii) sorghum/pearl millet rotation at Kovilpatti; (iii) groundnut at Anantapur; (iv) cotton + green gram (1:1) at Akola; (v) soybean at Indore are discussed in this paper.

## 2. Materials and methods

### 2.1. Experimental details

The permanent manorial experiments (PME) have been conducted on different crops with a set of organic and inorganic fertilizer treatments at different AICRPDA research centers for more than 20-25 years (Table 1). The PMEs were conducted on (i) finger millet at Bangalore (Karnataka); (ii) groundnut at Anantapur (Andhra Pradesh); (iii) soybean at Indore (Madhya Pradesh); (iv) cotton + green gram in 1:1 row ratio at Akola (Maharastra); (v) sorghum rotated with pearl millet (yearly) at Kovilpatti (Tamil Nadu); (vi) rice at Phulbani (Orissa); (vii) rice at Varanasi (Uttar Pradesh); (viii) rice at Ranchi (Jharkhand); (ix) pearl millet at Agra (Uttar Pradesh); (x) rabi sorghum at Solapur (Maharastra); (xi) rabi sorghum at Bijapur (Karnataka); (xii) pearl millet/castor/cluster bean rotation (yearly) at SK Nagar (Gujarat); (xiii) rice in kharif followed by wheat in rabi at Rewa (Madhya Pradesh). The treatments were replicated thrice and tested in a Randomized Block Design. The treatments were randomized only in the first year and were fixed and superimposed to the same plots every year. Before superimposing the fertilizer treatments, initial soil samples were collected from each plot at a soil depth of 0–30 cm and analyzed for soil organic carbon (Walkley and Black,

1934), available (easily oxidizable) N (Subbaiah and Asija, 1956), available P (Olsen et al., 1954) and available K (Jackson, 1973). Soil sulphur was estimated by turbidity method (Chesnin and Yien, 1950) in each season. Observations on daily rainfall, variety, date of sowing and harvest, crop growing period, length of dry spells and other related details were also recorded every year and used for analysis.

| Center | Climate | Soil type | Crop | Year | Treatments |
|--------|---------|-----------|------|------|-----------|
| Bangalore | Semi-arid | Alfisols | Finger millet | 1984-2008 | T1 : Control; T2 : FYM @ 10 t/ha; T3 : FYM @ 10 t/ha + 50% NPK; T4 : FYM @ 10 t/ha + 100% NPK; T5 : 100% recommended NPK |
| Anantapur | Arid | Alfisols | Ground nut | 1985-2006 | T1: Control; T2: 100% NPK (20–40–40 kg/ha); T3: 50% NPK (10–20–20 kg/ha); T4: 100% N (groundnut shells ~ 20 kg N/ha); T5: 50% N (FYM ~ 10 kg N/ha); T6: 100% N (groundnut shells ~ 20 kg N/ha) + 50% NPK (10–20–20 kg/ha); T7: 50% N (FYM~10 kg N/ha) + 50% NPK (10–20–20 kg/ha); T8: 100% NPK (20–40–40 kg/ha) + $ZnSO_4$ @ 25 kg/ha; T9: Farmers practice (FYM @ 5 t/ha) |
| Indore | Semi-arid | Vertisols | Soybean | 1992-2006 | T1 : Control; T2 : 20 kg N (urea) + 13 kg P/ha; T3 : 30 kg N (urea) + 20 kg P/ha; T4 : 40 kg N (urea) + 26 kg P/ha; T5 : 60 kg N (urea) + 35 kg P/ha; T6 : 20 kg N (urea) + 13 kg P + FYM @ 6 t/ha; T7 : 20 kg N (urea) + 13 kg P + FYM @ 5 t/ha; T8 : FYM @ 6  t/ha; T9 : Crop residue @ 5 t/ha |
| Kovilpatti | Semi-arid | Vertisols | Sorghum/ Pearl millet | 1987-2005 | T1 : Control; T2 : 40 kg N (urea) + 20 kg P/ha; T3 : 20 kg N (urea) + 10 kg P/ha; T4 : 20 kg N (crop residue)/ha; T5 : 20 kg N (FYM)/ha; T6 : 20 kg N (crop residue) + 20 kg N (urea)/ha; T7 : 10 kg N (FYM) + 10 kg N (urea)/ha; T8 : 40 kg N (urea) + 20 kg P + 25 kg $ZnSO_4$/ha; T9 : FYM @ 5 t/ha |
| Akola | Semi-arid | Vertisols | Cotton + green gram (1:1) | 1987-2006 | T1 : Control; T2 : 50 kg N + 25 kg P/ha; T3 : 25 kg N + 12.5 kg P/ha; T4 : 25 kg N/ha (*Leucaena*); T5 : 25 kg N/ha (FYM); T6 : 25 kg N (*Leucaena*) + 25 kg N (urea) + 25 kg P/ha; T7 : 25 kg N (FYM) + 25 kg N (urea) + 25 kg P/ha; T8 : 25 kg N (*Leucaena*) + 25 kg P/ha |

**Table 1.** Permanent manorial experiments conducted at different AICRPDA centers

At Bangalore, experiments on finger millet (*Eleusine coracana* L.) were conducted in a permanent site for 25 years from 1984 to 2008 in a semi-arid Alfisol. The experimental site is situated at latitude of 12.97° North, longitude of 77.58° East and an altitude of 930 m above mean sea level. The treatments were (i) Control; (ii) FYM @ 10 t/ha; (iii) FYM @ 10 t/ha + 50% NPK; (iv) FYM @ 10 t/ha + 100% NPK; and (v) 100% recommended NPK. The 100% recommended NPK dose comprised of 50 kg N, 50 kg $P_2O_5$ and 25 kg $K_2O$/ha. The experiment was conducted in a net plot size of 2.7 m x 11.0 m each with row spacing of 30 cm and plant spacing of 10 cm.

At Akola, PME on cotton + green gram (1:1) was conducted for 20 years in a fixed site during South-West monsoon of 1987 to 2006 (June to November) in a semi-arid Vertisol. The research center is located at a latitude of 12.97° North, longitude of 77.58° East and an altitude of 930 m above mean sea level. The fertilizer treatments were applied in the same plot every year and are T1 : Control; T2 : 50 kg N + 25 kg P/ha; T3 : 25 kg N + 12.5 kg P/ha; T4 : 25 kg N/ha (*Leucaena*); T5 : 25 kg N/ha (FYM); T6 : 25 kg N (*Leucaena*) + 25 kg N (urea) + 25 kg P/ha; T7 : 25 kg N (FYM) + 25 kg N (urea) + 25 kg P/ha; and T8 : 25 kg N (*Leucaena*) + 25 kg P/ha. The trials were conducted in a net plot size of 8.4 m x 9.4 m with spacing of 60 x 30 cm for cotton and 30 x 10 cm for green gram every year.

At Kovilpatti, 15 experiments on sorghum (*Sorghum bicolor* L.) and 9 experiments on pearl millet (*Pennisetum americanum* L.) were conducted in a permanent site during North-East monsoon season (October to January) of 1987 to 2005 in a semi-arid vertic Inceptisol. The center is located at a latitude of 9.12° North, longitude of 77.53° East and an altitude of 166.42 m above mean sea level. Sorghum was grown every year up to 1987, and was rotated with pearl millet. Nine treatments which are combinations of urea, FYM and crop residue were tested in a net plot size of 7.5 m x 3.6 m and row spacing of 45 cm for both crops. The treatments tested were (i) Control; (ii) 40 kg N (urea) + 20 kg P/ha; (iii) 20 kg N (urea) + 10 kg P/ha; (iv) 20 kg N (crop residue)/ha; (v) 20 kg N (FYM)/ha; (vi) 20 kg N (crop residue) + 20 kg N (urea)/ha; (vii) 10 kg N (FYM) + 10 kg N (urea)/ha; (viii) 40 kg N (urea) + 20 kg P + 25 kg $ZnSO_4$/ha; and (ix) FYM @ 5 t/ha. The crop residue contained 1.2% N, while FYM contained 0.5% N.

At Indore, 15 field experiments of soybean were conducted on a permanent site during *kharif* season (June to October) of 1992 to 2006 in a semi-arid Vertisol. The research center is located at a latitude of 20°43′ N and longitude of 76°54′ E. The experiments were conducted with a set of 9 fertilizer treatments which are combinations of urea, farmyard manure (FYM) and crop residue superimposed to the same plots in each season. The treatment combinations tested were (i) control (ii) 20 kg N (urea) + 13 kg P/ha (iii) 30 kg N (urea) + 20 kg P/ha (iv) 40 kg N (urea) + 26 kg P/ha (v) 60 kg N (urea) + 35 kg P/ha (vi) 20 kg N (urea) + 13 kg P + FYM @ 6 t/ha (vii) 20 kg N (urea) + 13 kg P + FYM @ 5 t/ha (viii) FYM @ 6 t/ha and (ix) Crop residue @ 5 t/ha. The crop residue contained 0.75% N, 0.045 P and 0.14% K, whereas FYM contained 0.66% N, 0.45% P and 0.50% K. The field experiments were conducted in a net plot size of 9.0 m x 6.4 m with row spacing of 30 cm. The fertilizer treatments were randomized and superimposed to plots in a Randomized Block Design with 4 replications. FYM was applied 10 days prior to sowing, while the crop residue was applied as surface mulch after emergence of the crop in the prescribed treatments.

At Anantapur, 22 experiments were conducted on groundnut (*Arachis hypogea*) in a fixed site in *kharif* (July to November) during 1985 to 2006 under arid Alfisols. Anantapur is located at a latitude of 14.68° North, longitude of 77.62° East and an altitude of 350 m above mean sea level. The permanent site where trials were conducted was a shallow Alfisol with soil depth of 30 cm. The earliest date of sowing of groundnut was on 1st July in 2000, while the farthest was on 12th September in 2006. The earliest date of harvest was on 23rd October in 2000, while the farthest was on 3rd January in 2006. The crop had a minimum duration of 107 days in 1996 and 1999 and maximum duration of 127 days in 1985 with mean duration of 116 days having variation of 5.3% over years. A general recommended fertilizer NPK dose of 20–40–40 kg/ha was used for rainfed groundnut under Alfisols in Andhra Pradesh. Nine fertilizer NPK treatments which are combinations of inorganic and organic sources were tested every year. Apart from inorganic N through urea, P through single super phosphate and K through muriate of potash, organic N through groundnut shells and FYM were included in the different treatment combinations. The trials were conducted based on Randomized Block Design with 3 replications. The fertilizer NPK and organic N treatments tested were : T1: Control; T2: 100% NPK (20–40–40 kg/ha); T3: 50% NPK (10–20–20 kg/ha); T4: 100% N (groundnut shells ~ 20 kg N/ha); T5: 50% N (FYM ~ 10 kg N/ha); T6: 100% N (groundnut shells ~ 20 kg N/ha) + 50% NPK (10–20–20 kg/ha); T7: 50% N (FYM ~ 10 kg N/ha) + 50% NPK (10–20–20 kg/ha); T8: 100% NPK (20–40–40 kg/ha) + ZnSO4 @ 25 kg/ha; T9: Farmers practice (FYM @ 5 t/ha). The FYM contained 0.5% N, 1% P and 0.75% K on dry weight basis. All the improved agronomic practices prescribed for groundnut were adopted while conducting the trials.

## 2.2. Statistical analysis

The differences in effects of treatments in influencing soil fertility of N, P and K nutrients and crop yield were tested based on the standard Analysis of Variance (ANOVA) procedure. The treatments with a significantly higher effect on soil nutrients and yield were identified based on Least Significant Difference (LSD) criteria (Gomez and Gomez, 1984). Based on correlation coefficients measured between pairs of variables, the type (positive or negative) and extent of relation between yield, crop seasonal rainfall, and soil N, P and K nutrients were assessed for each treatment over years. Regression models of yield attained by each treatment were calibrated for assessing the influence of crop seasonal rainfall, soil N, P and K nutrients on yield of a crop over years as suggested by Draper and Smith (1998). The regression model through crop seasonal rainfall, soil N, P and K could be postulated as

$$Y = \pm\alpha \pm \beta_1\,(\text{Jun})\ \pm\beta_2\,(\text{Jul})\ \pm\beta_3\,(\text{Aug})\ \pm\beta_4\,(\text{Sep})\pm$$
$$\pm\beta_5\,(\text{Oct})\ \pm\beta_6\,(\text{Nov})\ \pm\beta_7\,(\text{SN})\ \pm\beta_8(\text{SP})\ \pm\beta_9(\text{SK}) \tag{1}$$

In model (1), $\alpha$ is intercept and $\beta1$ to $\beta9$ are regression coefficients measuring effects of variables on yield. The variables of monthly rainfall are retained depending on the dates of sowing and harvest and crop growing period. Soil sulphur was also included in the model calibrated for soybean at Indore. The usefulness of a regression model for yield prediction could be assessed based on the coefficient of determination ($R^2$) and unexplained variation measured by the prediction error. The sustainability yield index (SYI) of a fertilizer treatment could be derived as a ratio of the 'difference between mean yield and prediction error' and 'maximum mean yield' attained by any treatment in the study period (Behera et al., 2007; Nema et al., 2008; Maruthi Sankar et al., 2011, 2012a, 2012b).

At Kovilpatti, observations were recorded on daily rainfall (mm) and Pan Evaporation ($E_P$, in mm) during 1987 to 2005. Accordingly, the daily soil water balance computational procedure of Rijtema and Aboukhaled (1975) was used to calculate the Water Requirement (WR, mm), Potential Evapotranspiration (PET, mm) and Actual Evapotranspiration (AET, mm) for sorghum and pearl millet. The Crop Water Stress (CWS) was estimated by using the procedure as discussed by Hiler and Clark (1971). The crop coefficient values were determined by interpolating the values given by Doorenbos and Kassam (1979). The CWS ranged from 0.1% in 1987, 1993 and 1997 to 60.5% in 1995 with mean of 15.6% and variation of 119.9% for sorghum. In pearl millet, it ranged from 0.1% in 1996 to 71.5% in 1994 with mean of 31.7% and variation of 71.8% for pearl millet.

At Kovilpatti, the treatment-wise regression models of yield were developed using different variables of soil N, P, and K, crop seasonal rainfall, crop growing period, crop water stress measured under each treatment (Maruthi Sankar, 1986). The regression model of yield could be postulated as

$$Y = \pm \alpha \pm \beta1 \, (CGP) \pm \beta2 \, (CRF) \pm \beta3 \, (CWS) \pm \beta4 \, (SN) \pm \beta5 \, (SP) \pm \beta6 \, (SK) \qquad (2)$$

In (2), $\alpha$ is intercept and $\beta1$ to $\beta6$ are regression coefficients of variables considered in the model.

## 3. Rainfall and its distribution in different years

### 3.1. Semi-arid alfisols at Bangalore

At Bangalore, the earliest date of sowing of finger millet was on 14th July in 2004, while the latest was on 30th September in 2002. The earliest date of harvest of the crop was on 25th October in 2004, while the latest was on 3rd January in 2003. The crop had a minimum duration of 96 days in 2002 and maximum of 155 days in 1994 with a mean of 126 days and variation of 9.2%. The crop seasonal rainfall received from June to November was in a range of 396.6 mm in 1990 to 1174.7 mm in 2005 with a mean of 756 mm and variation of 28.1%. Four crop seasonal rainfall situations viz., < 500, 500–750, 750–1000 and 1000–1250 mm were observed during 1984 to 2008. The crop seasonal rainfall was < 500 mm in 3 years, 500–750 mm in 11 years, 750–1000 mm in 8 years and 1000–1250 mm in 3 years. June received a mean rainfall of 81 mm with a variation of 77.4%; while July received 98 mm with variation of

59.1%. August received a mean rainfall of 139 mm with a variation of 61.2%, while September received a mean rainfall of 200 mm with variation of 50.3%. October received a mean rainfall of 188 mm with a variation of 66.7%, while November received 50 mm with variation of 95.5% over the 25 years of study. The mean rainfall in a month increased from < 500 mm to 1000–1250 mm crop seasonal rainfall group. Under < 500 mm crop seasonal rainfall situation occurred for 3 years (1990, 2002 and 2006), the mean monthly rainfall ranged from 54 mm with a variation of 51.1% in July to 105 mm with a variation of 62.9% in October. Under 500–750 mm crop seasonal rainfall situation for 11 years (1984, 1985, 1986, 1987, 1989, 1994, 1995, 1996, 2001, 2003 and 2007), the mean monthly rainfall ranged from 38 mm with a variation of 83.2% in November to 199 mm with a variation of 47.7% in September. Under 750–1000 mm crop seasonal rainfall situation for 8 years (1988, 1992, 1993, 1997, 1999, 2000, 2004 and 2008), the mean monthly rainfall ranged from 51 mm with a variation of 125.2% in November to 263 mm with a variation of 31.0% in September. Under 1000–1250 mm crop seasonal rainfall situation for 3 years (1991, 1998 and 2005), the mean monthly rainfall ranged from 77 mm with a variation of 84.8% in November to 435 mm with a variation of 38.6% in October. The mean crop growing period was 121 days with variation of 17.9% under < 500 mm; 131 days with variation of 8.8% under 500–750 mm rainfall; 122 days with variation of 7.1% under 750–1000 mm rainfall; and 125 days with variation of 2.9% under 1000–1250 mm rainfall situation. The details of crop growing period, rainfall, date of sowing and harvest of finger millet under different crop seasonal rainfall situations during 1984 to 2008 are given in Table 2.

### 3.2. Semi-arid vertisols at Akola

In the long term study, both cotton and green gram were sown on the same date every year, but had different dates of harvest. The earliest date of sowing of was on 11th June in 1993, while the latest was on 23rd July in 2004. The earliest date of harvest of green gram was on 22nd August in 2001, while the latest was on 15th October in 1997. In case of cotton, the earliest date of harvest was on 28th November in 1990, while the latest was on 26th March in 1997. The duration of green gram ranged from 62 days in 1989 to 103 days in 1997 with a mean of 75 days and variation of 14.5%. Cotton had a duration in the range of 155 days in 1991 to 265 days in 1997 with mean of 202 days and variation of 16.9%. The total crop seasonal rainfall received during June to November ranged from 351.7 mm in 2003 to 1307.8 mm in 1988. The monthly rainfall received was erratic and had a high variation. June rainfall ranged from 24.4 mm (1996) to 339 mm (1990), while July rainfall ranged from 53.4 mm (2002) to 392.5 mm (1988). August received a rainfall in the range of 12.6 mm (1995) to 393.8 mm (1992), while September received 'no' rainfall (1991) to a maximum of 301.2 mm (1988). October received 'no' rainfall (1991, 2000 and 2003) to a maximum of 183.6 mm (1990), while November received 'no' rainfall in 14 years to a maximum of 164.3 mm (1997). A mean rainfall of 145.2 mm (CV of 57%), 184.4 mm (CV of 56.9%), 195.9 mm (CV of 48.5%), 127.6 mm (CV of 63.9%), 65.3 mm (CV of 93.1%) and 17.9 mm (CV of 224%) was received in June, July, August, September, October and November respectively. The crop seasonal rainfall was found to be < 500 mm in 3 years (1991, 2003 and 2004); 500–750 mm in 8 years (1987,

1989, 1995, 1996, 2000, 2001, 2002 and 2005); 750–1000 mm in 7 years (1992, 1993, 1994, 1997, 1998, 1999 and 2006); and 1000–1250 mm in 2 years (1988 and 1990) (Table 2).

### 3.3. Semi-arid vertic inceptisols at Kovilpatti

At Kovilpatti, the earliest DOS of sorghum was on 29th September in 1995, while the farthest was on 27th October in 1984 and 1985. The earliest date of harvest of sorghum was on 7th January in 2004, while the farthest was on 25th February in 1986. The crop had a minimum growing period of 88 days in 1983 to a maximum of 138 days in 2005 with mean of 112 days and variation of 12.6%. The lowest crop seasonal rainfall of 96.4 mm occurred on 8 days in 1995, while the highest of 634.6 mm occurred on 21 days in 1989 with a mean of 380.3 mm and variation of 42.6%. However, maximum number of 39 rainy days occurred in 1997 with a crop seasonal rainfall of 585.7 mm. The crop received rainfall from mean of 21 rainy days with variation of 37.6%. The earliest DOS of pearl millet was on 30th September in 2000, while the farthest was on 10th November in 1998. The earliest date of harvest of pearl millet was on 3rd January in 2003, while the farthest was on 16th February in 2005. The crop had a growing period of 80 days in 1988 to 136 days in 2004 with a mean of 99 days and variation of 16.5%. Rainfall occurred on only 6 days in 1994, while it occurred on a maximum of 35 days in 2004 with a mean of 16 days and variation of 61.3%. The actual rainfall received during crop growing period ranged from 181.1 mm in 1988 to 789.2 mm in 1996 with a mean of 495.8 mm and variation of 34.9% (Table 2).

### 3.4. Semi-arid vertisols at Indore

The earliest date of sowing of soybean was on 17th June in 2004, whereas the farthest was on 20th July in 1996. The earliest date of harvest of soybean was on 1st October in 2001, as against the farthest on 29th October in 1996. The crop had a minimum growing period of 91 days in 1992 compared to a maximum of 117 days in 2004 and had a mean of 106 days with variation of 6.6% during 15 years. The lowest crop seasonal rainfall of 354.1 mm (64.9% of annual rainfall) occurred in 2002, whereas the highest of 1308.3 mm (98.3% of annual rainfall) occurred in 1996 with a mean of 840.9 mm and variation of 30.2% over years. The rainfall ranged from 54.7 mm in 1996 to 329.7 mm in 2001 in June, 50.3 mm in 2002 to 676.9 mm in 1996 in July, 91.1 in 1999 to 429.7 mm in 2006 in August, 9 mm in 2000 to 350.3 mm in 2003 in September and 'no rainfall' in 1994 and 2003 to 79.8 mm in 1996 in October. A mean rainfall of 132.1 mm with a variation of 57.5% in June, 294.7 mm with a variation of 53.1% in July, 243.7 mm with a variation of 44.4% in August, 140.6 mm with a variation of 70.3% in September and 29.9 mm with a variation of 96.1% in October was received (Table 2).

### 3.5. Arid alfisols at Anantapur

A wide range was observed in the monthly rainfall received from May to November during 1985 to 2006. The total rainfall of May to November ranged from 255.0 to 842.8 mm with a mean rainfall of 538.7 mm and variation of 33.2%. The rainfall received in May ranged from

0.2 to 155.6 mm; June from 5.2 to 212.2 mm; July from 0.6 to 453.7 mm; August from 4.4 to 343 mm; September from 13.2 to 354.6 mm; October from 22.2 to 211.8 mm; and November from 2.6 to 127.8 mm. A maximum mean rainfall of 117.5 mm was received in September; compared to 106.4 mm in October; 101.0 mm in August; 91.3 mm in July; 58.9 mm in June; 44.8 mm in May; and 31.1 mm in November in the study. It is observed that July received rainfall with a maximum variation of 119.9%; compared to November with 111.3%; August with 91.4%; June with 88.7%; May with 79.3%; September with 64.8%; and October with 56.3%.

| Statistic | CGP | Jun | Jul | Aug | Sep | Oct | Nov | CRF |
|---|---|---|---|---|---|---|---|---|
| **Bangalore : Finger millet : 1984-2008** | | | | | | | | |
| Minimum | 96 | 7.7 | 21.1 | 31.8 | 43.8 | 28.0 | 2.0 | 396.6 |
| Maximum | 155 | 230.6 | 272.0 | 352.2 | 388.1 | 540.9 | 193.8 | 1174.7 |
| Mean | 126 | 81.0 | 98.0 | 139.0 | 200.0 | 188.0 | 50.0 | 756.0 |
| CV (%) | 9.2 | 77.4 | 59.1 | 61.2 | 50.3 | 66.7 | 95.5 | 28.1 |
| **Akola : Cotton + green gram (1:1) : 1987-2006** | | | | | | | | |
| Minimum | 155 (62) | 24.4 | 53.4 | 12.6 | 0.0 | 0.0 | 0.0 | 351.7 |
| Maximum | 265 (103) | 339.0 | 392.5 | 393.8 | 301.2 | 183.6 | 164.3 | 1307.8 |
| Mean | 203 (74) | 145.8 | 189.0 | 194.9 | 127.9 | 62.2 | 18.0 | 737.9 |
| CV (%) | 16.5 (14.5) | 55.4 | 55.2 | 47.6 | 62.1 | 98.0 | 216.2 | 31.9 |
| **Kovilpatti : Sorghum : 1987-2005** | | **Sep** | **Oct** | **Nov** | **Dec** | **Jan** | | |
| Minimum | 96 | 0.0 | 71.6 | 14.0 | 0.0 | 0.0 | | 181.1 |
| Maximum | 138 | 181.0 | 503.1 | 262.2 | 212.3 | 212.0 | | 789.2 |
| Mean | 113 | 72.3 | 202.5 | 125.1 | 64.5 | 31.5 | | 495.8 |
| CV (%) | 12.7 | 85.3 | 60.6 | 73.6 | 102.3 | 204.7 | | 34.9 |
| **Kovilpatti : Pearl millet : 1987-2005** | | | | | | | | |
| Minimum | 80 | 24.5 | 20.8 | 63.4 | 4.2 | 0.0 | | 209.3 |
| Maximum | 136 | 178.6 | 412.2 | 268.8 | 162.2 | 14.6 | | 692.0 |
| Mean | 99 | 98.1 | 197.4 | 144.1 | 54.2 | 4.3 | | 498.1 |
| CV (%) | 16.5 | 56.6 | 67.3 | 52.6 | 100.6 | 138.0 | | 30.5 |
| **Indore : Soybean : 1992-2006** | | | | | | | | |
| Minimum | 91 | 54.7 | 50.3 | 91.1 | 9.0 | 0.0 | | 354.1 |
| Maximum | 117 | 329.7 | 676.9 | 429.7 | 350.3 | 79.8 | | 1308.3 |
| Mean | 106 | 132.1 | 294.7 | 243.7 | 140.6 | 29.9 | | 840.9 |
| CV (%) | 6.6 | 57.5 | 53.1 | 44.4 | 70.3 | 96.1 | | 30.2 |
| **Anantapur : Groundnut : 1985-2006** | | | | | | | | |
| Minimum | 107 | 5.2 | 0.6 | 4.4 | 13.2 | 22.2 | 2.6 | 255.0 |
| Maximum | 127 | 212.2 | 453.7 | 343.0 | 354.6 | 211.8 | 127.8 | 801.4 |
| Mean | 116 | 58.9 | 91.3 | 101.0 | 117.5 | 106.4 | 31.1 | 502.1 |
| CV (%) | 5.0 | 88.7 | 119.9 | 91.4 | 64.8 | 56.3 | 111.3 | 37.0 |

**Table 2.** Mean and variation of rainfall and crop growing period at different locations

## 4. Results and discussion

### 4.1. ANOVA of soil test values and yield in different seasons

The mean and coefficient of variation of soil fertility of nutrients and yield of crops attained under each rainfall situation at Bangalore, Akola, Kovilpatti and Indore are given in Table 3.

| TR | Yield (kg/ha) | | | | Soil N (kg/ha) | | | | Soil P (kg/ha) | | | | Soil K (kg/ha) | | | |
|---|---|---|---|---|---|---|---|---|---|---|---|---|---|---|---|---|
| | Min | Max | Mean | CV | Min | Max | Mean | CV | Min | Max | Mean | CV | Min | Max | Mean | CV |
| **Bangalore : Finger millet : 1984-2008** | | | | | | | | | | | | | | | | |
| T1 | 54 | 1356 | 537 | 79.7 | 87 | 210 | 163 | 12.4 | 4.5 | 27.9 | 9.7 | 46.1 | 37 | 88 | 59 | 19.5 |
| T2 | 1146 | 3125 | 2452 | 23.4 | 146 | 241 | 195 | 8.2 | 15.7 | 61.6 | 43.8 | 28.8 | 59 | 116 | 88 | 16.6 |
| T3 | 1432 | 3836 | 2891 | 21.9 | 170 | 217 | 196 | 4.7 | 24.0 | 90.4 | 59.4 | 27.2 | 61 | 152 | 95 | 18.3 |
| T4 | 1821 | 4552 | 3167 | 22.7 | 174 | 242 | 204 | 7.5 | 35.2 | 91.9 | 68.6 | 18.7 | 67 | 157 | 107 | 20.4 |
| T5 | 756 | 3429 | 1826 | 45.4 | 103 | 245 | 190 | 12.5 | 21.4 | 76.4 | 50.4 | 27.0 | 50 | 114 | 78 | 16.2 |
| LSD | | | 535 | | | | 13 | | | | 6.7 | | | | 13 | |
| **Akola : Green gram: 1987-2007** | | | | | | | | | | | | | | | | |
| T1 | 74 | 700 | 362 | 42.4 | 110 | 290 | 217 | 24.0 | 11.0 | 32.2 | 25.2 | 17.4 | 243 | 395 | 289 | 12.6 |
| T2 | 90 | 790 | 488 | 37.0 | 123 | 306 | 239 | 18.9 | 15.0 | 35.9 | 30.0 | 16.6 | 256 | 393 | 320 | 11.6 |
| T3 | 87 | 770 | 453 | 38.4 | 113 | 299 | 227 | 20.8 | 16.8 | 35.4 | 28.8 | 17.1 | 252 | 389 | 313 | 11.6 |
| T4 | 83 | 768 | 447 | 40.3 | 117 | 297 | 230 | 23.7 | 16.5 | 35.9 | 30.1 | 16.2 | 256 | 389 | 323 | 9.0 |
| T5 | 83 | 880 | 480 | 39.9 | 119 | 328 | 239 | 24.2 | 16.5 | 42.8 | 32.3 | 18.3 | 260 | 417 | 351 | 12.3 |
| T6 | 88 | 840 | 491 | 37.9 | 125 | 313 | 250 | 21.1 | 15.7 | 42.5 | 31.8 | 19.4 | 277 | 485 | 358 | 15.1 |
| T7 | 97 | 930 | 547 | 35.6 | 123 | 328 | 254 | 22.5 | 20.0 | 42.8 | 33.4 | 15.3 | 294 | 491 | 369 | 16.1 |
| T8 | 96 | 910 | 505 | 40.3 | 123 | 320 | 243 | 24.3 | 18.4 | 46.2 | 32.4 | 19.0 | 276 | 492 | 344 | 18.9 |
| LSD | 10 | 379 | 108 | 73.3 | | | | | | | | | | | | |
| **Akola : Cotton: 1987-2007** | | | | | | | | | | | | | | | | |
| T1 | 91 | 1016 | 492 | 51.2 | 110 | 290 | 217 | 24.0 | 11.0 | 32.2 | 25.2 | 17.4 | 243 | 395 | 289 | 12.6 |
| T2 | 171 | 1637 | 699 | 52.6 | 123 | 306 | 239 | 18.9 | 15.0 | 35.9 | 30.0 | 16.6 | 256 | 393 | 320 | 11.6 |
| T3 | 152 | 1338 | 637 | 53.6 | 113 | 299 | 227 | 20.8 | 16.8 | 35.4 | 28.8 | 17.1 | 252 | 389 | 313 | 11.6 |
| T4 | 112 | 1404 | 625 | 51.6 | 117 | 297 | 230 | 23.7 | 16.5 | 35.9 | 30.1 | 16.2 | 256 | 389 | 323 | 9.0 |
| T5 | 101 | 1637 | 678 | 55.3 | 119 | 328 | 239 | 24.2 | 16.5 | 42.8 | 32.3 | 18.3 | 260 | 417 | 351 | 12.3 |
| T6 | 174 | 1795 | 737 | 54.3 | 125 | 313 | 250 | 21.1 | 15.7 | 42.5 | 31.8 | 19.4 | 277 | 485 | 358 | 15.1 |
| T7 | 217 | 1910 | 805 | 54.8 | 123 | 328 | 254 | 22.5 | 20.0 | 42.8 | 33.4 | 15.3 | 294 | 491 | 369 | 16.1 |
| T8 | 126 | 1725 | 702 | 55.4 | 123 | 320 | 243 | 24.3 | 18.4 | 46.2 | 32.4 | 19.0 | 276 | 492 | 344 | 18.9 |
| LSD | 18 | 290 | 134 | 58.9 | | | | | | | | | | | | |
| **Kovilpatti : Sorghum: 1987-2005** | | | | | | | | | | | | | | | | |
| T1 | 157 | 1520 | 709 | 66.2 | 76 | 146 | 112 | 18.5 | 6.0 | 8.8 | 7.1 | 13.8 | 236 | 389 | 328 | 13.3 |
| T2 | 415 | 2122 | 1118 | 48.3 | 79 | 168 | 132 | 19.0 | 3.3 | 12.1 | 8.3 | 28.9 | 193 | 449 | 361 | 21.4 |
| T3 | 340 | 1840 | 917 | 53.3 | 76 | 159 | 124 | 18.2 | 6.5 | 11.9 | 8.9 | 19.3 | 270 | 435 | 355 | 15.7 |
| T4 | 130 | 1821 | 867 | 61.9 | 69 | 149 | 121 | 18.9 | 5.7 | 11.2 | 8.4 | 18.6 | 236 | 448 | 369 | 16.2 |
| T5 | 175 | 1919 | 867 | 60.4 | 76 | 145 | 125 | 17.1 | 7.5 | 10.3 | 8.8 | 10.0 | 237 | 450 | 360 | 16.3 |
| T6 | 147 | 2407 | 1031 | 68.5 | 70 | 159 | 126 | 21.9 | 3.3 | 11.3 | 9.4 | 24.1 | 271 | 475 | 375 | 16.4 |
| T7 | 325 | 2320 | 1163 | 55.5 | 80 | 187 | 142 | 20.8 | 9.0 | 12.9 | 10.7 | 10.8 | 215 | 528 | 394 | 21.4 |
| T8 | 342 | 2451 | 1246 | 50.2 | 84 | 163 | 136 | 16.6 | 6.3 | 12.3 | 10.1 | 17.9 | 252 | 481 | 382 | 19.9 |
| T9 | 195 | 2164 | 999 | 70.0 | 79 | 156 | 131 | 18.0 | 6.3 | 12.5 | 9.4 | 20.4 | 215 | 546 | 382 | 25.2 |

| TR | Yield (kg/ha) | | | | Soil N (kg/ha) | | | | Soil P (kg/ha) | | | | Soil K (kg/ha) | | | |
|---|---|---|---|---|---|---|---|---|---|---|---|---|---|---|---|---|
| | Min | Max | Mean | CV | Min | Max | Mean | CV | Min | Max | Mean | CV | Min | Max | Mean | CV |
| **Kovilpatti : Pearl millet : 1987-2005** | | | | | | | | | | | | | | | | |
| T1 | 163 | 936 | 453 | 51.8 | 77 | 107 | 87 | 12.4 | 6.0 | 8.2 | 6.5 | 9.9 | 243 | 357 | 319 | 12.2 |
| T2 | 230 | 1212 | 670 | 42.4 | 98 | 149 | 118 | 14.2 | 7.3 | 10.6 | 9.4 | 11.6 | 275 | 438 | 374 | 15.1 |
| T3 | 333 | 1083 | 626 | 33.2 | 82 | 123 | 104 | 14.1 | 6.5 | 10.0 | 7.7 | 12.7 | 256 | 437 | 360 | 18.2 |
| T4 | 253 | 1067 | 577 | 42.0 | 84 | 119 | 102 | 12.5 | 6.5 | 9.5 | 7.9 | 13.6 | 250 | 450 | 364 | 21.1 |
| T5 | 247 | 1135 | 561 | 45.2 | 80 | 116 | 96 | 12.7 | 7.0 | 10.0 | 8.4 | 11.8 | 270 | 468 | 385 | 19.1 |
| T6 | 265 | 1194 | 676 | 40.7 | 92 | 120 | 107 | 10.0 | 8.3 | 10.5 | 9.0 | 6.6 | 248 | 490 | 403 | 21.8 |
| T7 | 419 | 1220 | 774 | 32.1 | 90 | 122 | 107 | 10.9 | 8.0 | 12.1 | 9.7 | 13.4 | 248 | 460 | 382 | 18.7 |
| T8 | 502 | 1190 | 761 | 29.7 | 81 | 114 | 103 | 10.9 | 8.5 | 11.0 | 9.5 | 7.9 | 280 | 453 | 385 | 13.8 |
| T9 | 331 | 1046 | 552 | 41.0 | 87 | 125 | 100 | 12.5 | 6.0 | 10.3 | 7.9 | 15.9 | 290 | 493 | 407 | 16.9 |
| **Indore : Soybean : 1992-2006** | | | | | | | | | | | | | | | | |
| T1 | 691 | 2066 | 1275 | 31.1 | 161 | 209 | 178 | 8.4 | 5.0 | 21.5 | 11.4 | 57.5 | 325 | 830 | 540 | 22.2 |
| T2 | 987 | 2448 | 1620 | 29.4 | 185 | 232 | 202 | 7.0 | 5.8 | 24.9 | 14.1 | 53.8 | 340 | 743 | 552 | 18.1 |
| T3 | 1147 | 2720 | 1774 | 29.5 | 167 | 240 | 206 | 10.3 | 6.0 | 27.7 | 16.1 | 53.1 | 360 | 980 | 585 | 25.7 |
| T4 | 1250 | 2828 | 1886 | 27.7 | 204 | 280 | 225 | 8.4 | 5.0 | 28.2 | 16.9 | 46.6 | 320 | 721 | 545 | 19.0 |
| T5 | 1308 | 3050 | 1994 | 26.4 | 171 | 284 | 224 | 11.5 | 4.8 | 29.2 | 17.7 | 42.6 | 347 | 1042 | 629 | 25.6 |
| T6 | 1449 | 3247 | 2095 | 25.3 | 216 | 384 | 274 | 20.3 | 4.2 | 28.4 | 22.2 | 30.3 | 373 | 1132 | 741 | 24.3 |
| T7 | 997 | 3102 | 1790 | 34.0 | 185 | 337 | 254 | 18.9 | 4.5 | 27.8 | 15.0 | 53.1 | 340 | 943 | 642 | 23.2 |
| T8 | 1152 | 3061 | 1863 | 30.3 | 189 | 426 | 264 | 26.2 | 5.2 | 27.0 | 18.3 | 32.8 | 360 | 892 | 677 | 20.9 |
| T9 | 897 | 2504 | 1629 | 32.6 | 176 | 347 | 239 | 23.4 | 5.6 | 25.4 | 15.6 | 46.8 | 320 | 1215 | 629 | 31.7 |
| LSD | | | 241 | | | | 23 | 67.8 | | | 2.4 | 53.7 | | | 52.8 | 53.3 |

LSD: Least significant difference at p < 0.05, CV: Coefficient of variation (%)

**Table 3.** Descriptive statistics of finger millet grain yield, and available soil nutrients

## 4.1.1. Semi-arid alfisols at Bangalore

At Bangalore, the changes in soil N, P and K nutrients over years were assessed. The trends indicated that the soil N decreased in all treatments, however, the decrease was significant only in control. There was a build–up of soil P in all treatments, but, the increase was significant only in FYM @ 10 t/ha, FYM @ 10 t/ha + 50% NPK and 100% NPK treatments. There was a decrease of soil K over years, in all treatments, but the decrease was significant only in FYM @ 10 t/ha ha + 100% NPK application. Based on the predictability of changes in soil nutrient status over years ($R^2$), the prediction (%) of yield ranged from 1 to 26% for soil N; 2 to 44% for soil P; and 1 to 26% for soil K for different treatments. The standard error based on a regression model ranged from 9.3 to 23.1 kg/ha for soil N; 4.5 to 12.3 kg/ha for soil P; and 11.4 to 19.2 kg/ha for soil K over years. The trends of changes in yield, and soil nutrients as affected by treatments over years indicated that in general, the soil P tended to increase, while soil N reflected the decreasing tendency over years. However, soil K decreased over years. Thus, the trends of soil fertility changes were similar for soil N and P nutrients (except in control), while it was the opposite trend for soil K over years.

## 4.1.2. Semi-arid vertisols at Akola

At Akola, the ANOVA indicated that fertilizer treatments differed significantly in influencing soil fertility of nutrients and yield in all years. They were also significantly different when pooled over years under each rainfall situation (Gomez and Gomez, 1985). A minimum mean yield of 360 kg/ha (variation of 43.7%) and 492 kg/ha (variation of 52.5%) was attained under control, while a maximum of 527 kg/ha (variation of 33.9%) and 807 kg/ha (variation of 56.1%) was attained under 25 kg N (FYM) + 25 kg N (urea) + 25 kg P/ha in case of green gram and cotton respectively. Application of 25 kg N (FYM) + 25 kg N (urea) + 25 kg P/ha was also superior for enhancing soil fertility status by providing a maximum mean soil N of 251.8 kg/ha (variation of 22.8%), soil P of 33.5 kg/ha (variation of 19.5%) and soil K of 368.6 kg/ha (variation of 20.4%) over years. A build-up of soil N and a depletion of soil K were observed under all treatments over years. A build-up of soil P was observed under control and 25 kg N + 12.5 kg P/ha, while there was a depletion under all the remaining treatments. Comparison of pairs of treatments for differences in yield, soil N, P and K nutrients indicated that 25 kg N (FYM) + 25 kg N (urea) + 25 kg P/ha was superior to all other treatments by attaining a significantly higher yield of cotton and green gram and maintaining higher soil fertility status over years.

## 4.1.3. Semi-arid vertic inceptisols at Kovilpatti

At Kovilpatti, the ANOVA of sorghum data indicated a significant difference among treatments in individual years and also when pooled over years in influencing soil nutrients and grain yield. The mean sorghum yield ranged from 642 kg/ha with variation of 69.9% under control to 1190 kg/ha with variation of 50.9% under 40 kg N (urea) + 20 kg P + 25 kg ZnSO$_4$/ha. This treatment gave maximum potential yield of 2451 kg/ha in 1999. 40 kg N (urea) + 20 kg P/ha had a minimum variation of 47.8%, while FYM @ 5 t/ha had a maximum variation of 72.4% for sorghum yield. The mean soil N ranged from 97 kg/ha with 22.1% variation under control to 118 kg/ha with 16.3% variation under an application of 40 kg N (urea) + 20 kg P/ha. However, minimum soil N of 76 kg/ha was observed under 40 kg N (urea) + 20 kg P/ha, 20 kg N (urea) + 10 kg P/ha and 20 kg N (crop residue)/ha in 1986, while maximum of 154 kg/ha was observed under 20 kg N (FYM)/ha in 1985. Soil N had minimum variation of 14.1% under 20 kg N (crop residue) + 20 kg N (urea)/ha and maximum variation of 22.2% under 20 kg N (FYM)/ha. The mean soil P ranged from 8.8 kg/ha with 21.9% variation under 20 kg N (crop residue)/ha to 10.8 kg/ha with 79.2% under FYM @ 10 t/ha. However, minimum soil P of 6 kg/ha was observed under control and FYM @ 5 t/ha in 1987 and 1999, while maximum of 18.3 kg/ha was observed under 40 kg N (urea) + 20 kg P + 25 kg ZnSO$_4$/ha in 1983. Soil P had minimum variation of 14.5% under 40 kg N (urea) + 20 kg P/ha and maximum variation of 36.4% under control. The mean soil K ranged from 407 kg/ha with 43.3% variation under control to 473 kg/ha with 35.5% variation under 20 kg N (crop residue) + 20 kg N (urea)/ha. However, minimum soil K of 243 kg/ha was observed under control in 2005, while maximum of 854 kg/ha was under 10 kg N (FYM) + 10 kg N (urea)/ha in 1982. Minimum variation of 32.5% under 40 kg N (urea) + 20 kg P/ha and maximum variation of 43.3% under control was observed for soil K.

In case of pearl millet at Kovilpatti, the yield ranged from 399 kg/ha with 43% variation under control to 725 kg/ha with 28.3% variation under 10 kg N (FYM) + 10 kg N (urea). This treatment gave a maximum potential yield of 1062 kg/ha in 1994. 20 kg N (urea) + 10 kg P/ha had minimum variation of 24.3%, while control had maximum of 43% for yield. The mean soil N ranged from 110 kg/ha with 19.1% variation under control to 140 kg/ha with 21.7% variation under 10 kg N (FYM) + 10 kg N (urea) kg/ha. However, minimum soil N of 69 kg/ha was observed under 20 kg N (crop residue)/ha in 1988, while maximum of 187 kg/ha was observed under 10 kg N (FYM) + 10 kg N (urea)/ha in 1996. Soil N had minimum variation of 17% under 40 kg N (urea) + 20 kg P + 25 kg $ZnSO_4$/ha compared to maximum of 22.4% under 20 kg N (crop residue) + 20 kg N (urea)/ha.

The mean soil P ranged from 7 kg/ha with variation of 12.1% under control to 10.5 kg/ha with variation of 10.3% under 10 kg N (FYM) + 10 kg N (urea)/ha. However, minimum soil P of 3.3 kg/ha was observed under 40 kg N (urea) + 20 kg P/ha and 20 kg N (crop residue) + 20 kg N (urea)/ha in 2004, while maximum of 12.9 kg/ha was observed under 10 kg N (FYM) + 10 kg N (urea)/ha in 1994. The soil P had variation in the range of 10.3% under 10 kg N (FYM) + 10 kg N (urea)/ha to 26.7% under 40 kg N (urea) + 20 kg P/ha. The mean soil K ranged from 325 kg/ha with 14% variation under control to 389 kg/ha with 22.6% variation under 10 kg N (FYM) + 10 kg N (urea)/ha. However, minimum soil K of 193 kg/ha was observed under 40 kg N (urea) + 20 kg P/ha in 2004, while maximum of 546 kg/ha was observed under FYM @ 5 t/ha in 1992. Soil K had variation ranging from 14% under control to 26.8% under FYM @ 5 t/ha. A higher mean soil N was observed in sorghum trials, while higher mean soil P and K were observed in pearl millet trials under all treatments. A higher variation of yield was observed in sorghum compared to pearl millet in all treatments. In sorghum, soil K had maximum variation in 8 treatments compared to soil P in only one treatment, while soil N had minimum variation in all treatments. In pearl millet, soil P had maximum variation in 4 treatments, followed by soil K in 3 treatments and soil N in 2 treatments.

### 4.1.4. Semi-arid vertisols at Indore

The F-test indicated that the organic and inorganic treatment combinations were significantly different in both individual years and also when pooled over years in influencing the soybean yield and soil nutrients. The mean soybean yield ranged from 1275 kg/ha with variation of 31.1% under control to 2095 kg/ha with a variation of 25.3% under 20 kg N (urea) + 13 kg P + FYM @ 6 t/ha. The superior treatment also gave a maximum potential yield of 3247 kg/ha in 2006. Application of 20 kg N (urea) + 13 kg P + FYM @ 6 t/ha was also superior with a maximum mean soil N (274 kg/ha), soil P (22.2 kg/ha), soil K (741 kg/ha), and soil Sulphur (18.1 kg/ha). The control gave a minimum soil N of 178 kg/ha, soil P of 11.4 kg/ha, soil K of 540 kg/ha, and soil Sulphur of 13.6 kg/ha. 20 kg N (urea) + 13 kg P/ha had a minimum variation of 7% for soil N and 18.1% for soil K. The control had a maximum variation of 57.5% for soil P and 43.2% for soil S. FYM @ 6 t/ha had a maximum variation of 26.2% for soil N, while crop residue @ 5 t/ha had a maximum of 31.7% for soil K.

### 4.1.5. Arid alfisols at Anantapur

The distribution of organic carbon, soil P and K under each treatment as indicated by minimum, maximum, mean and variation are given in Table 4. A wide range of 0.13 to 0.69% in organic carbon; 9.7 to 171.8 kg/ha in soil P; and 89 to 454 kg/ha in soil K was observed. The mean organic carbon ranged from 0.23% in control to 0.38% in 50% N (FYM ~ 10 kg N/ha) and farmers practice (FYM @ 5 t/ha); soil P from 34.5 kg/ha in control to 100.6 kg/ha in 100% NPK (20–40–40 kg/ha) + ZnSO$_4$ @ 25 kg/ha; and soil K from 163 kg/ha in control to 297 kg/ha in 50% N (FYM ~ 10 kg N/ha) + 50% NPK (10–20–20 kg/ha) over years. 100% NPK (20–40–40 kg/ha) had lowest variation of 12.2%, while Farmers practice (FYM @ 5 t/ha) had highest variation of 38.6% for organic carbon. The variation ranged from 37.8% in 50% N (FYM ~ 10 kg N/ha) + 50% NPK (10–20–20 kg/ha) to 49.5% in Farmer's practice (FYM @ 5 t/ha) for soil P and 32.5% in 50% N (FYM ~ 10 kg N/ha) + 50% NPK (10–20–20 kg/ha) to 51.4% in 50% NPK (10–20–20 kg/ha) for soil K.

The groundnut pod yield attained by different treatments ranged from 171 kg/ha attained by 50% N (FYM ~ 10 kg N/ha) + 50% NPK (10–20–20 kg/ha) to 1546 kg/ha attained by 100% NPK (inorganic) application. Based on the ANOVA, the fertilizer treatments differed significantly from each other in all the years except 1992, 1996, 2000, 2002 and 2006. The crop failed to produce any pod yield in 1988 and 2001 due to insufficient soil moisture. The yield attained in different years along with LSD at p < 0.05 level are given in Table 3. Based on the LSD criteria, 100% NPK (20–40–40 kg/ha) gave significantly higher yield of 1367, 609, 745 and 1546 kg/ha in 1985, 1991, 1993 and 2004; while 100% N (groundnut shells ~ 20 kg N/ha) was superior with yield of 1300, 1518 and 1348 kg/ha in 1986, 1987 and 1990 respectively. Application of 100% N (groundnut shells ~ 20 kg N/ha) + 50% NPK (10–20–20 kg/ha) was superior with yield of 757, 388 and 588 kg/ha in 1994, 1997 and 2003 respectively; while 50% N (FYM ~ 10 kg N/ha) + 50% NPK (10–20–20 kg/ha) gave significantly higher yield of 1541 kg/ha in 1989 and 1123 kg/ha in 2005. Application of 100% NPK (20–40–40 kg/ha) + ZnSO$_4$ @ 25 kg/ha was superior with pod yield of 1131 kg/ha in 1995, 1329 kg/ha in 1998 and 1348 kg/ha in 1999. The study indicated that 100% NPK (20–40–40 kg/ha) + ZnSO$_4$ @ 25 kg/ha was superior with maximum mean pod yield of 926 kg/ha (variation of 46.6%), while control gave minimum yield of 741 kg/ha (variation of 44.6%). However, 100% N (groundnut shells ~ 20 kg N/ha) + 50% NPK (10–20–20 kg/ha) had minimum variation of 41.5%, while 100% NPK (20–40–40 kg/ha) had maximum variation of 47.4%. Highest yield increase of 24.9% was attained by 100% NPK (20–40–40 kg/ha) + ZnSO$_4$ @ 25 kg/ha, followed by 100% NPK (20–40–40 kg/ha) with 23.5%, 50% N (FYM ~ 10 kg N/ha) + 50% NPK (10–20–20 kg/ha) with 23.1%, while 100% N (groundnut shells ~ 20 kg N/ha) gave lowest yield increase of 16.5% over years (Table 4).

## 4.2. Relationship between yield, soil nutrients and rainfall over years

### 4.2.1. Finger millet experiments at Bangalore

The estimates of correlation between finger millet yield, soil fertility of nutrients and monthly rainfall are given in Table 5. At Bangalore, with application of 100% NPK over years, the grain yield had a significant negative correlation with soil P. It had a positive relationship with soil N in control, FYM @ 10 t/ha + 100% NPK and 100% NPK; soil K in all treatments except 100% NPK. The crop seasonal rainfall had a negative effect on finger millet yield in control and 100% NPK. The crop growing period had a positive correlation with grain yield attained by all treatments except FYM @ 10 t/ha and FYM @ 10 t/ha + 50% NPK. Among different treatments, the negative correlation of yield (in all treatments), soil N and soil K with time period indicated a decrease, while a positive correlation of soil P with time period indicated an increase with fertilizer application.

| Treat ment | Yield (kg/ha) | | | | Organic carbon (%) | | | | Soil P (kg/ha) | | | | Soil K (kg/ha) | | | |
|---|---|---|---|---|---|---|---|---|---|---|---|---|---|---|---|---|
| | | | | | Min | Max | Mean | CV | Min | Max | Mean | CV | Min | Max | Mean | CV |
| T1 | 237 | 1364 | 741 | 44.6 | 0.13 | 0.32 | 0.23 | 28.5 | 9.7 | 48.0 | 34.5 | 44.6 | 94 | 272 | 163 | 38.9 |
| T2 | 210 | 1546 | 916 | 47.4 | 0.24 | 0.33 | 0.26 | 12.2 | 57.0 | 162.1 | 97.1 | 41.1 | 133 | 393 | 234 | 37.0 |
| T3 | 247 | 1428 | 879 | 44.3 | 0.19 | 0.38 | 0.26 | 24.0 | 45.5 | 161.6 | 84.8 | 44.9 | 89 | 454 | 233 | 51.4 |
| T4 | 226 | 1518 | 863 | 42.0 | 0.21 | 0.45 | 0.33 | 21.2 | 19.4 | 79.0 | 55.2 | 39.5 | 121 | 391 | 241 | 36.9 |
| T5 | 219 | 1516 | 876 | 43.4 | 0.21 | 0.54 | 0.38 | 33.3 | 29.1 | 103.0 | 61.8 | 46.5 | 116 | 393 | 263 | 33.6 |
| T6 | 193 | 1478 | 889 | 41.5 | 0.26 | 0.49 | 0.36 | 19.1 | 34.3 | 143.6 | 89.1 | 44.1 | 104 | 415 | 253 | 40.1 |
| T7 | 171 | 1541 | 912 | 45.6 | 0.24 | 0.61 | 0.36 | 36.9 | 51.0 | 153.6 | 100.0 | 37.8 | 119 | 371 | 297 | 32.5 |
| T8 | 226 | 1473 | 926 | 46.6 | 0.18 | 0.38 | 0.28 | 23.1 | 40.0 | 171.8 | 100.6 | 40.1 | 111 | 394 | 258 | 38.4 |
| T9 | 205 | 1507 | 833 | 45.8 | 0.26 | 0.69 | 0.38 | 38.6 | 34.0 | 128.2 | 65.5 | 49.5 | 141 | 358 | 269 | 35.4 |

**Table 4.** Effect of fertilizer treatments on soil test values of organic carbon, P and K nutrients at Anantapur

| Variable1 | Variable2 | T1 | T2 | T3 | T4 | T5 |
|---|---|---|---|---|---|---|
| Bangalore | | | | | | |
| GY | SN | 0.17 | -0.02 | -0.02 | 0.07 | 0.11 |
| GY | SP | 0.13 | -0.13 | -0.01 | -0.03 | -0.46* |
| GY | SK | 0.12 | 0.07 | 0.15 | 0.34 | -0.02 |
| GY | CGP | 0.21 | -0.07 | -0.01 | 0.09 | 0.12 |
| GY | CRF | -0.34 | 0.04 | 0.09 | 0.09 | -0.22 |
| GY | Years | -0.78** | -0.34 | -0.23 | -0.54** | -0.67** |
| SN | Years | -0.51** | -0.17 | -0.02 | -0.16 | -0.32 |
| SP | Years | 0.13 | 0.65** | 0.67** | 0.34 | 0.48* |
| SK | Years | -0.05 | -0.35 | -0.37 | -0.51** | -0.47* |

**Table 5.** Relation between yield, soil nutrients and crop seasonal rainfall at Bangalore

### 4.2.2. Cotton and green gram experiments at Akola

The estimates of correlation of cotton and green gram yield with soil fertility of nutrients and monthly rainfall are given in Table 6. At Akola, June rainfall had a significant positive correlation with green gram yield of all treatments except 25 kg N/ha (FYM) compared to August rainfall with yield attained by control, 25 kg N + 12.5 kg P/ha, 25 kg N (*Leucaena*) + 25 kg N (urea) + 25 kg P/ha and 25 kg N (FYM) + 25 kg N (urea) + 25 kg P/ha. The yield had a significant negative correlation with soil N under all treatments except 25 kg N (FYM) + 25 kg N (urea) + 25 kg P/ha; while a significant positive correlation with soil K under 25 kg N (*Leucaena*) + 25 kg N (urea) + 25 kg P/ha. There was no significant correlation between any pair of variables in case of cotton. The analysis indicated that rainfall of June, July, August and September, soil P, and soil K had a positive correlation, while soil N have a negative correlation with green gram yield over years. Similarly, the monthly rainfall of June to October, and soil K have a positive correlation, while November rainfall, soil N and P have a negative correlation with cotton yield over years.

| Var 1 | Var 2 | T1 | T2 | T3 | T4 | T5 | T6 | T7 | T8 |
|---|---|---|---|---|---|---|---|---|---|
| **Green gram** | | | | | | | | | |
| Yield | Jun | 0.61** | 0.51* | 0.55** | 0.63** | 0.42 | 0.60** | 0.54** | 0.62** |
| Yield | Jul | 0.05 | 0.01 | 0.02 | 0.03 | 0.20 | 0.16 | 0.08 | 0.15 |
| Yield | Aug | 0.51* | 0.41 | 0.45* | 0.41 | 0.43 | 0.47* | 0.48* | 0.43 |
| Yield | Sep | 0.14 | 0.17 | 0.17 | 0.05 | 0.30 | 0.14 | 0.17 | 0.08 |
| Yield | Soil N | -0.64** | -0.50* | -0.58** | -0.55** | -0.46* | -0.54** | -0.38 | -0.58** |
| Yield | Soil P | 0.06 | 0.09 | -0.03 | 0.09 | -0.04 | 0.04 | 0.10 | 0.10 |
| Yield | Soil K | 0.08 | 0.05 | 0.04 | 0.00 | 0.37 | 0.46* | 0.30 | 0.33 |
| **Cotton** | | | | | | | | | |
| Yield | Jun | 0.15 | 0.15 | 0.10 | 0.18 | 0.09 | 0.10 | 0.01 | 0.12 |
| Yield | Jul | 0.42 | 0.42 | 0.42 | 0.39 | 0.40 | 0.39 | 0.40 | 0.39 |
| Yield | Aug | 0.25 | 0.21 | 0.27 | 0.23 | 0.23 | 0.22 | 0.24 | 0.24 |
| Yield | Sep | 0.08 | 0.01 | 0.08 | -0.01 | 0.08 | 0.07 | 0.15 | 0.05 |
| Yield | Oct | -0.02 | 0.07 | 0.01 | 0.02 | 0.06 | 0.08 | 0.06 | 0.05 |
| Yield | Nov | -0.24 | -0.23 | -0.25 | -0.24 | -0.27 | -0.30 | -0.26 | -0.29 |
| Yield | Soil N | -0.21 | -0.21 | -0.24 | -0.30 | -0.18 | -0.37 | -0.29 | -0.31 |
| Yield | Soil P | -0.38 | -0.20 | -0.29 | -0.29 | -0.26 | -0.22 | -0.29 | -0.18 |
| Yield | Soil K | 0.31 | 0.35 | 0.33 | 0.37 | 0.31 | 0.30 | 0.07 | 0.24 |
| **Soil nutrients** | | | | | | | | | |
| Soil N | Year | 0.74** | 0.44* | 0.59** | 0.64** | 0.65** | 0.53* | 0.56** | 0.59** |
| Soil P | Year | 0.16 | -0.17 | 0.02 | -0.20 | -0.25 | -0.14 | -0.10 | -0.27 |
| Soil K | Year | -0.11 | -0.34 | -0.51* | -0.18 | -0.40 | -0.71** | -0.71** | -0.71** |

\* and \*\* indicate significance at p < 0.05 and p < 0.01 level respectively

**Table 6.** Relation between yield, crop seasonal rainfall and soil nutrients at Akola

## 4.2.3. Sorghum and pearl millet experiments at Kovilpatti

The estimates of correlation of sorghum and pearl millet yield with soil fertility of nutrients and monthly rainfall are given in Table 7. At Kovilpatti, sorghum yield had a significant positive correlation with years under 20 kg N (crop residue) + 20 kg N (urea)/ha, 10 kg N (FYM) + 10 kg N (urea)/ha and FYM @ 5 t/ha and negative correlation with soil N under 20 kg N (crop residue) + 20 kg N (urea)/ha. It had positive correlation with the crop growing period (CGP) under application of FYM @ 5 t/ha and negative correlation with the crop water stress (CWS) under application of 40 kg N (urea) + 20 kg P + 25 kg ZnSO$_4$/ha. The soil N had a negative relation with years under control, 20 kg N (crop residue) + 20 kg N (urea)/ha, 10 kg N (FYM) + 10 kg N (urea)/ha and 40 kg N (urea) + 20 kg P + 25 kg ZnSO$_4$/ha, and positive correlation under FYM @ 5 t/ha. The soil P had negative correlation with years under control, 20 kg N (urea) + 10 kg P/ha, 20 kg N (crop residue)/ha and 20 kg N (FYM)/ha, while soil K was negatively correlated with years only under control.

| Var1 | Var2 | T1 | T2 | T3 | T4 | T5 | T6 | T7 | T8 | T9 |
|------|------|------|------|------|------|------|------|------|------|------|
| *Sorghum (1987-2005)* | | | | | | | | | | |
| GY | Year | 0.42 | 0.34 | 0.40 | 0.40 | 0.43 | 0.60* | 0.60* | 0.44 | 0.52* |
| | SN | -0.28 | -0.23 | -0.43 | -0.37 | -0.08 | -0.71** | -0.55* | -0.39 | 0.02 |
| | SP | -0.06 | 0.16 | 0.02 | -0.25 | -0.14 | -0.22 | -0.20 | -0.14 | 0.06 |
| | SK | -0.07 | 0.10 | 0.08 | -0.12 | -0.08 | -0.12 | -0.09 | -0.10 | 0.08 |
| | CGP | 0.29 | 0.24 | 0.29 | 0.38 | 0.36 | 0.14 | 0.23 | 0.34 | 0.53* |
| | CRF | -0.06 | 0.06 | -0.02 | 0.09 | 0.05 | 0.06 | 0.09 | 0.06 | -0.19 |
| | CWS | -0.26 | -0.43 | -0.29 | -0.33 | -0.38 | -0.34 | -0.41 | -0.53* | -0.01 |
| SN | Year | -0.62* | -0.47 | -0.46 | -0.40 | -0.39 | -0.89** | -0.71** | -0.92** | 0.53* |
| SP | Year | -0.65** | -0.26 | -0.60* | -0.69** | -0.54* | -0.47 | -0.44 | -0.43 | -0.44 |
| SK | Year | -0.57* | -0.40 | -0.43 | -0.34 | -0.35 | -0.33 | -0.36 | -0.44 | -0.32 |
| *Pearl millet (1987-2005)* | | | | | | | | | | |
| GY | Year | 0.77* | 0.35 | 0.01 | 0.52 | 0.68* | 0.40 | 0.59 | 0.19 | 0.82* |
| | SN | 0.46 | 0.17 | 0.68* | 0.60 | 0.58 | 0.65* | 0.62 | 0.69* | 0.53 |
| | SP | -0.39 | 0.53 | 0.29 | 0.01 | 0.14 | 0.56 | 0.41 | 0.66* | -0.36 |
| | SK | 0.25 | 0.57 | 0.20 | 0.66* | 0.68* | 0.39 | 0.39 | 0.52 | 0.26 |
| | CGP | 0.11 | 0.37 | 0.66* | 0.41 | 0.09 | 0.65* | 0.45 | 0.52 | 0.13 |
| | CRF | 0.41 | 0.50 | 0.51 | 0.86** | 0.72* | 0.70* | 0.61 | 0.65* | 0.65* |
| | CWS | -0.38 | 0.23 | -0.30 | -0.65* | -0.54 | -0.13 | 0.04 | 0.03 | -0.66* |
| SN | Year | 0.42 | 0.39 | 0.45 | 0.57 | 0.46 | 0.76* | 0.39 | 0.56 | 0.45 |
| SP | Year | -0.34 | 0.65* | -0.25 | -0.57 | -0.33 | -0.36 | -0.45 | -0.19 | -0.44 |
| SK | Year | 0.52 | 0.84** | 0.65* | 0.77* | 0.87** | 0.76* | 0.47 | 0.52 | 0.41 |

**Table 7.** Relation between yield, soil nutrients, crop seasonal rainfall, CGP and CWS at Kovilpatti

In case of pearl millet at Kovilpatti, the grain yield had significant positive correlation with years under control, 20 kg N (FYM)/ha and FYM @ 5 t/ha; soil N under 20 kg N (urea) + 10 kg P/ha, 20 kg N (crop residue) + 20 kg N (urea)/ha and 40 kg N (urea) + 20 kg P + 25 kg ZnSO₄/ha; soil P under 40 kg N (urea) + 20 kg P + 25 kg ZnSO₄/ha and soil K under 20 kg N (crop residue)/ha and 20 kg N (FYM)/ha. Pearl millet yield had significant correlation with CGP under 20 kg N (urea) + 10 kg P/ha and 20 kg N (crop residue) + 20 kg N (urea)/ha. It had significant positive correlation with crop seasonal rainfall under 20 kg N (crop residue)/ha, 20 kg N (FYM)/ha, 20 kg N (crop residue) + 20 kg N (urea)/ha, 40 kg N (urea) + 20 kg P + 25 kg ZnSO₄/ha and FYM @ 5 t/ha; and significant negative correlation with CWS under 20 kg N (crop residue)/ha and FYM @ 5 t/ha. Soil N had a significant positive correlation with years under 20 kg N (crop residue) + 20 kg N (urea)/ha compared to soil P under 40 kg N (urea) + 20 kg P/ha; and soil K under 40 kg N (urea) + 20 kg P/ha, 20 kg N (crop residue)/ha, 20 kg N (FYM)/ha and 20 kg N (crop residue) + 20 kg N (urea)/ha.

### 4.2.4. Soybean experiments at Indore

The estimates of correlation between soybean yield, soil fertility of nutrients and monthly rainfall are given in Table 8. The soybean yield had a significant and positive correlation with uptake N under all the 9 treatments. It ranged from 0.90** for application of 20 kg N (urea) + 13 kg P + FYM @ 6 t/ha to 0.97** under control plot. The soybean yield was found to have a significant negative correlation with soil N (-0.56*) observed under FYM @ 6 t/ha over years. Similarly, the control yield had a significant negative correlation with crop growing period (-0.54*).

| Var1 | Var2 | T1 | T2 | T3 | T4 | T5 | T6 | T7 | T8 | T9 |
|------|------|------|------|------|------|------|------|------|------|------|
| GY | Jun | 0.13 | 0.06 | 0.01 | -0.01 | -0.05 | 0.03 | -0.01 | 0.01 | 0.04 |
| GY | Jul | -0.27 | -0.04 | 0.02 | 0.01 | 0.01 | 0.06 | -0.07 | 0.01 | -0.11 |
| GY | Aug | -0.09 | 0.13 | 0.14 | 0.12 | 0.20 | 0.25 | 0.17 | 0.24 | 0.07 |
| GY | Sep | 0.07 | 0.28 | 0.30 | 0.32 | 0.34 | 0.38 | 0.36 | 0.36 | 0.34 |
| GY | Oct | 0.18 | 0.15 | 0.15 | 0.16 | 0.13 | 0.16 | 0.05 | 0.08 | 0.05 |
| GY | CGP | -0.54* | -0.40 | -0.43 | -0.42 | -0.37 | -0.25 | -0.33 | -0.28 | -0.37 |
| GY | SN | 0.40 | 0.20 | -0.01 | -0.16 | -0.10 | -0.48 | -0.47 | -0.56* | -0.49 |
| GY | SP | 0.05 | 0.17 | 0.21 | 0.18 | 0.18 | -0.23 | 0.30 | 0.17 | 0.23 |
| GY | SK | -0.24 | -0.20 | -0.25 | -0.49 | -0.28 | -0.11 | -0.21 | -0.26 | -0.09 |
| GY | SS | -0.16 | -0.11 | 0.01 | 0.06 | 0.01 | 0.08 | -0.02 | -0.03 | -0.23 |

**Table 8.** Relation between yield, soil nutrients, crop seasonal rainfall, and CGP at Indore

### 4.2.5. Groundnut experiments at Anantapur

The estimates of correlation between groundnut pod yield, soil fertility of nutrients and monthly rainfall are given in Table 9. The groundnut pod yield had a better correlation with July rainfall in the range of 0.22 to 0.53 compared to other months in different years. The pod yield attained by all the 9 fertilizer treatments was found to decrease over years as indicated by the negative correlation. The correlations were found to be non-significant over years although they indicated the likely positive or negative trends or effects on the yield. It is observed that yield had a better correlation with July and September rainfall in T9; August rainfall in T8; November rainfall in T3. Maximum pod yield decrease was observed under T4 over years.

| Var 1 | Var 2 | T1 | T2 | T3 | T4 | T5 | T6 | T7 | T8 | T9 |
|-------|-------|------|-------|-------|-------|-------|-------|-------|-------|-------|
| Yield | Jul | 0.34 | 0.42 | 0.41 | 0.22 | 0.48 | 0.28 | 0.45 | 0.43 | 0.53 |
| Yield | Aug | 0.06 | 0.10 | 0.08 | 0.03 | 0.03 | 0.10 | 0.11 | 0.15 | 0.07 |
| Yield | Sep | 0.17 | 0.19 | 0.19 | 0.18 | 0.26 | 0.18 | 0.21 | 0.21 | 0.26 |
| Yield | Oct | 0.04 | -0.04 | -0.03 | 0.08 | -0.09 | 0.01 | -0.04 | -0.05 | -0.07 |
| Yield | Nov | 0.18 | 0.17 | 0.21 | 0.20 | 0.14 | 0.17 | 0.15 | 0.20 | 0.16 |
| Yield | CGP | 0.04 | -0.03 | -0.02 | -0.03 | -0.05 | -0.07 | -0.07 | -0.08 | -0.04 |
| Yield | Year | -0.53 | -0.45 | -0.48 | -0.54 | -0.42 | -0.46 | -0.40 | -0.45 | -0.37 |

**Table 9.** Relation between yield, monthly rainfall, and CGP over years at Anantapur

## 4.3. Regression model of yield through soil nutrients and rainfall

Multiple regression models for yield attained by each treatment owing to simultaneous influence of crop seasonal rainfall, soil N, P and K nutrients were calibrated and the regression coefficients of variables along with coefficient of determination ($R^2$) and standard error (SE) are given in Table 10 for Bangalore, Table 11 for Akola and Table 12 for Kovilpatti.

### 4.3.1. Regression model of finger millet yield at Bangalore

At Bangalore, the yield predictability ($R^2$) was in the range of 11% for FYM @ 10 t/ha to 52% for the yield attained with application of 100% NPK over years based on the model. The standard error ranged from 435 to 717 kg/ha in the 25 year study. Based on the model, the crop seasonal rainfall had a positive effect under application of FYM @ 10 t/ha, FYM @ 10 t/ha + 50% NPK and FYM @ 10 t/ha + 100% NPK. The effect of soil N under control, FYM @ 10 t/ha + 50% NPK and 100% NPK; soil K under FYM @ 10 t/ha + 50% NPK and FYM @ 10 t/ha + 100% NPK were positive. The analysis indicated that soil P had a significant negative effect on finger millet yield attained by FYM @ 10 t/ha + 50% NPK. Similarly, soil K had a significant negative effect on the yield attained by 100% NPK application as given in Table 10.

| Treatment | Regression model | $R^2$ | SE (kg/ha) |
|---|---|---|---|
| T1 | $Y = -2579 - 0.48$ (CRF) $+ 65.89$ (SN) $- 0.19$ (SN)$^2$ $-$ $62.94$ (SP) $+ 4.08$ (SP)$^2 - 78.42$ (SK) $+ 0.69$ (SK)$^2$ | 0.27 | 435 |
| T2 | $Y = 13534 + 0.08$ (CRF) $- 82.60$ (SN) $+ 0.21$ (SN)$^2$ $-$ $3.83$ (SP) $+ 0.02$ (SP)$^2 - 73.68$ (SK) $+ 0.43$ (SK)$^2$ | 0.11 | 646 |
| T3 | $Y = 1341 + 0.21$ (CRF) $+ 4.38$ (SN) $+ 0.01$ (SN)$^2$ $-$ $96.25$ (SP) $+ 0.89^*$ (SP)$^2 + 47.40$ (SK) $- 0.21$ (SK)$^2$ | 0.23 | 660 |
| T4 | $Y = 26227 + 0.04$ (CRF) $- 213.30$ (SN) $+ 0.53$ (SN)$^2$ $-$ $102.16$ (SP) $+ 0.85$ (SP)$^2 + 12.26$ (SK) $- 0.01$ (SK)$^2$ | 0.30 | 717 |
| T5 | $Y = 12422^{**} - 0.47$ (CRF) $+ 54.92$ (SN) $- 0.18$ (SN)$^2$ $-$ $101.2$ (SP) $+ 0.74$ (SP)$^2 - 277.05^*$ (SK) $+ 1.72^*$ (SK)$^2$ | $0.52^*$ | 679 |

**Table 10.** Regression model of finger millet yield through rainfall and soil nutrients at Bangalore

## 4.3.2. Regression models of cotton and green gram yield at Akola

At Akola, the model of green gram yield through rainfall of June to September, crop duration, soil N, P and K gave a predictability in the range of 0.53 for 25 kg N (FYM) + 25 kg N (urea) + 25 kg P/ha to 0.73 for control (Table 11). The model gave a standard error of yield in the range of 108 kg/ha for control to 161 kg/ha for 25 kg N (FYM) + 25 kg N (urea) + 25 kg P/ha. June and September rainfall had a positive effect, while July had a negative effect on yield attained by all treatments except 25 kg N (*Leucaena*) + 25 kg N (urea) + 25 kg P/ha and 25 kg N (*Leucaena*) + 25 kg P/ha. August rainfall had a positive effect on yield attained by control, 25 kg N/ha (FYM), 25 kg N (*Leucaena*) + 25 kg N (urea) + 25 kg P/ha and 25 kg N (FYM) + 25 kg N (urea) + 25 kg P/ha. Soil N had a negative effect on yield of all treatments, while soil P had a positive effect for all treatments except 25 kg N + 12.5 kg P/ha and 25 kg N/ha (*Leucaena*). Soil K had a negative effect on yield of all treatments except 25 kg N/ha (FYM), 25 kg N (*Leucaena*) + 25 kg N (urea) + 25 kg P/ha and 25 kg N (FYM) + 25 kg N (urea) + 25 kg P/ha. The model indicated that effect of June rainfall on yield of all treatments except 25 kg N/ha (FYM), 25 kg N (*Leucaena*) + 25 kg N (urea) + 25 kg P/ha and 25 kg N (FYM) + 25 kg N (urea) + 25 kg P/ha; September rainfall and crop duration on yield of 25 kg N/ha (FYM); and soil N for control were significant.

In case of cotton at Akola (Table 11), the model of yield through rainfall of June to November, crop duration, soil N, P and K gave predictability of 0.36 for 25 kg N/ha (FYM) to 0.47 for 25 kg N/ha (*Leucaena*). The standard error ranged from 286 kg/ha for control to 516 kg/ha for 25 kg N (FYM) + 25 kg N (urea) + 25 kg P/ha. July rainfall had a positive effect, while November rainfall had negative effect on yield of all treatments. June rainfall had a positive effect on yield of control, 25 kg N + 12.5 kg P/ha, 25 kg N/ha (*Leucaena*) and 25 kg N/ha (FYM); August rainfall had a positive effect in all treatments except 25 kg N/ha (*Leucaena*); and October rainfall had a positive effect on yield attained by 50 kg N + 25 kg/ha, 25 kg N (*Leucaena*) + 25 kg N (urea) + 25 kg P/ha and 25 kg N (FYM) + 25 kg N (urea) + 25 kg P/ha. September rainfall had a negative effect on yield of all treatments except control and 25 kg N + 12.5 kg P/ha. Crop duration had a positive effect on yield attained by all treatments except control. Soil N had a positive effect on yield attained by all treatments except control, while soil P had a negative effect on yield attained by all treatments. Soil K had a positive effect on yield of 50 kg N + 25 kg/ha, 25 kg N + 12.5 kg P/ha, 25 kg N/ha (*Leucaena*) and 25 kg N/ha (FYM).

| Variable | T1 | T2 | T3 | T4 | T5 | T6 | T7 | T8 |
|---|---|---|---|---|---|---|---|---|
| *Green gram* | | | | | | | | |
| α | 223 | 321 | 526 | 606 | -379 | -64 | -142 | 256 |
| β1 (CGP) | 1.69 | 4.02 | 2.49 | 2.54 | 7.39* | 3.03 | 3.81 | 3.38 |
| β2 (Jun) | 1.07* | 1.49* | 1.41* | 1.68* | 1.11 | 1.30 | 1.24 | 1.55* |
| β3 (Jul) | -0.05 | -0.10 | -0.13 | -0.14 | -0.02 | 0.01 | -0.13 | 0.16 |
| β4 (Aug) | 0.16 | -0.09 | -0.12 | -0.29 | 0.07 | 0.05 | 0.22 | -0.12 |
| β5 (Sep) | 0.60 | 0.92 | 0.93 | 0.96 | 1.25* | 0.79 | 0.90 | 0.64 |
| β6 (SN) | -1.33* | -1.50 | -1.50 | -1.27 | -1.12 | -0.96 | -0.40 | -1.46 |
| β7 (SP) | 3.07 | 1.52 | -0.66 | -2.59 | 1.50 | 2.71 | 2.99 | 2.31 |
| β8 (SK) | -0.13 | -0.40 | -0.60 | -0.84 | 0.52 | 0.47 | 0.19 | -0.15 |
| R² | 0.73* | 0.62 | 0.67* | 0.65* | 0.68* | 0.64* | 0.53 | 0.67* |
| SE (kg/ha) | 108 | 147 | 131 | 142 | 142 | 145 | 161 | 152 |
| SYI (%) | 27.7 | 36.6 | 34.3 | 32.8 | 36.0 | 36.8 | 40.3 | 37.3 |
| *Cotton* | | | | | | | | |
| α | 853 | 172 | 38 | 435 | 725 | 1517 | 2337 | 1270 |
| β1 (CGP) | -1.58 | 2.44 | 0.83 | 2.40 | 1.64 | 2.18 | 1.35 | 3.53 |
| β2 (Jun) | 0.63 | -0.28 | 0.42 | 1.18 | 0.31 | -0.40 | -0.57 | -0.09 |
| β3 (Jul) | 1.00 | 1.44 | 1.10 | 0.97 | 1.29 | 1.55 | 2.14 | 1.32 |
| β4 (Aug) | 0.43 | 0.33 | 0.54 | -0.27 | 0.34 | 0.09 | 0.50 | 0.003 |
| β5 (Sep) | 0.97 | -1.64 | 0.19 | -0.60 | -0.71 | -1.62 | -1.27 | -1.63 |
| β6 (Oct) | -0.97 | 0.13 | -1.03 | -0.76 | -0.68 | 0.26 | 0.45 | -0.07 |
| β7 (Nov) | -0.50 | -1.90 | -1.62 | -2.95 | -2.62 | -2.61 | -1.16 | -3.55 |
| β8 (SN) | 0.03 | -2.61 | -0.36 | -1.19 | -0.99 | -3.26 | -3.17 | -2.80 |
| β9 (SP) | -14.40 | -3.05 | -13.42 | -23.58 | -18.27 | -11.08 | -15.63 | -15.31 |
| β10 (SK) | -0.29 | 2.16 | 1.92 | 1.75 | 0.79 | -0.19 | -2.01 | -0.18 |
| R² | 0.42 | 0.38 | 0.37 | 0.47 | 0.36 | 0.41 | 0.38 | 0.41 |
| SE (kg/ha) | 286 | 431 | 403 | 352 | 448 | 459 | 516 | 444 |
| SYI (%) | 10.8 | 14.0 | 12.4 | 14.3 | 12.0 | 14.6 | 15.2 | 13.7 |

**Table 11.** Regression models of yield through rainfall and soil nutrients at Akola

## 4.3.3. Regression models of sorghum and pearl millet yield at Kovilpatti

In sorghum at Kovilpatti, the model of 20 kg N (FYM)/ha had minimum R² of 0.26 with standard error of 573 kg/ha, while model of 20 kg N (crop residue) + 20 kg N (urea)/ha had maximum predictability of 0.77 with an error of 464 kg/ha. Soil N significantly influenced the yield attained under 20 kg N (crop residue) + 20 kg N (urea)/ha and 10 kg N (FYM) + 10 kg N (urea)/ha, while both soil N and CGP significantly influenced yield attained with 40 kg N (urea) + 20 kg P + 25 kg ZnSO₄/ha. The models indicated that CGP had a positive influence on yield of all treatments except control and 40 kg N (urea) + 20 kg P/ha. Both crop seasonal rainfall and CWS had negative influence on yield of all treatments except 20 kg N (crop residue)/ha. Soil N negatively influenced the yield attained by all treatments except 20 kg N (crop residue)/ha and FYM @ 5 t/ha, while soil P had positive influence on yield attained by all treatments except 20 kg N (crop residue)/ha, 20 kg N (FYM)/ha and 10 kg N (FYM) + 10 kg N (urea)/ha. Soil K had a positive influence on yield attained by all treatments except control and 40 kg N (urea) + 20 kg P/ha (Table 12).

| Treatment | Multiple regression model | $R^2$ | SE (kg/ha) | SYI (%) |
|---|---|---|---|---|
| *Sorghum (1987-2005)* | | | | |
| Control | Y = 3383 − 8.52 (CGP) − 1.90 (CRF) − 2194.27 (CWS) − 16.99 (SN) + 227.71 (SP) − 2.34 (SK) | 0.41 | 474 | 0.08 |
| 40 kg N (urea) + 20 kg P/ha | Y = 1499 − 2.62 (CGP) − 1.78 (CRF) − 2317.53 (CWS) − 1.31 (SN) + 108.14 (SP) − 0.05 (SK) | 0.28 | 570 | 0.25 |
| 20 kg N (urea) + 10 kg P/ha | Y = 1296 + 4.79 (CGP) − 0.61 (CRF) − 985.85 (CWS) − 7.48 (SN) + 18.80 (SP) + 0.07 (SK) | 0.32 | 506 | 0.17 |
| 20 kg N (crop residue)/ha | Y = −1618 + 26.39 (CGP) + 1.36 (CRF) + 530.11 (CWS) − 5.52 (SN) − 124.84 (SP) + 1.34 (SK) | 0.37 | 555 | 0.15 |
| 20 kg N (FYM)/ha | Y = 182 + 7.95 (CGP) − 0.81 (CRF) − 1373.72 (CWS) + 3.0 (SN) − 4.64 (SP) + 0.02 (SK) | 0.26 | 573 | 0.15 |
| 20 kg N (crop residue) + 20 kg N (urea)/ha | Y = 2767 + 14.52 (CGP) − 0.36 (CRF) − 1606.54 (CWS) − 35.81 ** (SN) + 97.49 (SP) + 0.47 (SK) | 0.77* | 464 | 0.21 |
| 10 kg N (FYM) + 10 kg N (urea)/ha | Y = 2753 + 13.83 (CGP) − 0.41 (CRF) − 1430.45 (CWS) − 38.21 ** (SN) − 39.22 (SP) + 2.62 (SK) | 0.74* | 444 | 0.27 |
| 40 kg N (urea) + 20 kg P + 25 kg ZnSO₄/ha | Y = −200 + 29.0 * (CGP) − 0.38 (CRF) − 1820.54 (CWS) − 28.28 ** (SN) + 83.34 (SP) + 2.22 (SK) | 0.71* | 458 | 0.30 |
| FYM @ 5 t/ha | Y = −23 + 9.46 (CGP) − 2.71 (CRF) − 2556.99 (CWS) + 9.1 (SN) + 14.73 (SP) + 0.7 (SK) | 0.38 | 757 | 0.20 |
| *Pearl millet (1987-2005)* | | | | |
| Control | Y = 2723 − 13.15 (CGP) + 0.21 (CRF) − 239.81 (CWS) + 6.22 (SN) − 212.62 * (SP) − 0.78 (SK) | 0.95* | 84 | 0.33 |
| 40 kg N (urea) + 20 kg P/ha | Y = − 4079 + 13.93 (CGP) − 0.78 (CRF) + 1373.54 (CWS) + 20.09 (SN) + 453.74 (SP) − 8.73 (SK) | 0.81 | 268 | 0.53 |
| 20 kg N (urea) + 10 kg P/ha | Y = − 653 + 2.64 (CGP) − 0.60 (CRF) − 580.17 (CWS) + 3.43 (SN) + 87.74 (SP) + 0.49 (SK) | 0.73 | 189 | 0.49 |
| 20 kg N (crop residue)/ha | Y = − 422 + 6.58 (CGP) + 0.41 (CRF) − 590.53 (CWS) − 6.69 (SN) + 22.39 (SP) + 2.66 * (SK) | 0.98* | 52 | 0.44 |
| 20 kg N (FYM)/ha | Y = 1787 − 6.26 (CGP) + 0.77 (CRF) + 4.18 (CWS) + 9.02 (SN) − 162.38 (SP) − 1.72 (SK) | 0.77 | 194 | 0.42 |
| 20 kg N (crop residue) + 20 kg N (urea)/ha | Y = − 2472 + 9.79 (CGP) + 0.74 (CRF) − 36.16 (CWS) − 2.15 (SN) + 140.34 (SP) + 2.14 (SK) | 0.97* | 101 | 0.53 |
| 10 kg N (FYM) + 10 kg N (urea)/ha | Y = − 1049 + 1.38 (CGP) + 1.30 (CRF) + 43.84 (CWS) − 4.47 (SN) + 100.62 (SP) + 1.95 (SK) | 0.91* | 170 | 0.63 |
| 40 kg N (urea) + 20 kg P + 25 kg ZnSO₄/ha | Y = − 869 + 11.16 (CGP) − 0.15 (CRF) + 869.91 (CWS) + 20.78 (SN) − 43.19 (SP) − 5.25 (SK) | 0.90* | 157 | 0.62 |
| FYM @ 5 t/ha | Y = 273 + 2.62 (CGP) + 0.07 (CRF) − 17.55 (CWS) + 6.36 (SN) − 74.04 (SP) − 0.41 (SK) | 0.92* | 103 | 0.42 |

\* and ** indicate significance at p < 0.05 and p < 0.01 level respectively

**Table 12.** Regression models of yield through rainfall, CGP, CWS and soil nutrients at Kovilpatti

In case of pearl millet at Kovilpatti, the model of 20 kg N (urea) + 10 kg P/ha had minimum $R^2$ of 0.73 with an error of 189 kg/ha, while the model of 20 kg N (crop residue)/ha had maximum $R^2$ of 0.98 with an error of 52 kg/ha. Soil P had a significant influence on control yield, while soil K had significant influence on yield attained by 20 kg N (crop residue)/ha. The models indicated that CGP positively influenced the yield of all treatments except control and 20 kg N (FYM)/ha. The crop seasonal rainfall had positive influence on yield attained by all treatments except 40 kg N (urea) + 20 kg P/ha, 20 kg N (urea) + 10 kg P/ha and 40 kg N (urea) + 20 kg P + 25 kg ZnSO₄/ha. The CWS negatively influenced the yield of all treatments except 40 kg N (urea) + 20 kg P/ha, 20 kg N (FYM)/ha, 10 kg N (FYM) + 10 kg N (urea)/ha and 40 kg N (urea) + 20 kg P + 25 kg ZnSO₄/ha. Soil N positively influenced the yield attained by all treatments except 20 kg N (crop residue)/ha, 20 kg N (crop residue) + 20 kg N (urea)/ha and 10 kg N (FYM) + 10 kg N (urea)/ha. Soil P positively influenced yield of all treatments except control, 20 kg N (FYM)/ha, 40 kg N (urea) + 20 kg P + 25 kg ZnSO₄/ha and FYM @ 5 t/ha, while soil K negatively influenced yield of all treatments except 20 kg N (urea) + 10 kg P/ha, 20 kg N (crop residue)/ha, 20 kg N (crop residue) + 20 kg N (urea)/ha and 10 kg N (FYM) + 10 kg N (urea)/ha (Table 12).

### 4.3.4. Regression model of soybean yield at Indore

Based on the regression model of soybean yield as a function of monthly rainfall received during June to October, crop growing period, soil N, P, K and S, crop growing period had a significant effect on yield attained by all treatments except control and crop residue @ 5 t/ha. July rainfall had a significant effect on yield attained by 20 kg N (urea) + 13 kg P + FYM @ 6 t/ha and 20 kg N (urea) + 13 kg P + FYM @ 5 t/ha, while August rainfall had a significant effect on yield attained by all treatments except control and crop residue @ 5 t/ha. September rainfall had a significant effect on yield attained by all treatments except control based on the model. Among soil nutrients, soil N and P had a significant influence on yield attained by 20 kg N (urea) + 13 kg P + FYM @ 5 t/ha. The $R^2$ ranged from 0.81 for control to 0.98 for 20 kg N (urea) + 13 kg P + FYM @ 5 t/ha, while standard error ranged from 130 kg/ha for 20 kg N (urea) + 13 kg P + FYM @ 5 t/ha to 320 kg/ha for control. The regression model indicated that maximum rate of change in yield of 31.07 for a unit change in soil N and 1.81 for soil K occurred in control compared to a minimum of −21.23 in 30 kg N (urea) + 20 kg P/ha and −2.16 in 40 kg N (urea) + 26 kg P/ha for the two soil nutrients respectively. The yield attained by 20 kg N (urea) + 13 kg P + FYM @ 6 t/ha had a minimum rate of change for soil P and maximum rate of change for soil Sulphur, while 20 kg N (urea) + 13 kg P/ha had maximum rate of change for soil P and control had a minimum rate of change for soil S (Table 13).

### 4.3.5. Regression model of groundnut pod yield at Anantapur

The regression models of yield through monthly rainfall gave $R^2$ in the range of 0.18 for 100% N (groundnut shells ~ 20 kg N/ha) and 100% N (groundnut shells ~ 20 kg N/ha) + 50% NPK (10–20–20 kg/ha) to 0.43 for Farmers practice (FYM @ 5 t/ha). The estimate of standard error based on regression models was in the range of 315 kg/ha for control and Farmers practice (FYM @ 5 t/ha) to 397 kg/ha for 100% NPK (20–40–40 kg/ha). Among the effects of rainfall on yield, the rainfall received in July, September, October and November had a positive influence, while rainfall received in August had negative influence on the yield attained by all treatments. July rainfall had significant influence on yield attained by all treatments except 100% N (groundnut shells ~ 20 kg N/ha) and 100% N (groundnut shells ~ 20 kg N/ha) + 50% NPK (10–20–20 kg/ha); while November rainfall had significant influence on yield attained by 100% NPK (20–40–40 kg/ha), 50% NPK (10–20–20 kg/ha), 50% N (FYM ~ 10 kg N/ha) + 50% NPK (10–20–20 kg/ha), 100% NPK (20–40–40 kg/ha) + $ZnSO_4$ @ 25 kg/ha and Farmers practice (FYM @ 5 t/ha). The rate of change in yield of all treatments was positive and maximum for an unit change in November rainfall, followed by July, October and September, while it was negative for August. The analysis indicated that the model of farmers practice (FYM @ 5 t/ha) had maximum $R^2$ and minimum standard error, while the models of 100% NPK (20–40–40 kg/ha) and 100% N (groundnut shells ~ 20 kg N/ha) + 50% NPK (10–20–20 kg/ha) had minimum $R^2$ and maximum standard error (Table 14).

| Variable | T1 | T2 | T3 | T4 | T5 | T6 | T7 | T8 | T9 |
|---|---|---|---|---|---|---|---|---|---|
| α | -1926 | 4065 | 12409* | 9645** | 7728 | 5223* | 9248** | 8225** | 8794** |
| β1(CGP) | -14.6 | -77.9** | -77.8** | -61.1* | -86.5* | -77.0** | -94.5** | -69.9** | -53.5 |
| β2 (Jun) | -6.83 | 3.68 | 2.04 | -0.02 | 5.24 | 4.87 | 2.75 | 3.54 | -2.19 |
| β3 (Jul) | 0.15 | -1.63 | -2.49 | -0.78 | -2.98 | -3.48* | -2.81** | -1.47 | -0.81 |
| β4 (Aug) | 0.35 | 4.22* | 6.72* | 4.02* | 7.33* | 7.36** | 6.71** | 5.29* | 3.44 |
| β5 (Sep) | 1.65 | 5.07** | 7.01** | 5.61** | 7.93* | 7.68** | 8.12** | 6.11** | 5.34* |
| β6 (Oct) | 17.90 | -1.24 | -0.34 | 4.95 | -10.65 | -7.30 | -3.21 | -3.71 | 7.98 |
| β7 (SN) | 31.07 | 13.81 | -21.23 | -2.79 | -2.36 | 4.45 | -3.44* | -1.54 | -3.33 |
| β8 (SP) | -7.17 | 13.30 | -33.95 | -27.40 | -20.25 | -40.98 | -28.15* | -18.94 | -38.17 |
| β9 (SK) | 1.81 | 1.64 | -0.48 | -2.16 | 1.17 | 1.27 | 1.64 | -0.47 | -0.38 |
| β10 (SS) | -125.04 | 16.54 | 46.23 | -41.05 | 76.68 | 87.68 | 37.85 | 5.45 | -73.22 |
| $R^2$ | 0.81 | 0.94* | 0.95* | 0.94* | 0.96** | 0.96** | 0.98** | 0.95* | 0.91 |
| SE (kg/ha) | 320 | 215 | 216 | 238 | 191 | 187 | 130 | 225 | 290 |
| SYI (kg/ha) | 29.4 | 43.3 | 48.0 | 50.8 | 55.5 | 58.8 | 51.1 | 50.4 | 41.2 |

**Table 13.** Regression model of soybean yield through rainfall, CGP and soil nutrients

| Treatment | Regression model | $R^2$ | SE (kg/ha) | SYI (%) |
|---|---|---|---|---|
| Control | Y = 376 + 1.30* (Jul) − 0.09 (Aug) + 0.58 (Sep) + 1.02 (Oct) + 2.97 (Nov) | 0.24 | 315 | 27.6 |
| 100% NPK (20–40–40 kg/ha) | Y = 433 + 1.95* (Jul) − 0.04 (Aug) + 0.82 (Sep) + 0.99 (Oct) + 4.01* (Nov) | 0.30 | 397 | 33.6 |
| 50% NPK (10–20–20 kg/ha) | Y = 423 + 1.76* (Jul) − 0.09 (Aug) + 0.80 (Sep) + 0.96 (Oct) + 4.04* (Nov) | 0.32 | 351 | 34.2 |
| 100% N (GS ~ 20 kg N/ha) | Y = 473 + 1.03 (Jul) − 0.11 (Aug) + 0.80 (Sep) + 1.16 (Oct) + 3.30 (Nov) | 0.18 | 357 | 32.7 |
| 50% N (FYM ~ 10 kg N/ha) | Y = 477* + 1.86* (Jul) − 0.46 (Aug) + 1.13 (Sep) + 0.54 (Oct) + 3.23 (Nov) | 0.36 | 330 | 35.3 |
| 100% N (GS ~ 20 kg N/ha) + 50% NPK (10–20–20 kg/ha) | Y = 527* + 1.11 (Jul) + 0.14 (Aug) + 0.73 (Sep) + 0.72 (Oct) + 3.11 (Nov) | 0.18 | 363 | 34.0 |
| 50% N (FYM ~ 10 kg N/ha) + 50% NPK (10–20–20 kg/ha) | Y = 432 + 1.97* (Jul) − 0.03 (Aug) + 0.82 (Sep) + 0.99 (Oct) + 3.75* (Nov) | 0.32 | 373 | 34.9 |
| 100% NPK (20–40–40 kg/ha) + ZnSO4 @ 25 kg/ha | Y = 399 + 1.94* (Jul) + 0.24 (Aug) + 0.88 (Sep) + 0.94 (Oct) + 4.55* (Nov) | 0.34 | 383 | 35.1 |
| Farmers practice (FYM @ 5 t/ha) | Y = 366 + 2.08** (Jul) − 0.32 (Aug) + 1.02 (Sep) + 0.84 (Oct) + 3.69* (Nov) | 0.43* | 315 | 33.5 |

**Table 14.** Regression models of groundnut pod yield through monthly rainfall at Anantapur

## 4.4. Sustainability yield index of treatments under different rainfall situations

Using the mean yield of treatments over years under different crop seasonal rainfall situations, standard error, and maximum mean yield attained by any treatment over years, the estimates of sustainability yield index of treatments were derived for different crop seasonal rainfall situations and are given in Table 15.

| Treatment | Mean yield (kg/ha) under different rainfall situations | | | | SE (kg/ha) | Sustainability Yield Index (%) | | | |
|---|---|---|---|---|---|---|---|---|---|
| Rainfall (mm) | <500 | 500-750 | 750-1000 | 1000-1250 | | <500 | 500-750 | 750-1000 | 1000-1250 |
| *Bangalore : Finger millet : 1984 – 2008* | | | | | | | | | |
| T1 | 532 | 706 | 444 | 172 | 435 | 3.1 | 8.6 | 0.3 | -8.3 |
| T2 | 1883 | 2567 | 2572 | 2283 | 646 | 39.1 | 60.6 | 60.8 | 51.7 |
| T3 | 2206 | 2941 | 3082 | 2880 | 660 | 48.8 | 72.0 | 76.5 | 70.1 |
| T4 | 2239 | 3409 | 3176 | 3183 | 717 | 48.1 | 85.0 | 77.6 | 77.9 |
| T5 | 1415 | 2197 | 1662 | 1315 | 679 | 23.2 | 47.9 | 31.0 | 20.1 |
| *Akola : Green gram : 1987-2007* | | | | | | | | | |
| T1 | 258 | 316 | 392 | 580 | 108 | 16.5 | 22.8 | 31.2 | 51.9 |
| T2 | 330 | 435 | 558 | 610 | 147 | 20.1 | 31.6 | 45.2 | 50.9 |
| T3 | 303 | 407 | 500 | 600 | 131 | 18.9 | 30.3 | 40.6 | 51.5 |
| T4 | 318 | 392 | 488 | 650 | 142 | 19.3 | 27.4 | 38.1 | 55.8 |
| T5 | 328 | 398 | 553 | 680 | 142 | 20.4 | 28.1 | 45.1 | 59.1 |
| T6 | 329 | 427 | 539 | 710 | 145 | 20.2 | 31.0 | 43.3 | 62.1 |
| T7 | 365 | 469 | 623 | 670 | 161 | 22.4 | 33.8 | 50.8 | 55.9 |
| T8 | 356 | 435 | 534 | 775 | 152 | 22.4 | 31.0 | 42.0 | 68.5 |
| *Akola : Cotton : 1987-2007* | | | | | | | | | |
| T1 | 233 | 524 | 551 | 545 | 286 | -2.8 | 12.5 | 13.9 | 13.6 |
| T2 | 309 | 765 | 760 | 800 | 431 | -6.4 | 17.5 | 17.2 | 19.3 |
| T3 | 268 | 709 | 693 | 725 | 403 | -7.1 | 16.0 | 15.2 | 16.9 |
| T4 | 289 | 681 | 695 | 670 | 352 | -3.3 | 17.2 | 17.9 | 16.6 |
| T5 | 326 | 706 | 785 | 700 | 448 | -6.4 | 13.5 | 17.7 | 13.2 |
| T6 | 349 | 795 | 821 | 800 | 459 | -5.8 | 17.6 | 18.9 | 17.9 |
| T7 | 371 | 851 | 948 | 795 | 516 | -7.6 | 17.5 | 22.6 | 14.6 |
| T8 | 345 | 745 | 795 | 775 | 444 | -5.2 | 15.8 | 18.4 | 17.3 |
| *Kovilpatti : Sorghum : 1987-2005* | | | | | | | | | |
| Rainfall (mm) | <250 | 250-500 | 500-750 | | | <250 | 250-500 | 500-750 | |
| T1 | 250 | 785 | 493 | | 474 | -36.4 | 12.7 | 1.4 | |
| T2 | 488 | 1195 | 997 | | 570 | -13.4 | 25.5 | 31.8 | |
| T3 | 448 | 965 | 765 | | 506 | -9.5 | 18.7 | 19.3 | |
| T4 | 354 | 879 | 870 | | 555 | -32.6 | 13.2 | 23.4 | |
| T5 | 290 | 923 | 756 | | 573 | -45.9 | 14.3 | 13.6 | |
| T6 | 317 | 1102 | 944 | | 464 | -23.9 | 26.0 | 35.7 | |
| T7 | 394 | 1246 | 1072 | | 444 | -8.1 | 32.7 | 46.8 | |
| T8 | 480 | 1339 | 1168 | | 458 | 3.5 | 35.9 | 52.9 | |
| T9 | 436 | 1136 | 619 | | 757 | -52.1 | 15.4 | -10.3 | |

| Treatment | Mean yield (kg/ha) under different rainfall situations | | | | SE (kg/ha) | Sustainability Yield Index (%) | | | |
|---|---|---|---|---|---|---|---|---|---|
| *Kovilpatti : Pearl millet : 1987-2005* | | | | | | | | | |
| T1 | 290 | 531 | 454 | | 84 | 41.0 | 51.7 | 30.3 | |
| T2 | 408 | 622 | 721 | | 268 | 27.9 | 41.0 | 37.1 | |
| T3 | 463 | 543 | 673 | | 189 | 54.6 | 40.9 | 39.7 | |
| T4 | 253 | 575 | 624 | | 52 | 40.0 | 60.5 | 46.9 | |
| T5 | 358 | 578 | 585 | | 194 | 32.7 | 44.4 | 32.1 | |
| T6 | 265 | 646 | 743 | | 101 | 32.7 | 63.1 | 52.6 | |
| T7 | 419 | 837 | 807 | | 170 | 49.6 | 77.2 | 52.2 | |
| T8 | 502 | 667 | 825 | | 157 | 68.7 | 59.0 | 54.7 | |
| T9 | 336 | 619 | 564 | | 103 | 46.4 | 59.7 | 37.8 | |
| *Indore : Soybean : 1992-2006* | | | | | | | | | |
| T1 | 985 | 1617 | 1292 | 1137 | 320 | 39.3 | 54.8 | 40.3 | 25.2 |
| T2 | 1181 | 1835 | 1600 | 1708 | 215 | 57.2 | 68.4 | 57.3 | 46.0 |
| T3 | 1365 | 1902 | 1730 | 1950 | 216 | 68.0 | 71.2 | 62.7 | 53.4 |
| T4 | 1460 | 2041 | 1830 | 2065 | 238 | 72.3 | 76.1 | 65.9 | 56.3 |
| T5 | 1573 | 2175 | 1889 | 2225 | 191 | 81.8 | 83.8 | 70.3 | 62.6 |
| T6 | 1608 | 2257 | 1983 | 2386 | 187 | 84.1 | 87.4 | 74.4 | 67.7 |
| T7 | 1290 | 2122 | 1674 | 1964 | 130 | 68.6 | 84.1 | 63.9 | 56.5 |
| T8 | 1379 | 1980 | 1787 | 2133 | 225 | 68.3 | 74.1 | 64.7 | 58.8 |
| T9 | 1178 | 1906 | 1599 | 1693 | 290 | 52.5 | 68.2 | 54.2 | 43.2 |
| *Anantapur : Groundnut : 1985-2006* | | | | | | | | | |
| T1 | 726 | 738 | 966 | | 315 | 26.6 | 27.4 | 49.0 | |
| T2 | 879 | 941 | 1250 | | 397 | 31.2 | 35.3 | 64.2 | |
| T3 | 846 | 910 | 1112 | | 351 | 32.0 | 36.3 | 57.3 | |
| T4 | 857 | 854 | 1013 | | 357 | 32.3 | 32.2 | 49.4 | |
| T5 | 833 | 932 | 1088 | | 330 | 32.5 | 39.1 | 57.0 | |
| T6 | 883 | 848 | 1256 | | 363 | 33.6 | 31.5 | 67.2 | |
| T7 | 863 | 971 | 1182 | | 373 | 31.7 | 38.8 | 60.9 | |
| T8 | 879 | 960 | 1329 | | 383 | 32.1 | 37.5 | 71.2 | |
| T9 | 776 | 912 | 1087 | | 315 | 29.8 | 38.7 | 58.1 | |

**Table 15.** Sustainability of fertilizer treatments under different crop seasonal rainfall situations

## 4.4.1. Bangalore

At Bangalore, the SYI was in a range of 3.1 to 48.8% under < 500 mm; 8.6 to 85.0% under 500–750 mm; 0.3 to 77.6% under 750–1000 mm; and –8.3 to 77.9% under 1000–1250 mm of rainfall based on the standard error. The study indicated the superiority of FYM @ 10 t/ha under < 500 mm rainfall; while FYM @ 10 t/ha + 100% NPK was superior under 500–750, 750–1000 and 1000–1250 mm rainfall situations for attaining maximum mean yield and sustainability over years. Thus, based on the long term study, an efficient fertilizer treatment having a high sustainability has been identified for attaining maximum productivity. Conclusively, the results obtained from this long term study incurring huge expenditure provide very good conjunctive nutrient use options with good conformity for different rainfall conditions of rainfed semi-arid Alfisol for ensuring higher finger millet yield, maintaining higher SYI, and maintaining improved soil fertility.

### 4.4.2. Akola

At Akola, the SYI of each treatment was determined for green gram and cotton using mean yield of a treatment, standard error determined from the treatment-wise regression models and maximum yield of 910 kg/ha of green gram and 1910 kg/ha of cotton attained in the study. The SYI was determined under each of the 4 rainfall situations of < 500, 500–750, 750–1000 and 1000–1250 mm. Based on the regression model of green gram yield, 25 kg N (FYM) + 25 kg N (urea) + 25 kg P/ha had a maximum SYI of 22.4% under < 500 mm, 33.8% under 500–750 mm, 50.8% under 750–1000 mm, while 25 kg N (*Leucaena*) + 25 kg P/ha had a maximum of 68.5% under 1000–1250 mm rainfall. An increase in rainfall significantly increased the sustainability of treatments in green gram. This is evident from SYI range of 16.5 to 22.4% under < 500 mm; 22.8 to 33.8% under 500–750 mm; 31.2 to 50.8% under 750–1000 mm for control and 25 kg N (FYM) + 25 kg N (urea) + 25 kg P/ha respectively; and 50.9% for 50 kg N + 25 kg/ha to 68.5% for 25 kg N (*Leucaena*) + 25 kg P/ha under 1000–1250 mm rainfall situation.

At Akola, based on regression model of cotton yield through rainfall and soil nutrient variables, control had a maximum SYI of –2.8% under < 500 mm; while 25 kg N (*Leucaena*) + 25 kg N (urea) + 25 kg P/ha had 17.6% under 500–750; 25 kg N (FYM) + 25 kg N (urea) + 25 kg P/ha had 22.6% under 750–1000 mm; and 50 kg N + 25 kg/ha had 19.3% under 1000–1250 mm rainfall situation. In case of cotton also, an increase in rainfall increased the sustainability of treatments. This is evident from a SYI range of –7.6 (25 kg N (FYM) + 25 kg N (urea) + 25 kg P/ha) to –2.8% (control) under < 500 mm; 12.5 (control) to 17.6% (25 kg N (*Leucaena*) + 25 kg N (urea) + 25 kg P/ha) under 500–750 mm; 13.9 (control) to 22.6% (25 kg N (FYM) + 25 kg N (urea) + 25 kg P/ha) under 750–1000 mm; and 13.2% (25 kg N/ha (FYM)) to 19.3% (50 kg N + 25 kg/ha) under 1000–1250 mm rainfall. Thus fertilizer treatments have a better sustainability for green gram compared to cotton. The study indicated that cotton is unsustainable under < 500 mm rainfall situation.

### 4.4.3. Kovilpatti

At Kovilpatti, using mean yield of a treatment over 9 years each for sorghum and pearl millet; standard error based on the models; and maximum yield potential of sorghum of 617, 2451 and 1342 kg/ha under < 250, 250-500 and 500-750 mm rainfall situations respectively, SYI of each treatment was derived. In sorghum, T8 had SYI of 3.5%; while the other treatments had a negative SYI under < 250 mm. The SYI ranged from 12.7% for T1 to 35.9% for T8 under 250-500 mm rainfall situation. It ranged from 1.4% for T1 to 52.9% for T8 under 500-750 mm rainfall situation. In pearl millet, the maximum potential yield was 502, 864 and 1220 under < 250, 250-500 and 500-750 mm rainfall situations respectively. In pearl millet, the SYI ranged from 27.9% for T2 to 68.7% for T8 under < 250 mm. The SYI ranged from 40.9% for T3 to 77.2% for T7 under 250-500 mm rainfall situation. It ranged from 30.3% for T1 to 54.7% for T8 under 500-750 mm rainfall situation.

### 4.4.4. Indore

At Indore, the sustainable yield index of fertilizer treatments was measured by using the mean yield of treatments over years, standard error of the respective treatment under regression model; and maximum yield attained by treatments under different rainfall situations. The maximum yield attained was 1690 kg/ha under < 500 mm; 2368 kg/ha under 500-750 mm; 2415 kg/ha under 750-1000 mm; and 3247 kg/ha under 1000-1250 mm rainfall situation. The SYI ranged from 39.3 to 84.1% under < 500 mm; 54.8 to 87.4% under 500-750 mm; 40.3 to 74.4% under 750-1000 mm; 25.2 to 67.7% under 1000-1250 mm rainfall situations attained by control and 20 kg N (urea) + 13 kg P + FYM @ 6 t/ha respectively. 20 kg N (urea) + 13 kg P + FYM @ 6 t/ha was found to be most efficient treatment for attaining sustainable soybean yield over years. The treatment was also superior with a maximum available mean soil N, P, K and S at the end of 15 years. Based on the study, 20 kg N (urea) + 13 kg P + FYM @ 6 t/ha could be recommended for large scale adoption under farmer's fields for attaining maximum sustainable yields and maintenance of soil fertility of N, P, K and S under semi-arid Vertisols.

### 4.4.5. Anantapur

Using mean yield of each treatment over years; standard error based on regression model of yield; and maximum pod yield ($Y_{max}$) of 1546 kg/ha attained by 100% NPK (20–40–40 kg/ha) in 2004, SYI was derived for each treatment. The index ranged from 27.6% for control to 35.3% for 50% N (FYM~10 kg N/ha) based on the model through monthly rainfall. The SYI indicated that 50% N (FYM~10 kg N/ha) was superior for attaining sustainable yield based on model of monthly rainfall. 100% NPK (20–40–40 kg/ha) + $ZnSO_4$ @ 25 kg/ha was the 2nd best treatment based on the study. The maximum pod yield was 1546 kg/ha under < 500 mm; 1541 kg/ha under 500-750 mm; 1329 kg/ha under 750-1000 mm rainfall situation. The SYI ranged from 26.6 for control to 33.6% for T6 under < 500 mm; 27.4% for control to 39.1% for T5 under 500-750 mm; 49.0 for control to 71.2% for T8 under 750-1000 mm rainfall situation.

## Author details

G.R. Maruthi Sankar*, K.L. Sharma, N. Ashok Kumar, B. Venkateswarlu A. Girija and P. Ravi
*All India Coordinated Research Project for Dryland Agriculture (AICRPDA), CRIDA, Santoshnagar, Hyderabad, Andhra Pradesh, India*

Y. Padmalatha, K. Bhargavi, M.V.S. Babu and P. Naga Sravani
*AICRPDA, Acharya NG Ranga Agricultural University, Anantapur, Andhra Pradesh, India*

B.K. Ramachandrappa and G. Dhanapal
*AICRPDA, University of Agricultural Sciences, Bangalore, Karnataka, India*

Sanjay Sharma and H.S. Thakur
*AICRPDA, College of Agriculture, Indore, Madhya Pradesh, India*

* Corresponding Author

A. Renuka Devi and D. Jawahar
*AICRPDA, Tamil Nadu Agricultural University, Kovilpatti, Tamil Nadu, India*

V.V. Ghabane
*AICRPDA, Punjabrao Deshmukh Krishi Vidyapeeth, Akola, Maharastra, India*

Vikas Abrol, Brinder Singh and Peeyush Sharma
*AICRPDA, Sher-e-Kashmir University of Agriculture and Technology, Rakh Dhiansar, Jammu &*
*Kashmir, India*

A.K. Singh
*Krishi Anusandhan Bhavan, ICAR, Pusa, New Delhi, India*

## Acknowledgement

The PMEs have been conducted for more than 20-25 years at different AICRPDA centers. The experiments were conducted by many scientists at Bangalore, Kovilpatti, Anantapur, Akola and Indore centers. The authors are grateful to all the scientists who have directly or indirectly associated with the PMEs conducted over years. The authors would also express their gratitude to the Indian Council of Agricultural Research for funding the PMEs conducted at different centers over years.

## 5. References

Bansal, R.K., N.K.Awadhwal and V.M.Mayande (1987). Implement development for SAT Alfisols. In Alfisols in the semi-arid tropics : Proceedings of the Consultants' Workshop on the State of the Art and Management Alternatives for Optimizing the Productivity of SAT Alfisols and Related Soils, 97 – 107. India : ICRISAT Center. Patancheru, Andhra Pradesh 502324, India.

Behera, B., G.R. Maruthi Sankar., S.K.Mohanty., A.K.Pal ., G.Ravindra Chary., G.Subba Reddy and Y.S.Ramakrishna (2007). Sustainable fertilizer practices for upland rice from permanent manorial trials under sub-humid Alfisols. *Indian Journal of Agronomy*, 52 (2): 33-38.

Bhat, A.K., V.Beri and B.S.Sindhu (1991). Effect of long term recycling of crop residue on soil properties. *Journal of Indian Society of Soil Science*, 39: 380–382.

Charreau, C. (1977). Controversial points in dryland farming practices in semi –arid West Africa. In Symposium on *Rainfed Agricultural Semi – arid Regions*, ed. G.H.Cannell, 313 – 360. Riverside, Calif. : University of California, Riverside.

Chesnin, L. and Yien, C.H. (1950). Turbedimetric determination of available sulphur. *Soil Science Society of America Proceeding* 15, 149-151.

Cocheme, J., and P.Franquin. (1967). *An agroclimatology survey of a semiarid area in Africa south of Sahara* (FAO/Unesco/WMO interagency Project on Agroclimatology Technical Note 86 – 160). Rome, Italy: FAO.

Dalal, R.C. and R.J.Mayer (1986). Long term trends in fertility of soils under continuous cultivation and cereal cropping in Southern Queensland: In Overall changes in soil properties and trends in winter cereal yields. *Australian Journal of Soil Research,* 24: 265-279.

Doorenbos, J. and Kassam, A.H. (1979). Yield response to water. FAO Irrigation and Drainage. Paper 33. FAO, Rome.

Draper, N.R. and H.Smith (1998). *Applied Regression Analysis.* John Wiley Inc., New York.

El-Swaify, S.A., S.Singh, and P.Pathak (1983). Physical and conservation constraints and management components for SAT Alfisols. In *ALFISOLS in the semi-arid tropics : Proceedings of the Consultants' Workshop on the State of the Art and Management Alternatives for Optimizing the Productivity of SAT Alfisols and Related Soils.* ICRISAT Center. Patancheru, Andhra Pradesh 502324, India.

Gomez, K.A. and A.A.Gomez (1984). *Statistical Procedures for Agricultural Research.* John Wiley Inc., New York.

Hiler, E.A. and Clark, R.N. (1971). Stress day index to characterize effects of water stress on crop yields. *Transactions American Society of Agricultural Engineers,* 14 : 757 – 761.

Jackson, M.L. (1973). *Soil Chemical Analysis.* Prentice Hall of India, New Delhi.

Kampen, J., and J.Burford (1980). Production systems, soil related constraints, and potential in the semi-arid tropics, with special reference to India. In *Priorities for alleviating soil-related constraints to food production in the tropics.* International Rice Research Institute, Los Banos, Laguna, Phillippines : 141-165.

Maruthi Sankar, G.R. (1986). On screening of regression models for selection of optimal variable subsets. *Journal of Indian Society of Agricultural Statistics,* 38 (2): 161-168.

Maruthi Sankar, G.R., G.Ravindra Chary., G.Subba Reddy., G.G.S.N.Rao., Y.S.Ramakrishna and A.Girija (2008). Statistical modeling of rainfall effects for assessing efficiency of fertilizer treatments for sustainable crop productivity under different agro-eco sub-regions. Paper presented in the International symposium on "Agro-meteorology and food security" at CRIDA, Hyderabad during 18–21: February.

Maruthi Sankar, G.R., Vittal, K.P.R., Ravindra Chary, G., Ramakrishna, Y.S. and Girija, A. (2006). Sustainability of tillage practices for rainfed crops under different soil and climatic situations in India. *Indian Journal of Dryland Agricultural Research and Development,* 21 (1), 60–73.

Maruthi Sankar, G.R., Sonar, K.R. and Reddy K.C.K. (1988). Pooling of experimental data for predicting fertilizer requirements of *Rabi* sorghum for varying soil test values. *Journal of Maharastra Agricultural Universities,* 13 (1) : 59 – 62.

Maruthi Sankar, G.R. (1992). Application of Statistical Model-Building and Optimization techniques in Fertiliser Use Research. Chapter 28 in the book on Dryland Agriculture: State of Art of Research in India. pp: 653-671.

Maruthi Sankar, G.R. and Vanaja, M. (2003). Crop growth prediction in sunflower using weather variables in a rainfed alfisol. *Helia,* 26 (39): 125 – 140.

Maruthi Sankar, G.R. and Raghuram Reddy, P. (2005). Identification of maize (*Zea mays* L.) genotypes for rainfed condition based on modelling of plant traits. *Indian Journal of Genetics and Plant Breeding,* 65 (2): 88–92.

Mathur, G.M. (1997). Effects of long term application of fertilizers and manures on soil properties under cotton–wheat rotation in North West Rajasthan. *Journal of Indian Society of Soil Science,* 42 (2): 288–292.

Mohanty, S.K., G.R. Maruthi Sankar., B.Behera., A.Mishra., A.K.Pal and C.R.Subudhi. (2008). Statistical evaluation and optimization of fertilizer requirement of upland rice (*Oryza sativa*) genotypes at varying levels of crop seasonal rainfall under moist sub-humid Alfisols. *Indian Journal of Agricultural Science,* 78 (3): 18–23.

Nema, A.K., G.R.Maruthi Sankar and S.P.S.Chauhan. (2008). Selection of superior tillage and fertilizer practices based on rainfall and soil moisture effects on pearl millet yield under semi-arid inceptisols. *Journal of Irrigation and Drainage Engineering,* 134 (3): 361–371.

Olsen, S.R., C.V.Cole., F.S.Watanabe and L.A.Dean. (1954). Estimation of available phosphorus in soils by extraction with sodium bicarbonate. Circular of US Department of Agriculture, 939.

Pal, A.K., Behera, B., and Mohanty, S.K. (2006). Long term effect of chemical fertilizers and organic manures on sustainability of rice (*Oryza sativa* L.)–horse gram (*Microtyloma uniflorum*) cropping sequence under rainfed upland soil. *Indian Journal of Agricultural Sciences,* 76 (4) : 211–274.

Prasad, R. and N.N.Goswami. (1992). Soil fertility restoration and management for sustainable agriculture in South Asia. *Advances in Soil Science.* Soil Sci. 17, 37–77.

Prihar, S.S. and Gajri, P.R. (1988). Fertilization of dryland crops. *Indian Journal of Dryland Agricultural Research and Development,* 3 (1), 1–33.

Rijtema, P.E. and Aboukhaled, A. (1975). Crop water use. In. Research on Crop water use in salt affected soils and drainage in the Arab Republic of Egypt. Edited by Aboukhaled, A., Arara. Balba, A.M., Bishay, B.G., Kadry, L.T., Rijtema, P.E. and Taher, A. FAO Regional Office for the near East. pp: 5–61.

Singh, N.P., Sachan, R.S., Pandey, P.C., and Bisht, P.S. (1999). Effect of a decade long fertilizer and manure application on soil fertility and productivity of rice – wheat system in a mollisol. Journal of *Indian Society of Soil Science,* 47 (1) : 72 – 80.

Singh, R., Singh, Y., Prihar, S.S. and Singh, P. (1975). Effect of N fertilization on yield and water use efficiency of dryland winter wheat as affected by stored water and rainfall. *Agronomy Journal,* 67, 599–603.

Snedecor, G.W. and Cochran, W.G. (1967). Statistical methods. Iowa State University Press, Ames, Iowa, USA.

Subbaiah, B.V. and G.L.Asija. (1956). A rapid procedure for determination of available nitrogen in soil. *Current Science,* 25 : 259–260.

Velayutham, M., Reddy, K.C.K. and Maruthi Sankar, G.R. (1985). All India Coordinated Research Project on Soil Test Crop Response Correlation and its impact on Agricultural Production. *Fertilizer News,* 30 (4): 81–85.

Venkateswarlu, J. and Singh, R. (1982). Crop response to applied nutrients under limited water conditions. Review of Soils Research in India. Transactions of 12th International Soil Science Congress, New Delhi.

Vikas Abrol, G.R.Maruthi Sankar., Mahinder Singh and J.S.Jamwal. (2007). Optimization of fertilizer requirement of maize based on yield and rainfall variations from permanent manorial trials under dry-sub humid Inceptisols. *Indian Journal of Dryland Agricultural Research and Development*, 22 (1) : 15–21.

Vittal, K.P.R., Maruthi Sankar, G.R., Singh, H.P. and Samra, J.S. (2002). Sustainability of Practices of Dryland Agriculture – Methodology and Assessment. Research Bulletin of AICRP for Dryland Agriculture, CRIDA, Hyderabad, 100 pages.

Vittal, K.P.R., G.R.Maruthi Sankar., H.P.Singh., D.Balaguravaiah., Y.Padamalatha and T.Yellamanda Reddy.( 2003). Modeling sustainability of crop yield on rainfed groundnut based on rainfall and land degradation. *Indian Journal of Dryland Agricultural Research and Development*, 18 (1):7–13.

Vittal, K.P.R., Basu, M.S., Ravindra Chary, G., Maruthi Sankar, G.R., Srijaya, T., Ramakrishna, Y.S., Samra, J.S., and Gurbachan Singh. (2004). District wise promising technologies for rainfed groundnut based production system in India. Research Bulletin – An AICRPDA contribution, CRIDA, Hyderabad. (92 pages).

Walkley, A.J., and Black, C.A. (1934) Estimation of organic carbon by chromic acid titration method. *Soil Science*, 37: 29-38.

# Tillage and Soil Properties

# Chapter 9

# Impact of Agricultural Traffic and Tillage Technologies on the Properties of Soil

Ioan Tenu, Petru Carlescu, Petru Cojocariu and Radu Rosca

Additional information is available at the end of the chapter

## 1. Introduction

Soil has an essential part in preserving life on earth. The main function of soil lies in the fact that it represents the support for the agriculture practice, aiming to insure the peoples' alimentary security and safety, due to its physical and biological properties, to its fertility, to its capacity to provide plants with the water and nutrients needed for their growth.

Taking into account the intensive development of agriculture, the concept of "sustainable development" is a new and complex one, imposed to humanity by the need for the preservation of the soil functions and which is connected not only to agriculture but also to other knowledge fields.

Within the frame of this concept, a special place belongs to sustainable agriculture, which constitutes a system of technologies and good practices aiming not only towards better productions but also to achieve conservative goals.

The main objectives of sustainable agriculture are:

- alimentary safety, by providing the human needs with the necessary food and fiber;
- preservation of the quality of the environment and of the natural resources vital to the agriculture;
- more efficient use of the renewable and non-renewable resources.

In essence, sustainable agriculture must harmoniously combine the three main dimensions: economical, social and ecological.

In order to achieve the goals of sustainable agriculture, the way in which soil tillage works are fulfilled is of an extreme importance, because the induced physical and mechanical changes affect the soil's physical, chemical and biological processes. This is the reason why the concept of "conservative agriculture" (CA) was developed, as a part of the sustainable agriculture systems, which is based on the use of natural renewable resources, especially soil, and on real time soil regeneration; some complementary agricultural practices are included herein [9, 17]:

- minimum soil disturbance (through reduced soil tillage works and stubble seeding), in order to preserve the soil's structure, fauna and organic matter;
- permanent soil coating (covering crops, residue, mulch), in order to protect it and contribute to the weeds removal;
- different crops rotation and combination schemes, in order to stimulate soil microorganisms and to remove weeds, diseases and pests.

Hence, conservative agriculture is based on an unconventional soil tillage system, named "Soil conservation tillage system" (SCTS). Within this system, moldboard plowing is deferred (completely or partially), the number of agricultural operations is limited and at least 15...30% of the vegetable debris is kept on the soil surface. Worldwide, this system is used on nearly 45% of the farmland, and an increase to 60% is appraised for the next twenty years [9, 17]. The unconventional soil tillage system consists of very different methods, starting with the seeding in the untilled soil and ending with the deep loosening without furrow overturning. Between these two extreme methods, many other variants are possible: reduced tillage, minimum tillage (when up to 30% of the vegetable debris is left on the soil surface), minimum tillage with vegetable mulch (more than 30% of the vegetable debris is left on the soil surface), ridge seeding, partial or strip tillage etc. [9]

As a result, three directions were outlined in order to define the unconventional soil tillage systems:

- direct sowing, when seeds are inserted into the practically non-tilled soil. In this case, soil is tilled only in order to create very small gutter, using small knives mounted on the seeding machine.
- minimum tillage or reduced tillage. In this case, the fact that different types of soil must be differently loosened, in order to favor the normal plant growth, is taken into account. The minimum tillage system includes either the base soil loosening, without furrow overturn, or the superficial tillage, followed by seeding. Sometimes, a minimum of mechanical works is required in order to destroy the weeds, to favor some biologic processes and to support the development of the roots. This system allows the reduction of the energy consumption and working time.
- "rotation" tillage represents another possibility to diminish the intensity of soil loosening. In this case, tillage should be very well correlated with crop rotation. This system is characterized by some peculiarities: the different plants, which are grown in a crop rotation system, have different requirements regarding the soil tillage system; soil requirements towards the tillage system are different from one crop to another; soil characteristics are gradually changing (in a favorable or unfavorable way). It is important to notice that the soil tillage system should be modified according to requirements of the respective crop.

## 2. Aggressiveness of wheels and active parts of the agricultural units towards soil

Soil degradation is one of the most important problems to be faced nowadays. It is appraised that 5-7 mha of soil are degrading worldwide each year, with a tendency to attend 10 mil. ha per year in the near future [8]. Soil degradation may be physical, chemical and biological. In the case of the physical soil degradation, two important features are affected: bulk density and structure. This means that bulk density increases (the soil is compacted) and the structural elements are damaged (deformed, crushed, sheared, broken, fragmented) [4, 12, 15].

### 2.1. Effect of agricultural traffic over soil

Mechanization of the agricultural processes and the use of heavy productive units, with high working width, resulted, over time, in soil compaction and reduction of harvest. The use of the organic fertilizers and deep tillage finally results in the diminishing of the content of organic matter and in an increased sensibility to compaction [20].

**Soil compaction processes**. Soil compaction may be defined as "the compaction of soil mass in a smaller volume". The increase of the bulk density is accompanied by structural changes, changes in the thermal and hydraulic conductivity and in the gas transfer characteristics, all of them affecting the chemical and biological equilibrium of soil.

The artificial (anthropic) soil compaction is the result of exaggerated traffic imposed by agricultural operations, transport operations etc. The intensity of anthropic compaction depends on different factors. Some of them belong to soil, namely to its susceptibility to compaction: uneven grain size distribution, unstable structure, reduced humus content etc. Other factors are influenced by the characteristics of the agricultural equipment; this compaction is favored by the heavy machineries exerting high pressure over soil, by the increased number of passings, by the increased tire air pressure, by the agricultural traffic performed over wet soil etc.

**The effects of soil compaction.** Soil compaction leads to the reduction of harvest by 50%, compared to the non-compacted soil, while fuel consumption is expected to be increased by 35% [2]. Soil compaction is one of the main causes of surface flow and erosion. In the meantime, compacted soils require higher costs of the irrigation arrangements and exploitation, due to the poor infiltration of water and to the intensified evapotranspiration. Soil compaction also causes the reduction of the water holding capacity and of the permeability, reduces soil aeration, significantly increases the penetration resistance and plowing resistance, inhibits the development of the plant roots and the quality of the ploughland deteriorates. It was established that not only the active parts of the tillage equipment deteriorate the soil structure, but also the tractors' wheels and the tracking wheels of the agricultural machinery; the use of heavier equipments leads to contact pressures of 0.2...1.8 MPa, while the specific resistance of the soil's structure elements is lower than 0.1 MPa (usually 0.02...0.006 MPa); as a result, soil compaction occurs up to 30-50 cm under the tracking wheel and on an area with a width four times the wheel width [6].

**Plant and response to compaction.** In compacted soils or compacted layers the penetration depth of the roots and their density are restricted, the effect being a slow development of the root system. As a result, the plant's access to water and nutrients is limited, while the capacity of the root system to counteract the noxious effect of pathogenic agents is diminished. This is why plant species with deep roots are less sensible to compaction. Compaction induces changes in the soil's water and air regime, also affecting the activity of microorganisms. Soil compaction also favors the ammonia nitrogen in the detriment of nitric nitrogen, with unfavorable effects over the harvest [10].

**Effect of soil tillage over compaction.** The agricultural operations related to soil are: ploughing, land preparation, seeding and some of the maintenance operations. In most of the cases, the aim of these operations is to loosen the compacted soil layers. Where soil compaction is a problem, tillage has an ameliorative effect. Soils are usually subjected to two types of traffic: one that produces compaction (wheel traffic) and another one that produces loosening (tillage traffic). Soil reaction to compaction depends on traffic characteristics, soil properties and humidity when the traffic takes place; soil compaction is usually expressed by the means of bulk density, porosity or penetration resistance. Because of wheeled traffic the bulk density of soil increases; the magnitude of the density change depends on the soil texture, its humidity, the wheel-soil contact pressure and the number of passages. The maximum compaction effect is reached when wheel slip reaches 15...25%, due to the fragmentation of the structure elements under the effect of shear stress [18].

## 2.2. Aggressiveness of the tillage active parts towards soil

Unlike the wheels (tracking wheels, driving wheels etc.), which compact the soil, the active parts of the tillage equipment (rotary cultivator tines, plow shares, cultivator tines etc.) loosen the soil; the rollers, the combined seedbed preparation devices and the combined cultivator are exceptions [6].

In the case of the active parts used for seedbed preparation, their destructive action over the soil structure elements is of an utmost importance. The active parts destroy, to a greater or a lesser extent, the structure elements through deformation, crumbling, cutting, breaking-up. The destruction of the soil's structure is a general phenomenon, occurring at any tillage operation, but it gets large proportions in soils with a rough texture or average-rough texture (sandy soils, sand-loamy soils, clay-loamy soils) for which the mechanical stability of the structure elements is lower, due to the lower clay content. The recently tilled soils, the humid soils or the dry clay soils on which loosening is obtained by the means of rotating active parts (disc harrows, rotary cultivators) are also vulnerable [4].

It should be emphasized that not only the active parts of the tillage equipments destroy the soil's structure, but also the wheels of the tractors or the tracking wheels of the agricultural machinery. In this case, the wheel-soil contact pressure produces compaction as a result of the deformation and breaking-up of the structure elements. The division of the structure elements into fragments results in the increase of the bulk density because of a more stuffed settlement. Consequently, the soil's capacity to drain and store water is diminished, the thermal regime worsens, the accessibility of plants to nutrients is diminished and the activity of the anaerobic microorganisms is reduced; on sloping lands, the erosion due to water is intensified, and the plants have difficulties in developing the root system, loose their stability and harvest is diminished [14].

**Effect of the moldboard plough over soil.** In the working process, the plough's share cuts the furrow horizontally and begins its detachment, overturning and lateral displacement, the process being finalized by the moldboard. These actions cause complex deformations of the soil slice, resulting in it's the fragmentation and crumbling. It should be emphasized that, under the action of the share's cutting edge, a number of soil's structural elements are also cut. In the meantime, the twisting of the furrow, both vertically and horizontally, as well as the friction between soil and the parts of the plough body results in the destruction of a number of the structural elements through deformation, crumbling, breaking, fragmentation. The process of destruction of the structure elements is strengthened by the working speed and by the use of worn shares or aggressive moldboards (cylindrical or helical moldboards).

**Effect of the rotary cultivators over soil.** In the working process, due to the advance movement of the equipment and the rotation movement of the rotor, the active parts penetrate into the soil and cut slices with a particular shape. Under the action of centrifugal force, the soil slices are thrown over the inner surface of the housing and louver. As a result, a supplementary crumbling is achieved, the soil being left behind the rotary harrow in a loosened and fine layer.

The cutting edges of the tines of the rotary harrow cut some of the soil's structural elements, which are also destroyed through deformation, breaking, fragmentation, crumbling, due to the peripheral speed of the tines, to the friction between tines and soil and to the impact between the slices and the housing of the rotary harrow. Pulverization of the structural elements may also occur, when some of them are fragmented to the maximum extent, resulting in particles of clay, silt and sand [19].

When using rotary harrows for the seedbed preparation care should be taken so that the width of the soil slices should not be less than 25 mm. In the working process, the number of destroyed structural elements increases when the slices get thinner. Therefore, the peripheral speed of the tines should no exceed 6 m/s, while the speed of the advance movement should not exceed 1 m/s.

**Effect of the disc harrows over soil.** The active parts of these agricultural equipments are spherical discs. In the tillage process, the disc is displaced forward, following the movement of the equipment; in the same time, the disc is rotating, due to the contact with the soil. Over the effect of weight, the disc penetrates into the soil and cuts a soil layer, which is raised over the interior concave surface of the disc, is crumbled, displaced laterally and partially overturned. The aggressiveness of the discs depends on their shape, as well as on the cutting angle (disk angle), which can be adjusted between 15 and 30 degrees; crumbling increases when the cutting angle increases.

During the working process, the cutting edge of the disc cuts some of the soil's structural elements; the number of cut structural elements is higher when the disk's cutting edge is indented. In the meantime, the bending of the soil layer, vertically and horizontally, its twisting, as well as the friction between soil and disk, cause the destruction of some structural elements, through deformation, breaking, fragmentation. The proportion of damaged structural elements increases with speed [22].

**Effect of the cultivators over soil.** Depending on their destination, cultivators are equipped with different types of active parts, the more important being the ones used for weed cutting and for soil loosening.

The weed cutting active parts are aimed to cut the weeds and loosen the soil. During the working process of the arrow type tines, soil is cut horizontally, at a certain depth; in the meantime, the weeds are cut and soil is loosened and crumbled.

The straight, chisel, diamond pointed and narrow arrow type loosening active parts are mounted on rigid or elastic holders. When elastic holders are used, the active part vibrates, exerting an energetic action over soil. As a result of the displacement of these active parts, the superficial soil layer is crumbled and loosened.

The weed cutting type active parts cut some of the soil's structural elements. All the types of active parts (for both weed cutting and soil loosening) destroy the structural elements due to the advance movement, friction, breaking of soil layers, vibration; the structural elements are destroyed through deformation, crumbling, breaking, fragmentation. In order to diminish these adverse effects, speed should be limited to 10-12 km/h.

## 3. Simulation of the tire wheel-soil and active parts-soil interaction, in laboratory conditions

### 3.1. Design and testing of a laboratory soil channel test rig

Physical degradation of the soil, caused by the interaction with the active parts and support wheels of the agricultural equipment, refers especially to structural deterioration and its compaction. In order to quantify these issues studies must be performed in order to establish the critical values of the working parameters (of the active parts and wheels) leading to the physical degradation of the soil. In the meantime, correlations should be established between the above-mentioned parameters and the ones characterizing soil structure and compaction.

With the purpose to perform these studies a laboratory test rig for the study of the interaction between the active parts and wheels of agricultural equipment, on one hand, and soil, on the other hand, was designed and built. The designing process has taken into account the similarity laws, in order to reproduce in laboratory conditions the complex processes occurring at the contact interface between the working parts and soil and between the wheels of the agricultural equipments and soil.

The test rig (Figure 1) consists of a rigid frame (1), the soil bin (2), the carriage (3), on which the active part (plough body) for soil working is mounted (6), the wheel with tire (7) and the drum for leveling - settling (8); at the end of laboratory test rig a winch is fixed, which is for trolley carriage with the cable (5). Due to its length (10240 mm), the soil bin is composed of five sections, joined together with screws. The bin was coated, at the inside, with a plastic sheet.

An electric motor (9) and a cylindrical gear (10) are used to drive the drum (11); the drum drives the cable (5), thus towing the cart (3). The ends of the traction cable are attached to the carriage by the means of the load cells (4).

The carriage is also fitted with an electric motor (12) and a gear transmission in order to drive the tire wheel (7). The working depth of the active body can be adjusted by the means of the screw mechanism (13). The screw mechanism (14) is used to adjust the vertical position of the tire wheel; the screw mechanism (15) is used to change the vertical position of the soil leveling and compaction drum. The soil leveling and compaction roller (8), mounted on the carriage, is used to achieve a certain soil compaction, before it is processed by the active body or performing various experiments with the tire wheel.

Four upper trundles – two in the front and two in the back – and four lower trundles - also two in the front and two in the back - are mounted on the carriage frame; the trundles are rolling on rails, mounted on each side of the soil channel frame.

When the carriage is towed by the means of the cable, the wheel (7) rotates due to the interaction with the soil and thus the conditions for a driven tractor wheel are simulated. When the carriage is not towed, the wheel (7) becomes a driving one, being driven by the electrical motor (12) by the means of a cylindrical gear drive and of a trapezoidal belt drive. Thus, the driving wheel of the tractor is simulated.

The towing cable is connected to the carriage by the means of two 1000 daN strain gauge load cells, allowing the measurement of the traction force needed to displace the carriage; the transducers are fitted with spherical joints at both ends and are connected to a programmable weighing controller, which displays the mean values of the corresponding sampling signal and eliminates the load spikes that could occur due to vibrations and soil unevenness.

A control panel (18) is used for the power supply of the two electric motors; the electric cables are guided by the steel cable (16), fixed between the pillars (17). The electrical motors are controlled by the means of a frequency converter, allowing the adjustment of the rotation speed when the frequency is modified between 3 and 50 Hz. The dynamic braking principle is used in order to stop the carriage at the end of travel. Switches on the control panel allow the selection of the electric motor (the carriage towing motor or the tire wheel driving motor), as well as its forward or reverse motion.

A bill chernozem type of soil was used to fill the soil channel bin, having a loam-clay texture, the aggregate size of 0.02 ... 50 mm and 17-19% humidity.

## 3.2. Laboratory tests concerning some active parts and tire wheels

The test rig was tested in two phases. In the first phase, the laboratory test rig (soil channel) was tested to see if the prescribed constructive-functional parameters are attained. It was found that the test rig has the following constructive-functional parameters: depth of active enforcement machinery of working the soil: 0 ... 300 mm; adjustment of working angle of the active part relative to the soil surface: (-) 25⁰ ... (+) 25⁰; the speed of the carriage when it is hauled by the 5,5 kW electric motor: 0,5 ... 1,55 m/s (1,8 ... 5,58 km/h); the maximum pull-down force for the with tire wheel and the soil leveling and settlement drum: 500 daN; maximum traction force of the cable (of the carriage), at a speed of the trolley of 0,55 m/s: 800 daN; maximum traction force of the cable (of the carriage) at a speed of the carriage of 1,55 m/s: 280 daN; cable breakdown point: 40,83 kN. It was concluded that there were no significant differences between the design parameters and the achieved ones.

In the second phase, the laboratory test rig was used in order to study the soil - moldboard plough body interaction and the tire wheel-soil interaction (Figure 2).

The studies concerning the active part-soil interaction were performed over a semi helical moldboard plough body, with a working width of 200 mm. The plough body was mounted on the carriage of the test rig. The influence of working depth, soil resistance to penetration and travel speed on the traction resistance and specific power consumption were evaluated. The results presented in the Table 1 show that increasing the traveling speed of the plough body results in an increased traction resistance. In the meantime, as the plough body speed increases, the specific power consumption significantly increases.

When soil penetration resistance increases, the traction force and specific power consumption also increase.

As far as the working depth is concerned, it was concluded that the increase of the penetration depth of the plough body resulted in a significant increase of traction resistance.

It was also noticed that increasing the working depth of the plough body resulted in an uneven change of the specific power consumption: increasing the working depth from 100 mm to 150 mm caused a slight decrease of the specific power consumption; when the working depth was further increased from 150 mm to 200 mm, the specific power consumption increased. These variations of the power specific consumption could be explained as follows: for small working depths (below 15 cm) the soil slice was not yet deployed as furrow and was not overturned, so that the specific power consumption was low; for depths over 15 cm the conditions to overturn the furrow were better, so that supplementary power was consumed in order to overturn it.

In the series of experiments regarding the driving wheel mounted on the test rig, the tire wheel was used to drive the carriage along the soil channel, the wheel being driven by the 3 kW electric motor. The wheel was equipped with a 5.00-12/4PR TA60 traction tire (with angled lugs).

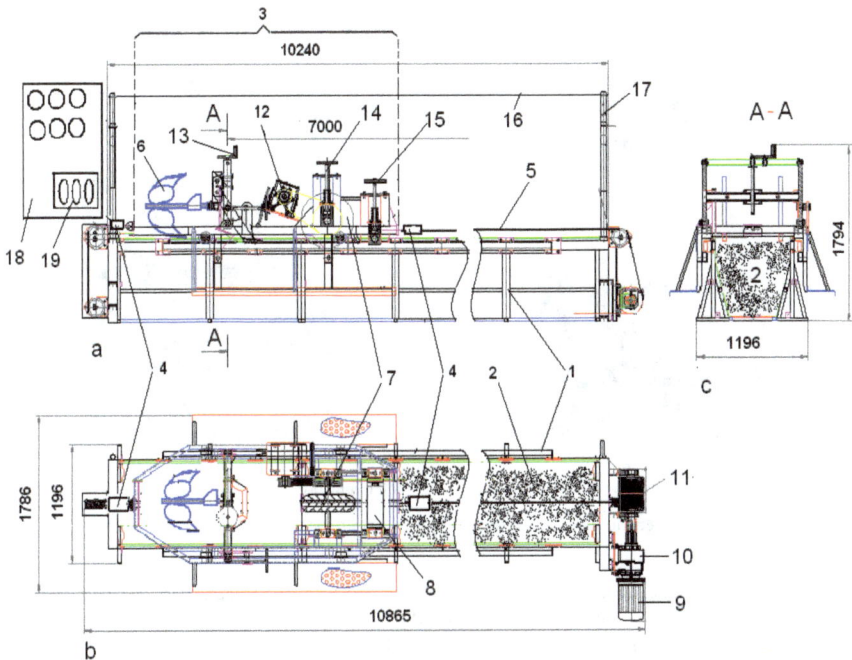

*a – side view; b – top view;* 1 – laboratory test rig frame; 2 – soil bin; 3 – carriage 4 – load cell; 5 – carriage cable trolley; 6 – active part (plow body); 7 – wheel with tire; 8 – drum for soil leveling and compaction; 9 – electric motor winch; 10 – cylindrical gear of the winch; 11 – traction cord reel; 12 – electric motor (and cylindrical gear) to drive the wheel with tire; 13 – screw mechanism for adjusting the working depth of the active body; 14 – screw mechanism for adjusting the vertical position of the wheel with tire; 15 – screw mechanism for adjusting the vertical position of the drum for soil leveling and compaction; 16 – steel cable to support the electric conductors; 17 – supporting pillars for the steel cable; 18 – electric power and control panel; 19 – controller to measure the traction force.

**Figure 1.** Laboratory test rig with soil channel for the study of the interaction between the active parts or agricultural wheels and the soil

The tests were aimed to evaluate the main operating parameters of the driving wheel test rig. The effects of the wheel speed, soil penetration resistance and wheel pushdown force over the carriage speed, wheel slip and wheel traction force were studied. The results obtained in these experiments are presented in Table 2.

Based on the experimental results it was concluded that wheel slip decreased when its speed was increased; in the meantime, the traction force decreased when its speed was increased.

As far as the soil resistance to penetration was concerned, it was concluded that its increase led to the increase of the wheel speed, because a more compact soil increases the adherence of the driving wheel; the increased adherence diminishes wheel slip, increases the carriage travel speed and also increases the traction force of the wheel.

(a)                                                (b)

a – study of wheel – soil interaction; b –study of plough body – soil interaction.

**Figure 2.** Laboratory experiments in order to study of the interaction of tire wheel and active part with the soil

| Working depth of plough body (mm) | Soil resistance to penetration (MPa) | Speed of plough body (m/s) | Traction force (N) | Specific power consumption (W/cm²) |
|---|---|---|---|---|
| 100 | 0,2 | 0,75 | 705 | 2,65 |
|  |  | 1,00 | 720 | 3,60 |
|  |  | 1,25 | 735 | 4,59 |
|  | 0,4 | 0,75 | 925 | 3,47 |
|  |  | 1,00 | 940 | 4,70 |
|  |  | 1,25 | 960 | 6,00 |
| 150 | 0,2 | 0,75 | 1055 | 2,64 |
|  |  | 1,00 | 1070 | 3,57 |
|  |  | 1,25 | 1080 | 4,50 |
|  | 0,4 | 0,75 | 1380 | 3,45 |
|  |  | 1,00 | 1400 | 4,67 |
|  |  | 1,25 | 1420 | 5,92 |
| 200 | 0,2 | 0,75 | 1450 | 2,72 |
|  |  | 1,00 | 1470 | 3,67 |
|  |  | 1,25 | 1485 | 4,64 |
|  | 0,4 | 0,75 | 1859 | 3,47 |
|  |  | 1,00 | 1890 | 4,72 |
|  |  | 1,25 | 1930 | 6,03 |

**Table 1.** Operating parameters of the laboratory test rig by modeling the plough body –soil interaction

| Wheel speed (rot/min) | Soil resistance to penetration (MPa) | Down force over the wheel (N) | Speed of the trolley (m/s) | Driving wheel slip (%) | Traction force of the driving wheel (N) |
|---|---|---|---|---|---|
| 20 | 0,2 | 500 | 0,50 | 17 | 230 |
| | | 750 | 0,51 | 15 | 360 |
| | | 1000 | 0,53 | 14 | 580 |
| | 0,4 | 500 | 0,51 | 15 | 290 |
| | | 750 | 0,52 | 15 | 442 |
| | | 1000 | 0,54 | 13 | 610 |
| 30 | 0,2 | 500 | 0,75 | 16 | 220 |
| | | 750 | 0,76 | 15 | 350 |
| | | 1000 | 0,79 | 12 | 570 |
| | 0,4 | 500 | 0,76 | 15 | 285 |
| | | 750 | 0,77 | 14 | 438 |
| | | 1000 | 0,79 | 12 | 600 |
| 40 | 0,2 | 500 | 0,99 | 15 | 220 |
| | | 750 | 1,01 | 14 | 340 |
| | | 1000 | 1,02 | 11 | 560 |
| | 0,4 | 500 | 1,01 | 14 | 280 |
| | | 750 | 1,01 | 13 | 410 |
| | | 1000 | 1,04 | 10 | 580 |

**Table 2.** Evaluation of the test rig's key operating parameters with motored wheel

Regarding the effect of the wheel down force, it was established that an increased force resulted in an increased wheel speed and, consequently, an increased travel speed of the carriage, due to the diminished wheel slip. The explanation is that an increased down force leads to a lower wheel slip, due to the higher wheel adherence; for the same reason, traction force increased when the wheel down force was increased.

Based on the experimental results precise correlations could be established between soil compaction and wheel slip, for a given vertical force acting over the wheel. In the same manner, the relationship between the vertical force and the traction force of the wheel was established, for a certain state of the soil.

## 4. Physical and mathematical modeling of the interaction between the active parts and soil

### 4.1. Introduction to numerical simulation

The finite volumes method (FVM) was used in order to model the interaction between the active parts of the agricultural equipment and soil. The equations specific to fluid flow are partial derivatives equations. Moreover, these equations are non-linear and interconnected. Under these circumstances, solving the equations is possible only by the means of numerical methods [1, 5].

There are several types of numerical methods for solving the flow equations; the finite volumes method was preferred because it was considered the most appropriate for the simulation of flow and other soil related phenomena, for the particular case of soil-agricultural active part interaction. Within this method, the main stages for solving a problem regarding the soil-active part interaction are as follows:

• division of the soil model into control volumes (finite volumes), based on a computational mesh;
• integration of the equations for each control volume in order to obtain the algebraic equations which are characteristic to the unknowns of the problem;
• solving of the algebraic equations.

Regarding the first stage, the computational mesh may be "structured" (the geometrical elements are 2D triangles or 3D parallelepipeds) or "unstructured" (the geometrical elements are 2D quadrilaterals or 3D tetrahedrons).

The numerical simulation of the physical phenomena characteristic to fluid flow by the means of the CFD (computer fluid dynamics) method is entirely based on the fundamental laws that describe the flow of a fluid: mass conservation, momentum conservation, energy conservation [11].

Five methods may be used in order to solve the problems regarding the interaction between soil and the active parts: empirical and semi empirical methods, dimensional analysis, finite elements method (FEM), discrete elements method (DEM) and the neuronal networks method – ANN (Figure 3) [13].

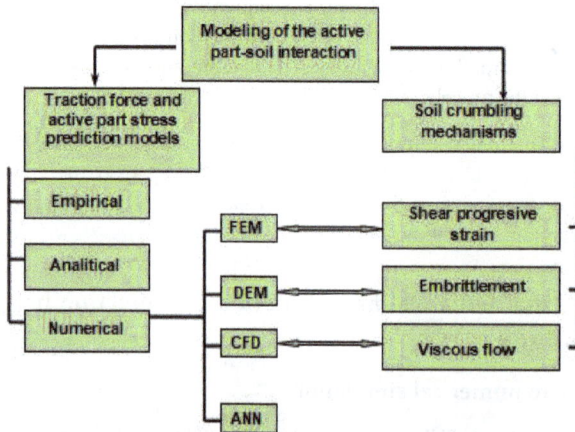

**Figure 3.** Methods for the numerical simulation of the active part-soil interaction

The mechanism of the active part-soil interaction may also be described using the rheological behavior of soil, taking into account its dynamic characteristics. The use of CFD method should take into account the dynamics of the active part-soil interaction, considering the irreversible deformations of the soil (see Figure 3) [21].

## 4.2. Case study regarding the CFD numerical simulation of the active part-soil interaction

The design criteria taken into account when building a tillage active part are traction resistance, the volume of the tilled soil and the overall energy requirements. The traction force depends on the pressure exerted by soil over the surface of the active part; soil pressure and its distribution are important items that should be considered when designing the size and shape of the active part.

The condition of the tilled soil depends on the soil's mechanical behavior and initial condition and on the characteristics of the active part. Soil is a complex material and its behavior is not yet fully understood. The complexity increases even further when taking into account different locations, with different climate conditions and different bulk densities. Several models for predicting the mechanical behavior of soil were considered over time, combining springs, dampers, slide-blocks, and from an elastic, plastic or viscous perspective. The behavior of most soils is non-linear and they can be considered as non-linear plastic or viscous-plastic materials. Thus, soil deformation may be described by a simple viscous-plastic model, the Bingham rheologic model.

The suggested shape of the winged share is a typical one (Figure 4a), but was adapted in order to increase its tillage performances. In the numerical simulation, the active part is mounted on a vertical holder (Figure 4b).

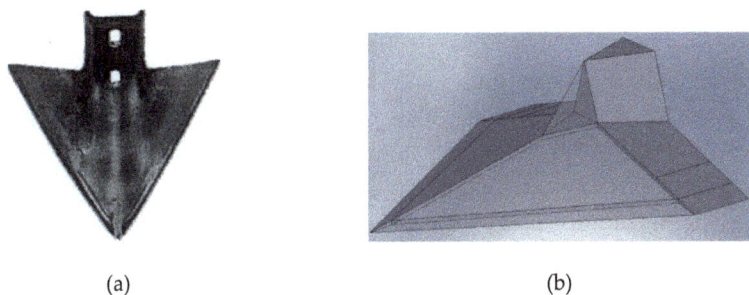

(a)                                                        (b)

a – existent; b - 3D model

**Figure 4.** Winged share

The numerical simulation assumes that the active part is stationary, while the viscous-elastic soil is flowing around the tillage tool (Figure 5).

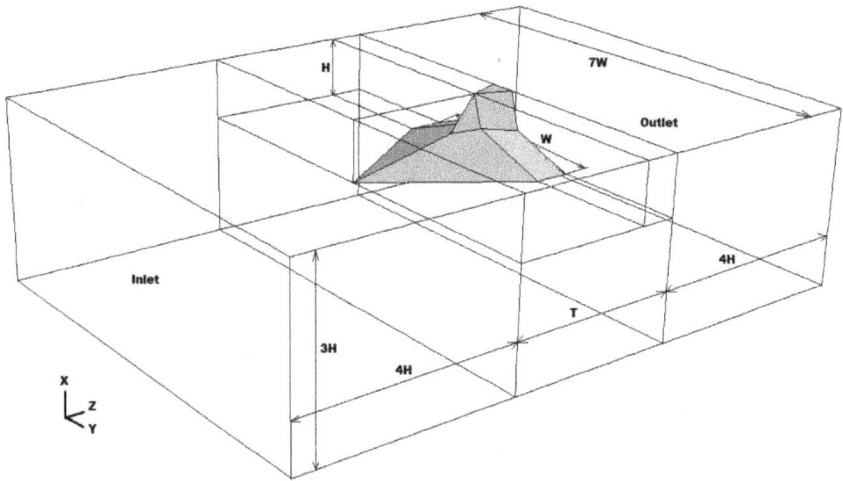

**Figure 5.** Schematics of the simulated domain (soil bin) and position of the active part

The active part acts as a resistance in the flow field and has the following dimensions: length T=140 mm, width W = 200 mm and height (equal to the working depth) H = 100 mm. The flow field is considered a parallelepiped with a length (8H+T) of 940 mm and a height (3H) of 300 mm. The sidewalls and the lower wall represent boundary conditions for the simulation field and their flow characteristics were neglected.

The CFD simulation showed that, for a soil bin seven times wider than the width of the active part, the wall effect of the bin disappears and an idealized flow pattern around the active part may be considered.

A mixed mesh was used in order to cover the computing domain, structured at the beginning and at the end of the bin and unstructured near the active part. At the tillage body level, the computational mesh was unstructured in the front and in the holder area, with a high density of the mesh, and structured behind the active part, with a lower mesh density (fig. 6). In order to cover the entire model 658000 elements were defined.

**General and boundary conditions.** The 3D CFD simulations were performed in isothermal conditions, for a loam-clay soil (38% clay, 32% silt and 30% sand – Figure 7), characteristic for the northeast area of Moldavia; these conditions imposed the use of the finite volumes method (FVM) for the CFD simulation.

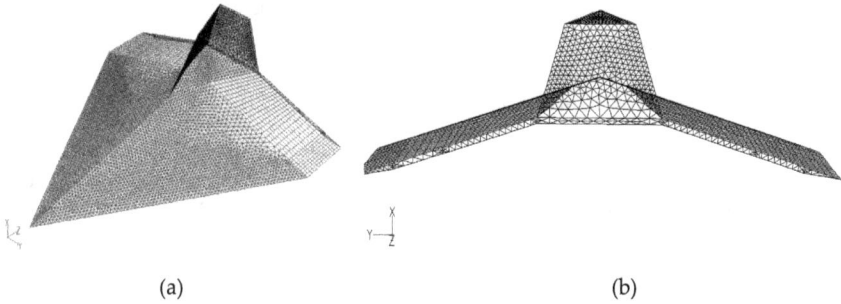

(a)            (b)

a – view in the horizontal plane; b – view from behind in the vertical plane

**Figure 6.** Computational mesh and elements for the arrow type tillage body:

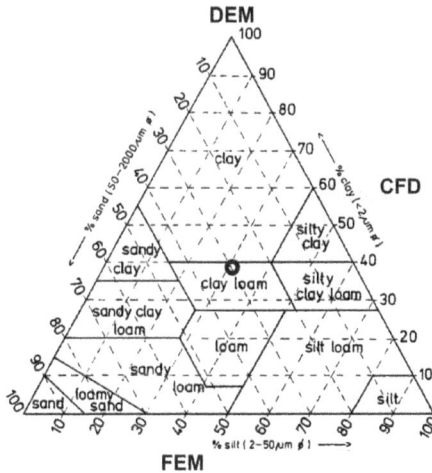

**Figure 7.** Soil classification according to sand, silt and clay content (texture triangle) and models for numerical simulation (**O** – present case study for soil with 38% clay, 32% silt and 30% sand) [15]

The viscous-plastic parameters of soil, such as the dynamic viscosity $\eta$ and the shear yield stress $\tau_0$, were evaluated at a constant speed, with a rheometer. Within the simulation the soil was considered at a humidity of 19% and a resistance to penetration of 400 kPa; other parameters taken into account were as follows: bulk density $\rho$ = 1250 kg/m³, dynamic viscosity $\eta$ = 900 Pa·s and shear yield stress $\tau_0$ = 12 kPa. The soil flow speed was w = 3 m/s, this being the traveling speed of the active part.

The following hypothesis were considered in the numerical simulation: a constant working depth of the tillage body; the active part was considered to be rigid; the soil flow was considered laminar and axisymetrical relative to the vertical section; 3D soil crumbling; the soil was considered an continuous, homogenous and anisotropic medium; soil behaves like a viscous-plastic Bingham material; the movement of the active part is considered with a finite yield tension and the displacement of the active part is considered as interaction between the submerged body and the fluid; the interspace between the clods of soil is neglected and soil is considered to be incompressible; cracking of soil occurs when the shear yield for the soil in front of the active part is exceeded.

**Generation of the model.** The Navier-Stokes equations are the basic equations for solving any problem related to fluid flow, by taking into account the momentum conservation equations, applied to an element of volume. The assumption that soil is flowing as a fluid allows the analysis of different types of interactions between soil and the active part, like the soil pressure over the surface of the active part, the traction resistance etc.

According to the second law of Newton, the acceleration of the volume of fluid (elementary volume) is directly related to the force acting upon it by the means of the momentum conservation equation:

$$\rho \cdot \frac{du_i}{dt} = \rho \cdot g - \frac{\partial p}{\partial x_i} + \frac{\partial \tau_{ij}}{\partial x_j} \tag{1}$$

where $d/dt$ is the full derivative, $g$ is the acceleration of the gravity field (m/s²), $p$ is the hydrostatic pressure (Pa), $\tau_{ij}$ is the tensor of the shear stress (Pa), $\rho$ is the density (kg/m³), $u_i$ is the travel speed (m/s) and $x$ is the distance (m).

The full derivative of the fluid volume variation is a time and space dependent function:

$$\frac{du_i}{dt} = \frac{\partial u_i}{\partial t} + u_j \cdot \frac{\partial u_i}{\partial x_j} \tag{2}$$

Equation (2) shows that the gravitational force, the hydrostatic pressure and the viscous stress (hydrodynamic stress) balance the acceleration of the fluid element. According to this equation soil flowing may be processed taking into account several types of dynamic interactions between the active part and soil:

- forces produced by the speed and acceleration of the active part;
- soil pressure over the surface of the active part, taking into account the weight of soil;
- soil cracking due to the viscous-plastic deformations.

For the Bingham type of plastic fluids, the strain-stress relationship may be written as:

$$\tau_{ij} = \sigma_i = 2 \cdot \mu \cdot \frac{\partial u}{\partial x} \tag{3}$$

$$\tau_{ij} = \tau_y + \mu \cdot \left( \frac{\partial u_i}{\partial x_j} + \frac{\partial u_j}{\partial x_i} \right), \text{for } \left| \tau_{ij} \right| > \tau_y \tag{4}$$

$$\gamma = 0, \text{for } \left| \tau_{ij} \right| \le \tau_y \tag{5}$$

where $\tau_y = \tau_0$ is the yield stress (Pa), $\mu = \eta$ is the plastic or dynamic viscosity ((Pa·s), $\gamma$ is the shearing speed (u/h being the flowing speed gradient) (s$^{-1}$) and h is the thickness of the sheared layer (m).

In a cartesian system of coordinates (x, y, z) the x component of the equation (1) may be expressed, based on the equations (2 - 4), in the form:

$$\rho \cdot \left( \frac{\partial u}{\partial t} + u \cdot \frac{\partial u}{\partial x} + v \cdot \frac{\partial u}{\partial y} + w \cdot \frac{\partial u}{\partial z} \right) = -\frac{\partial p}{\partial x} + \frac{\partial}{\partial x} \sigma_x + \frac{\partial}{\partial y} \tau_{xy} + \frac{\partial}{\partial z} \tau_{xz} \qquad (6)$$

where $\tau_{xy}$ is the shear stress in the xy plane (Pa), $\tau_{xz}$ is the shear stress in the xz plane (Pa), $\sigma_x$ is the normal stress (Pa) and u, v, w are characteristic coefficients of the x, y, z axes.

In order to satisfy the assumption that the viscous-plastic fluid is homogenous and isotropic, the simulation assumes that the yield stress is not dependent on the position or the orientation of the fluid particles.

In the tillage process, the active body has to overcome soil rigidity and cracking appears only when the tangential stress exceeds the yield stress. When the applied force exceeds the yield stress limit, soil flows as a viscous-plastic fluid, due to shear cracking.

The traction force is equal to or higher than the drag force; in the present simulation the active part is considered to be submerged into soil; hence, the drag force F is given by the relation:

$$F = \frac{1}{2} \cdot C_D \cdot \rho \cdot u_z^2 \cdot A \qquad (7)$$

where $C_D$ is the drag coefficient, $\rho$ is the density of soil, (kg/m$^3$); $u_z$ is the speed along the z axis (m/s), A is the characteristic area of the active part (m$^2$).

The drag coefficient has two components: the drag coefficient due to pressure $C_{Dp}$ and the drag coefficient due to friction, $C_{Df}$:

$$C_D = C_{Dp} + C_{Df} \qquad (8)$$

The weight of each drag component (pressure and friction) depends on the geometrical shape of the active part. The aim of the simulation is to obtain the geometrical shape leading to a minimum drag of the active part with.

A value of $10^{-4}$ was considered as the convergence criterion for each iteration and for each of the equations of the flowing process. A relaxation coefficient of 0.3 was chosen in order to maintain a stable convergence (a higher relaxation coefficient increases the convergence time). The results of the simulations were than interpreted considering that the active part is traveling at a constant speed.

Simulations were performed over three types of active parts, with the same geometrical shape, but with different angles between the cutting edges: $2\gamma = 60^0$; $2\gamma = 66^0$; $2\gamma = 70^0$.

Pressure distribution over the surface of the arrow type active body depends on the position of the cutting surfaces and on the characteristics of the soil. The simulations show that the maximum pressure is recorded on the cutting edges.

Table 3 presents the effect of the $2\gamma$ angle over the variation of the average normal pressure, for the same type of soil. The distribution of the normal pressure on the surface of the active part is shown in Figure 8.

| $2\gamma$ (degrees) | 60 | 66 | 70 |
|---|---|---|---|
| Average pressure (kPa) | 4.4 ... 45.2 | 4.5 ... 53.5 | 4.45 ... 53.35 |

**Table 3.** Variation of the average normal pressure along the cutting edges

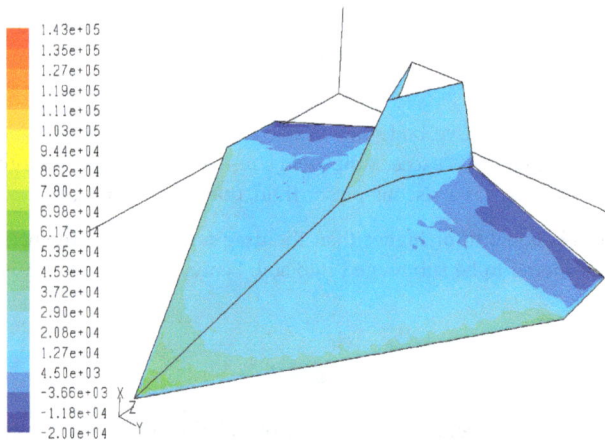

**Figure 8.** Distribution of the normal pressure (Pa) on the surface of the active part

As previously mentioned, drag has two components, one due to pressure and the other due to friction; therefore, the same two components (pressure and viscosity) are present within the traction force (Table 4).

The tangential stress distribution over the surface of the active part (the viscosity or friction component) reaches a maximum on the cutting edges and diminishes towards the inner part of the active body (Figure 9).

Figure 10 presents the variation of the soil speed; the soil is disturbed in front of the active body and the distribution pattern of the flowing speed clearly indicates the region where the cracks appear. Speed variation along the longitudinal (z) axis presents two distinct flowing regions (Figure 10b): one corresponds to the viscous-plastic flow, in the vicinity of the active part, when the tangent stress exceeds the yield point, and the second one corresponds to the area where the tangent stress does not exceed the yield point, while the second implies a flowing resistance ("solid flow"), when the tangent stress does not exceed the yield point; the soil is an elastic state.

Because the simulation assumes that soil flows while the active part is considered stationary, the soil speed on its surface is zero; speed is also zero in the contact area with the soil bin walls. Moving away from the surface of the active part, the speed increases; as a result, near the active part, the tangent stress records very high levels and the yield point is exceeded, leading to the formation of cracks and to a viscous-plastic strain.

| Region | Force due to static pressure (N) | | | Force due to viscosity (N) | | | Overall force (N) | | |
|---|---|---|---|---|---|---|---|---|---|
| | $2\gamma = 60^0$ | $2\gamma = 66^0$ | $2\gamma = 70^0$ | $2\gamma = 60^0$ | $2\gamma = 66^0$ | $2\gamma = 70^0$ | $2\gamma = 60^0$ | $2\gamma = 66^0$ | $2\gamma = 70^0$ |
| Surface of the active part | 86.1 | 95.6 | 102.4 | 415.6 | 520.4 | 642.3 | 501.7 | 616.0 | 744.7 |
| Surface of the active part holder | 40.07 | 36.3 | 39.1 | 46.2 | 42.5 | 32.8 | 86.27 | 78.8 | 71.9 |
| TOTAL | 126.17 | 131.9 | 141.5 | 461.8 | 562.9 | 675.1 | 587.97 | 694.8 | 816.6 |

**Table 4.** Traction force for the winged share type active part

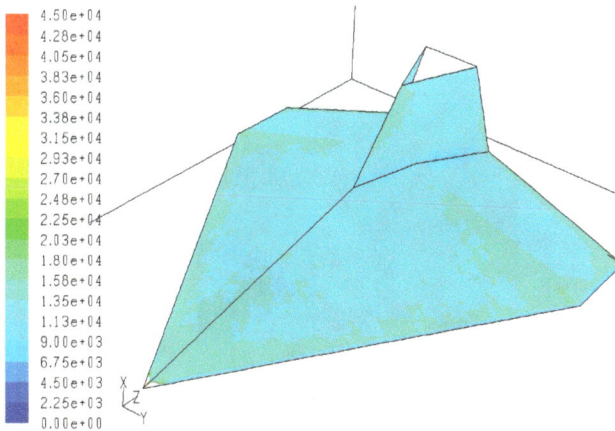

**Figure 9.** Distribution of the tangential stress (Pa)

# 5. Physical and mathematical modeling of the tire wheel-soil interaction

## 5.1. Simulation principles applied within the finite elements method (FEM)

The finite elements method (FEM) is used in order to study the bodies with a complex shape, providing numerical solutions for different physical characteristics when analytical solutions are impossible or very difficult to obtain. The finite element analysis (FEA) is used within this method [6].

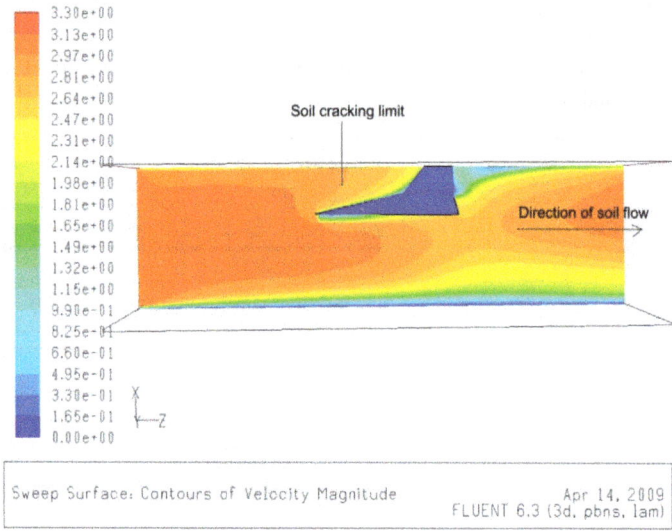

(a)

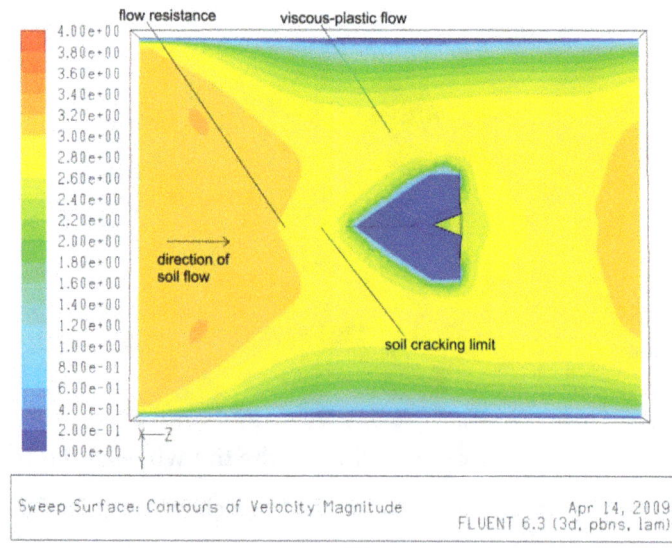

(b)

**Figure 10.** Soil speed profile (m/s) in the longitudinal vertical plane XOZ (a) and horizontal plane YOZ (b), for $2\gamma = 66^0$.

The finite elements method is based on the principle of the overall potential energy, which states that a structure or a body is deformed or displaced in a position that minimizes the potential energy (overall potential). This is in correspondence with the second law of thermodynamics, which states that the entropy of an isolated system can only increase towards a maximum value, meaning that the capacity to produce mechanical work can only decrease.

The principle of the minimum potential energy has many applications in the mechanics of the solid bodies and in the analysis of structures. In these cases, the principle of the minimum overall potential is a special case of the principle of virtual mechanical work applied to systems being under the action of conservative forces. The principle of the virtual mechanical work states that the virtual mechanical work of the exterior forces is equal and opposed to the virtual mechanical work of the interior forces (normal stress, shear stress, torsion and bending stress). It is assumed that forces and stresses remain constant and only the variations of strains are taken into account; only the strains that satisfy the internal compatibility of the body and the boundary conditions (resulting from the connections to other bodies) are accepted.

The finite elements method was imposed by the need to solve complex problems regarding the mechanics of deformable bodies. The method may be also applied to the problems referring to the flow of fluids, heat transfer, magnetic fields etc. [16].

## 5.2. Numerical simulation of the wheel-soil interaction

The first applications of the finite elements method – FEM – were aimed to simulate the linear elastic materials, but now this method is also used for non-linear, non –elastic materials like soil. Unlike metals, soil has a very low tensile strength; when compressed, the deformation reaches the plastic domain and soil has the tendency to remain in the deformed state. The wheel-soil interaction takes place in the non-linear domain, which means that a greater number of finite elements should be used in the simulation; the computation time increases accordingly.

In agriculture, the 3D simulations of the wheel-soil interaction aim to produce solutions leading to the limitation of soil compaction. A certain number of tests must be performed, with different types of soils and the results of the tests are than used in the numerical simulation in order to obtain a realistic prediction. These results allow the reduction of the design costs and the design time of the agricultural equipment according to the principles of sustainable agriculture.

The 3D numerical simulation based on FEM assumes some simplification hypothesis: the wheel is considered to be under the action of a dynamic vertical load; the wheel is assumed rigid; the number of vertices is limited by the computing power available; the soil characteristics are estimated ones, due to its heterogeneous characteristics.

**Description of the model**. The FEM software ABAQUS was used in order to build the 3D model for the analysis of soil compaction. Taking into account the complexity of the model the "explicit" version of the software was used, instead of the "standard" one; this approach allows the further development of the model by considering a deformable wheel instead of a rigid one.

The model is composed of two distinct 3D elements: a portion of the soil bin and the rigid wheel. In order to diminish the computation time, a vertical plane of symmetry, parallel to the direction of travel, was considered, dividing the contact patch into two symmetrical portions; simulation was therefore performed on only one half of the 3D soil-wheel model. The cartesian reference system was defined as follows (Figure 11): the negative z axis was considered to be along the travel direction, the positive x axis was perpendicular to the soil surface and the y axis was perpendicular to the travel direction and parallel to the wheel axis.

**Static soil model**. The 3D model of the soil was divided into five sections (Figure 11), each with a different density of the dicretization mesh; the contact surfaces between the sections were considered to be bound to each other. A higher density of the mesh was used for the wheel-soil contact area, on a depth of 0.05 m; a lower mesh density was used for the rest of the model.

The finite soil elements are 3D solid parallelepipeds, with 8 vertices and only three degrees of freedom, corresponding to the translations along the directions x, y and z. The computation time is thus significantly reduced and the stability of the model is increased.

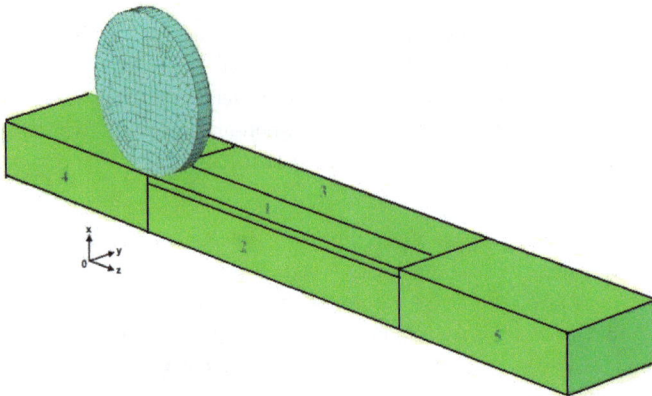

**Figure 11.** The 3D wheel-soil model

In order to extend the length and width of the simulation model to infinity, infinite solid elements were attached; they were considered 3D parallelepipeds, with eight vertices and one infinite face. These elements were used for the elastic linear deformation domain and were attached to the boundaries of the solid element.

The strain-stress curve was considered to be composed of linear portions, as shown in Table 5. The physical and mechanical properties of soil were those presented by Block [3]. The density of soil was $\rho$ = 1255.2 kg/m³, the modulus of elasticity was E = 0.3262 MPa and the initial Poisson's ratio was assumed to be $\nu$ = 0.3.

| Yield strength (MPa) | Volumetric strain | Yield strength (MPa) | Volumetric strain |
|---|---|---|---|
| 0 | 0 | 0.08 | 0.149679 |
| 0.005 | 0.014661 | 0.09 | 0.160028 |
| 0.01 | 0.028334 | 0.10 | 0.169280 |
| 0.02 | 0.053024 | 0.12 | 0.185036 |
| 0.03 | 0.074619 | 0.16 | 0.208422 |
| 0.04 | 0.093572 | 0.21 | 0.228045 |
| 0.05 | 0.110262 | 0.29 | 0.248232 |
| 0.06 | 0.125006 | 0.40 | 0.266976 |
| 0.07 | 0.138069 | 0.50 | 0.280999 |

**Table 5.** The strain/stress curve

The infinite elements use the same material properties; the initial elasticity modulus increases with compaction, due to the increase of the rigidity. The Poisson's ratio is modified during the simulation. The soil was modeled with ABAQUS/Explicit as "cap plasticity" and optionally as rigid "cap hardening". These options allow the plastic strain to start from a prescribed level, which is included in the Drucker-Prager rigidity model. Each layer of soil has an initial volumetric strain corresponding to the hydrostatic pressure induced by the weight of the above layers. An initial compaction of the soil was taken into account. The "cap plasticity" option imposes the definition of three supplementary parameters (for the sandy clay soil): the cohesion coefficient c = 350 Pa, the friction angle $\beta$ = 57.8⁰ and the "cap eccentricity" parameter R = 0.0005. The values "cap plasticity" and "cap eccentricity" were evaluated in the process of optimizing the Block model.

The yield strength due to the hydrostatic pressure was limited to values between 0.005 and 0.5 MPa.

**Dynamic model of the rigid wheel.** The rigid wheel was considered a cylinder, composed of 3D discrete rigid elements. The simulation in ABAQUS/Explicit was performed for two types of loads: concentrated load, applied in the center of the wheel, and distributed load, applied on the circumference of the wheel.

**Boundary conditions.** Specific boundary and loading conditions were considered within the simulation. The boundary conditions were defined for the soil model, for the rigid wheel model and for the wheel-soil interaction. The base of the soil model was restrained from moving along the x, y and z axis; the initial condition was defined considering an untilled soil with a certain degree of compaction. No constraint was defined for the soil-wheel interface.

Two constraints were defined for the wheel, referring to the acceleration of gravity and wheel rotation.

The interaction between wheel and soil was simulated by the means of two surfaces, corresponding to the cylindrical surface of the wheel and exterior soil surface. Thus, the wheel gradually loaded the soil. No slip was assumed during wheel rolling.

**Loading conditions.** The maximum wheel load of 5.8 kNa was reached in 12 time steps. The duration of each of the first five steps was set to one second; during these steps, each element was progressively loaded from zero to 9.81 $m/s^2$. After the first five seconds, the wheel acceleration was constant. During the following seven time steps (each with a duration of two seconds), the rotation speed of the wheel was increased from zero to 0.244 rad/s (16.74 cm/s travel speed).

**Simulation results.** The simulation was aimed to evaluate soil compaction, the indentation depth of the wheel and the normal stress. The 3D simulation emphasized the residual soil stress, leading to soil compaction and destruction of the structure in the contact area.

Three locations were chosen for the evaluation of the normal stress. The first one was placed at a depth of 0.3 m under the center of the wheel-soil contact patch, the second one was placed at a depth of 0.3 m under the outer edge of the wheel (y = 0,16 m) and the third one was placed at 0.15 m beneath the center of the contact patch. Table 6 presents the values of the normal stress, after 12 seconds, in the above-mentioned locations.

| Wheel load (kN) | Indentation depth (m) | Location | Normal stress (kPa) |
|---|---|---|---|
| 5.8 | 0.101 | 1 | 14.5 |
|  |  | 2 | 7.0 |
|  |  | 3 | 17.0 |

**Table 6.** Simulation results

# 6. The effect of tillage operations and of the agricultural traffic over soil

## 6.1. Materials and methods

The task of tillage is to prepare soils for productive use; it is performed in order to bring the seedlings into the soil and procure for them a good environment for further development. Tillage operations may have negative effects over the physical and mechanical properties of soil, over plant development and agricultural yield.

The experiments were aimed to investigate the effect of soil tillage technologies and soil compaction over the penetration resistance, apparent soil density, weighted average diameter of the soil's structural elements and water stability of the aggregates.

The experiments were performed on a mezocalcaric cambic chernozem, with a clay-loam texture and eastern exposure.

**Experimental variants**. The experimental variants are summarized in table 7.    Soil compaction was achieved before tillage, by running the Valtra T190 tractor once over the field (compacted once) or twice (compacted twice), at a travel speed of 2.5 km/h.

The experimental field was seeded with the Glosa variety of autumn wheat (approved in Hungary, in 2005); the row spacing was 125 mm, and the seeding depth was 5 cm.

| Agricultural equipments and operations | Soil compaction | Experimental variants |
|---|---|---|
| Plowing with T190[1] + Opal 140 | non-compacted | V$_1$ - control |
| Secondary tillage with U-650[2] + GD-3,2 | compacted once | V$_2$ |
| Sowing with U-650 + SUP-29 | compacted twice | V$_3$ |
| Plowing with T190 + Opal 140 | non-compacted | V$_4$ |
| Secondary tillage with T190 – BS 400 A | compacted once | V$_5$ |
| Sowing with U-650 + SUP-29 | compacted twice | V$_6$ |
| Direct sowing, in untilled soil, with T190 + MCR-2,5 | non-compacted | V$_7$ |
|  | compacted once | V$_8$ |
|  | compacted twice | V$_9$ |
| Loosening of the unplough soil (14 cm), seedbed preparation and sowing with T190+OA+AGPS-24DR[3] | non-compacted | V$_{10}$ |
|  | compacted once | V$_{11}$ |
|  | compacted twice | V$_{12}$ |
| Plowing with T190 + Opal 140 Seedbed preparation and sowing with T190+ AGPS-24DR[4] | non-compacted | V$_{13}$ |
|  | compacted once | V$_{14}$ |
|  | compacted twice | V$_{15}$ |

**Table 7.** Experimental variants

Evaluated soil degradation indices

**The specific resistance to penetration** was evaluated ten days after sowing, using the Eijelkamp (Holland) static penetrometer. The specific penetration resistance is ranked as:

- very low (lower than 1.08 MPa);
- low (1.08 – 2.45 MPa);
- average (2.45 – 4.9 MPa);
- high (4.9 – 9.81 MPa).

A specific penetration resistance lower than 2.45 MPa allows the normal development of plant roots; for values between 2.45 and 9.81 MPa the development of roots is partially limited; for values higher that 9.81 MPa, the roots development is stalled.

---

[1] T190 – Valtra tractor, 190 HP.
[2] U650 – Romanian tractor, 65 HP.
[3] OA+AGPS-24DR is composed of a rotary harrow and a seeder; OA are winged type shares mounted in the front of the rotary harrow.
[4] AGPS-24DR is composed of a rotary harrow and a seeder.

**Bulk density of soil**

In order to evaluate this index untilled soil samples were taken ten days after sowing, from the following depths: 0 – 10 cm, 10 – 20 cm, 20 – 30 cm and 30 – 40 cm.

For the clay-loam soil, the bulk density is ranked as:

- extremely low (loosened soil) – under 1.05 g/cm³;
- very low (moderately loosened soil) - 1.05 – 1.18 g/cm³;
- low (slightly loosened soil) - 1.19 – 1.31 g/cm³;
- average (slightly compacted soil) - 1.32 – 1.45 g/cm³;
- high (moderately compacted soil) - 1.46 – 1.58 g/cm³;
- very high (very compacted soil) – over 1.58 g/cm³.

**The weighted average diameter of the structural elements**. The structure of the arable layer is characterized by the grain size distribution. A device with sieves of different sizes (10; 5; 3; 2; 1; 0,5 and 0.25 mm) is used to evaluate this index, by sieving the soil samples.

The weighted average diameter is calculated with the formula:

$$WAD = \frac{\sum(p_i \cdot d_i)}{100}, \text{mm}$$

where $p_i$ is the share of each size fraction [%] and $d_i$ is the mean diameter of each size fraction [mm].

The Tiulin- Erikson method ranks the weighted average diameter as:

- very good (WAD = 2 – 5 mm);
- good (WAD = 1 -2 mm and 5 – 7 mm);
- acceptable (WAD = 0.25 – 1 and 7 – 10 mm);
- poor (WAD under 0.25 and over 10 mm).

**Water stability of the aggregates.** Represents the property of soil aggregates to withstand the dispersion action of water. The soil sample (20 g) is dispersed, entrained and sieved in a flow of water; in the Tiulin-Erikson apparatus, the sieves have apertures of 0.25; 0.5; 1; 2; 3; 5 mm.

The aggregates retained on each sieve are weighted and the $I_1$ water stability index is calculated with the formula:

$$I_1 = \frac{I + II + III}{IV + V}$$

where I is the percent of aggregates bigger than 5 mm, II is the percent of aggregates of 3 – 5 mm, III is the percent of aggregates of 2 – 3 mm, IV is the percent of aggregates with 1 – 2 mm and V is the percent of aggregates of 0.5 - 1 mm. According to $I_1$, the water stability of aggregates is ranked as follows:

- 3.00 – 5.00 – very good structure;
- 0.61 – 3.00 – good structure;
- 0.30 – 0.61 – medium structure;
- 0.18 – 0.30 – poor structure;
- lower than 0.18 – bad structure.

## 6.2. Results and discussions regarding the effect of soil tillage technologies and soil compaction over the physical and mechanical properties of soil

The experiments were aimed to investigate the effect of soil tillage technologies and soil compaction on the penetration resistance, apparent soil density, weighted average diameter of the soil's structural elements and water stability of the aggregates.

**Resistance to penetration.** The experiments were performed over three agricultural years (2008/2009, 2009/2010, 2010/2011) and the penetration resistance was evaluated for four depths (0 – 10 cm, 10 – 20 cm, 20 – 30 cm, 30 – 40 cm); the averaged values for the agricultural years and 0 – 40 cm depth are presented in table 8.

The results show very low values of the penetration resistance (0.438); according to the agro technical requirements, values lower than 1.08 MPa are considered "very low" and roots may develop normally.

Based on the penetration resistance, the five variants referring to the non-compacted soil were ranked as follows: $V_{13}$ (the best), $V_4$, $V_1$, $V_{10}$ and $V_7$ (the worst). Because soil was not compacted, these values were considered as reference in order to establish the effect of the tillage equipments.

| Compaction | Penetration resistance [MPa] | | | | |
|---|---|---|---|---|---|
| Non-compacted | $V_1$ | $V_4$ | $V_7$ | $V_{10}$ | $V_{13}$ |
| | 0.247 | 0.236 | 0.286 | 0.260 | 0.215 |
| Compacted once | $V_2$ | $V_5$ | $V_8$ | $V_{11}$ | $V_{14,}$ |
| | 0.326 | 0.315 | 0.356 | 0.347 | 0.306 |
| Compacted twice | $V_3$ | $V_6$ | $V_9$ | $V_{12}$ | $V_{15}$ |
| | 0.406 | 0.390 | 0.438 | 0.410 | 0.387 |

**Table 8.** The effect of the tillage technologies over the resistance to penetration

The experimental data show that the compaction of the soil (once and respectively twice) has significantly increased the penetration resistance.

The lowest values of the penetration resistance were recorded for the variants $V_{13}$ (non-compacted soil), $V_{14}$ (compacted once) and $V_{15}$ (compacted twice), when plowing was performed with the T190 + Opal 140 unit and the seedbed preparation and sowing were performed with the T190+ AGPS-24DR equipment.

Low values of the penetration resistance were also recorded for the variants $V_4$ (non-compacted), $V_5$ (compacted once) and $V_6$ (compacted twice), when the same equipment was used for plowing (T190 + Opal 140) and the BS 400 combined equipment was used for seedbed preparation.

The increase of the penetration resistance (compared to variants $V_{13}$, $V_{14}$ and $V_{15}$) was due to the presence of cage rollers and cross kill roller (mounted in the back of the equipment).

For the following variants ($V_{10}$, $V_{11}$ and $V_{12}$), classical plowing was replaced by a tillage operation performed with the OA + AGPS-24DR complex equipment; the soil was loosened over a depth of 15 cm by the OA winged type shares, mounted in front of the rotary harrow. As a result, the penetration resistance increased significantly.

Strip tillage was performed for the variants $V_7$, $V_8$ and $V_9$, by the means of the MCR 2.5 combined equipment; in this case, isolated bands of soil were tilled (only one third of the equipment's working width), to a depth of 8 cm. This solution led to the higher values of the penetration resistance.

**Soil bulk density**. Table 9 presents the experimental results referring to the bulk density of soil. The values are averaged ones, for three agricultural years and four depths.

| Compaction | Bulk density [g/cm³] | | | | |
|---|---|---|---|---|---|
| Non-compacted | $V_1$ | $V_4$ | $V_7$ | $V_{10}$ | $V_{13}$ |
| | 1.31 | 1.28 | 1.33 | 1.32 | 1.26 |
| Compacted once | $V_2$ | $V_5$ | $V_8$ | $V_{11}$ | $V_{14,}$ |
| | 1.42 | 1.40 | 1.45 | 1.43 | 1.39 |
| Compacted twice | $V_3$ | $V_6$ | $V_9$ | $V_{12}$ | $V_{15}$ |
| | 1.51 | 1.49 | 1.58 | 1.54 | 1.46 |

**Table 9.** Effect of tillage technologies over the bulk density of soil

For variants $V_{13}$, $V_4$ and $V_1$ (non-compacted soil) the bulk density was low (poorly loosened soil), while average values were recorded for variants $V_{10}$ and $V_7$ (slightly compacted soil). Average values were also recorded for all the "compacted once" variants (slightly compacted soil); the high values of the bulk density reported for the "compacted twice" variants ranked them as "moderately compacted".

Taking into account that lower values of the bulk density are desirable, the "non-compacted" variants were ranked as follows: $V_{13}$ (the best), $V_4$, $V_1$, $V_{10}$ and $V_7$ (the worst).

The experimental results showed that bulk density increases when soil compaction increases. The "compacted once" and "compacted twice" variants were ranked exactly in the same order as the "non-compacted" variants, in terms of bulk density, because their ranking is only due to the type of tillage equipment. Moreover, the same ranking was recorded in terms of penetration resistance, because the both indices are characterizing the compaction state of soil and the increase of one index implies the increase of the other one, too.

**Average weighted diameter (WAD) of the structure elements**. Table 10 presents the experimental results referring to the average weighted diameter of the soil's structure elements. The values are averaged ones, for three agricultural years and three depths.

| Compaction | Weighted average diameter [mm] | | | | |
|---|---|---|---|---|---|
| Non-compacted | $V_1$ | $V_4$ | $V_7$ | $V_{10}$ | $V_{13}$ |
| | 3,70 | 3,68 | 3,56 | 3,40 | 3,35 |
| Compacted once | $V_2$ | $V_5$ | $V_8$ | $V_{11}$ | $V_{14,}$ |
| | 3,05 | 3,09 | 3,34 | 3,20 | 3,19 |
| Compacted twice | $V_3$ | $V_6$ | $V_9$ | $V_{12}$ | $V_{15}$ |
| | 2,84 | 2,88 | 3,29 | 2,97 | 2,90 |

**Table 10.** Effect of tillage technologies over the weighted average diameter.

WAD varied between narrow limits – 2.845 mm to 3.702 mm. The results include the weighted average diameter in the 2 – 5 mm class (very good), for all the variants. Within this class, the best weighted average diameter is the one that is closest to 3.5 mm.

According to this index, the "non-compacted" variants were ranked as follows: $V_7$ (the best), $V_{10}$, $V_{13}$, $V_4$ and $V_1$ (the worst).

The "compacted once" and "compacted twice" variants were ranked exactly in the same order as the "non-compacted" variants, in terms of weighted average diameter, due to the facts previously mentioned.

Increasing the compaction of soil resulted in a marked destruction process, the weighted average diameter being diminished.

Variant $V_7$ was ranked the first due to the limited tillage process, during which the soil aggregates were less affected; when using the MCR-2.5 equipment, only isolated bands of soil were tilled (on one third of the equipment's working width), to a lower depth of 8 cm and thus the structural elements are preserved.

Variant $V_{10}$ was ranked second because soil plowing was replaced by loosening to a depth of 15 cm. The fact that seedbed preparation and sowing were performed simultaneously also contributed to the preservation of the soil's structure. In the same time, the more intense tillage (performed by the OA winged type shares and the rotary harrow) applied within this variant led to lower weighted average diameters of the aggregates.

In the case of variant $V_{13}$, which was ranked the third, classic moldboard plowing led to the destruction of the structure elements; seedbed preparation by the means of the FRB-3 rotary harrow (part of the AGPS-24DR complex equipment) also contributed to the diminishing of the aggregates' diameter.

Variant $V_4$ was ranked the fourth due to conventional plowing, followed by the secondary tillage performed with the BS 400A combined equipment. The great number of tillage equipments within its structure, of which three are rollers, had an unfavorable effect over the soil structure.

**Water stability of the aggregates.** Table 11 presents the experimental results referring to the water stability of the soil's aggregates. The values are averaged ones, for three agricultural years and three depths.

| Compaction | $I_1$ index | | | | |
|---|---|---|---|---|---|
| Non-compacted | $V_1$ | $V_4$ | $V_7$ | $V_{10}$ | $V_{13}$ |
|  | 3.60 | 3.61 | 4.18 | 3.70 | 3.68 |
| Compacted once | $V_2$ | $V_5$ | $V_8$ | $V_{11}$ | $V_{14,}$ |
|  | 2.84 | 2.96 | 3.43 | 3.18 | 3.02 |
| Compacted twice | $V_3$ | $V_6$ | $V_9$ | $V_{12}$ | $V_{15}$ |
|  | 2.64 | 2.68 | 2.93 | 2.77 | 2.78 |

**Table 11.** Effect of tillage technologies over the water stability of aggregates ($I_1$ index)

The values of the $I_1$ index were comprised between 2.64 and 4.18, which means that two classes were included: very good structure (3.00 – 5.00) and good structure (0.61 – 3.00). The higher the $I_1$ index, the better the soil structure. Within the 3.00 – 5.00 class, the best soil structure is the one that is closest to 4.00.

For the "non-compacted" variants, their ranking was as follows: $V_7$ (first place), $V_{10}$, $V_{13}$, $V_4$ and $V_1$ (last place).

The "compacted once" and "compacted twice" variants were ranked exactly in the same order as the "non-compacted" variants in terms of weighted average diameter, due to the previously mentioned facts regarding soil compaction.

The ranking of the variants respected the same order as in the case of the weighted average diameter.

It should be mentioned that an increased soil compaction results in a lower $I_1$ index; this tendency becomes more important when passing from the "non-compacted" variants to the "compacted once" variants.

## 6.3. Results regarding the effect of tillage technologies and soil compaction over the yield of seed for the autumn wheat crop.

The experimental variants are the ones already presented in the discussion referring to the effect of soil tillage and compaction over the physical and mechanical properties of soil.

The results regarding the seed yield are presented in table 12, being the averaged values for three agricultural years.

The experimental variants on the first five places ("non-compacted" soil) are ranked as follows: $V_{10}$ (best), $V_7$, $V_{13}$, $V_4$ and $V_1$ (5$^{th}$ place).

The next five places belong to the "compacted once" variants, in the following order: $V_{11}$ (6$^{th}$ place), $V_8$, $V_{14}$, $V_5$ and $V_2$ (10$^{th}$ place).

The last five places were taken by the "compacted twice" variants: $V_{12}$ (11$^{th}$ place), $V_9$, $V_{15}$, $V_6$ and $V_3$ (15$^{th}$ place).

As far as the tillage and sowing technology are concerned, the order is similar for the "non-compacted", "compacted once" and "compacted twice" variants, as only the type of tillage and sowing equipment affected ranking.

It should be noticed that yield decreases when soil compaction increases; this tendency is more significant when passing from the "non-compacted" variants to the "compacted once" variants.

| Tillage and sowing equipments | Compaction | Experimental variants | Seed yield, kg/ha |
|---|---|---|---|
| Valtra T190 + Opal 140; | non-compacted | $V_{1 - witness}$ | 5765 |
| U-650 + GD-3,2; | compacted once | $V_2$ | 4453 |
| U-650 + SUP-29 | compacted twice | $V_3$ | 4025 |
| Valtra T190 + Opal 140; | non-compacted | $V_4$ | 5800 |
| Valtra T190 + BS 400 A; | compacted once | $V_5$ | 4518 |
| U-650 + SUP-29 | compacted twice | $V_6$ | 4104 |
| | non-compacted | $V_7$ | 6016 |
| Valtra T190 + MCR-2,5 | compacted once | $V_8$ | 4789 |
| | compacted twice | $V_9$ | 4258 |
| | non-compacted | $V_{10}$ | 6268 |
| Valtra T190 + OA + AGPS-24DR | compacted once | $V_{11}$ | 5461 |
| | compacted twice | $V_{12}$ | 4297 |
| | non-compacted | $V_{13}$ | 5834 |
| Valtra T190 + Opal 140 Valtra T190 + AGPS-24DR | compacted once | $V_{14}$ | 4557 |
| | compacted twice | $V_{15}$ | 4182 |

**Table 12.** Seed yield for the autumn wheat crop

# 7. Conclusions and recomendations

Soil has an essential part in maintaining life on earth because it represents the support for the agriculture practice, creating the necessary conditions for obtaining the food products, due to its physical and biological properties, to its fertility, to its capacity to provide plants with the water and nutrients needed for their growth.

The intensification of the agricultural processes - mechanization, fertilization - led to a continuous degradation of soil, affecting five to seven million hectares each year. Sustainable agriculture could be a solution to this problem.

Sustainable agriculture is based on the soil conservation tillage system (SCTS). Within the unconventional soil tillage system moldboard plowing is deferred (completely or partially), the tillage works are limited and at least 15...30% of the vegetable debris is kept on the soil surface. The unconventional soil tillage system consists of very different methods: seeding in the untilled soil, reduced soil tillage, minimum soil tillage (when up to 30% of the vegetable debris is left on the soil surface), minimum tillage with vegetable mulch (more than 30% of the vegetable debris is left on the soil surface), ridge seeding, partial or strip tillage, deep loosening without furrow overturning etc.

The intensity of the anthropic soil compaction is affected by the type of agricultural machinery; compaction is promoted by the use of heavy machinery, with high wheel-soil contact pressures, by the increased number of passings, by the increased tire air pressure, by the agricultural traffic performed over wet soil. As far as the active parts for seedbed preparation are concerned, their destructive action over the structure elements of soil is of an utmost importance. The structure elements are destroyed through deformation, breaking, fragmentation, and cutting; in order to preserve soil structure one should comply with the technical recommendations of the equipment's manufacturer regarding the working speed and peripheral speed of the active parts.

In order to reproduce, in laboratory conditions, the working process of the active parts of the tillage equipment, a test rig was designed and built. The rig was used in order to study the soil - moldboard plough body interaction and the tire wheel-soil interaction; the results were then used to simulate the respective processes. The mathematical and physical simulations were performed by the means of CFD method. It was concluded that, in the case of the winged share type of active part, the maximum pressure would be recorded on the cutting edges. Based on the results of the simulations and on further field tests, new types of active parts will be developed; within the frame of conservative agriculture, this method allows the reduction of the time needed to produce new types of active parts.

The results of the mathematical and physical simulation were focused on soil, aiming to evaluate the compaction depth, the wheel indentation depth and the normal soil stress. The dynamic 3D simulation of a wheel rolling over an isotropic and non-linear soil produced truthful results regarding soil deformation and stresses at different depths, emphasizing the remanent soil stresses which lead to soil compaction and to the deterioration of its structure.

The laboratory results were validated by the field experiments, in which the effects of different agricultural equipments and of soil compaction over the penetration resistance, soil structure and yield were investigated.

Based on the results regarding the penetration resistance and bulk density, the variants (as presented in Table 7) were ranked as follows: $V_{13}$ (1st place), $V_4$ (2nd place) and $V_{1\text{-witness}}$ (3rd place).

The analysis of the weighted average diameter and water stability of the aggregates led to the following rating: $V_7$ (1st place), $V_{10}$ (2nd place) and $V_{13}$ (3rd place).

The analysis of the seed yield led to the following rating: $V_{10}$ (1st place), $V_7$ (2nd place) and $V_{13}$ (3rd place).

When the indices referring to soil structure were considered, the best results were obtained by the variant $V_7$ (non compacted soil; tillage and sowing performed with the T190 + MCR-2.5 equipment); the second place was taken by variant $V_{10}$ (non-compacted soil; tillage and sowing performed with the T190 + OA + AGPS-24DR equipment); variant $V_{13}$ (plowing performed with the T190 + Opal 140 equipment, sowing performed with the T190 + AGPS-24DR equipment) was ranked the third.

Taking into account all the facts we consider that all decision makers, who are connected in anyway with soil (farmers, equipment producers, chemical products manufacturers, researchers), should be focused on the degradation of the arable layer and encourage agricultural technologies aiming to its preservation.

## Author details

Ioan Tenu, Petru Carlescu, Petru Cojocariu and Radu Rosca
*University of Agricultural Sciences and Veterinary Medicine, Iasi, Romania*

## 8. References

[1] Abo-Elnor M., Hamilton, R., Boyle, J. T. (2004) Simulation of soil-blade interactions for sandy soil using advanced 3D finite element analysis. Soil&Tillage Research, 75: 61-73.

[2] Agoegwu S.N. (1994) Egusi-melon response to compaction due to traffic on a sandy loam soil. Proceedings of the 13th International Conference ISTRO, Denmark (1): 1-6.

[3] Block W.A. (1991) Analysis of soil stress under rigid wheel loading. Unpublished PhD Dissertation, Auburn University, Auburn AL.

[4] Canarache A. (1990) Physics of agricultural soils (in romanian). CERES Publishing House, Bucharest, Romania.

[5] Chi L., Kushwaha R.L. (1989) Finite element analysis of forces on a plane soil blade. Canadian Agricultural Engineering 31 (2): 135-140

[6] Eriksson J., Danfors B., Hakansson I. (1974) The effect of soil compaction on soil structure and crop yields. Bulletin 354, Swed. Inst. Agr. Eng., Uppsala, Sweden.

[7] Fervers C.W (2004) Improved FEM simulation model for a tire-soil interaction. Journal of Terramechanics 41(2-3): 87-100.

[8] Florea N. (2003) Degradation, protection and reclamation of soils and fields (in romanian). Universal Publishing House, Bucharest, Romania.

[9] Guş P., Rusu T. (2005) Sustainable development of agriculture (in romanian). Risoprint Publishing House, Cluj-Napoca, Romania.

[10] Hadas A., Larson W.E., Allmaras R.R. (1988) Advances in modeling machine-soil-plant interaction. Soil&Tillage Research, 11 (4): 349-372.

[11] Liu Y., Hou Z.M. (1985) Three-dimensional non-linear finite element analysis of soil cutting by narrow blades. Proceedings of the international conference on soil dynamics, Auburn, AL (2): 338-347.

[12] Shen J., Kushawa R.L. (1998) Soil-machine interaction: a finite perspective. Dekker, New York.

[13] Shmulevitch I., Asaf, Z., Rubinstein, D. (2007) Interaction between soil and a wide cutting blade using the discrete element method. Soil&Tillage Research, 97 (1): 37-50.

[14] Soane B.D., Bonne F.R (1986) The effect of tillage and traffic on soil structure. Soil&Tillage Research, 8: 303-306.

[15] Stout B.A. (editor) (1999). CIGH Handbook of agricultural engineering; vol. III – Plant production engineering. ASAE, St. Joseph, MI.

[16] Tanaka H., Momozu M., Oida A., Yamazaki M. (2000) Simulation of soil deformation and resistance at bar penetration by the distinct element method. Journal of Terramechanics, 37 (1): 41-56.

[17] Tolba M.L. (1987). Sustainable development-constraints and opportunities., Butterworth, London.

[18] Ţenu I., Cojocariu P., Cârlescu P., Roşca R., Leon D. (2010) Soil interaction with the active parts of the agricultural equipments (in Romanian). "Ion Ionescu de la Brad" Publishing House, Iaşi, Romania.

[19] Van Doren D., Unger P.W. (editors) (1982) Predicting tillage effects on soil physical properties and processes. ASA Special Publication 44, American Society of Agronomy and Soil Science Society of America Inc., Madison, WI, 198 pp.

[20] Wood R.K., Reeder, R.C., Morgan, M.T., Holmes, R.G. (1993) Soil physical properties as affected by grain cart traffic. Trans. of ASAE, 36 (1): 11-15.

[21] Yong R.N, Hanna A.W. (1977) Finite element analysis of plane soil cutting. Journal of Terramechanics, 14 (3):103-125.

[22] Zhang G.S., Chan K.Y., Oates A., Heenan D.P., Huang G.B. (2007) Relationship between soil structure and runoff/soil loss after 24 years of conservation tillage. Soil&Tillage Research, 92 (1-2):122-128.

# Chapter 10

# An Appraisal of Conservation Tillage on the Soil Properties and C Sequestration

Vikas Abrol and Peeyush Sharma

Additional information is available at the end of the chapter

## 1. Introduction

Soil is a fundamental natural resource on which civilization depends. Agricultural production is directly related to quality of soil. In view of the rapidly expanding global population and its pressure on the finite amount of land available for agricultural production; maintaining soil quality is essential not only for agricultural sustainability, but also for environmental protection. Maintenance of soil quality would reduce the problems of land degradation, decreasing soil fertility and rapidly declining production levels that occur in many parts of the world which lack the basic principles of good farming practices. Intensification of agricultural production has been an important factor influencing GHG emission and affecting the water balance. Currently, agriculture accounts for approximately 13% of total global anthropogenic emissions and is responsible for about 47% of total anthropogenic emissions of methane ($CH_4$) and 58% nitrous oxide ($N_2O$).

Soil tillage is one of the very important factors in agriculture that would affect soil physical properties and yield (Keshavarzpour and Rashidi, 2008). Among different operations, the soil tillage is considered one of the most important practices in agricultural production due to its influence on physical, chemical, and biological properties of the soil environment. The tillage would aim to create a soil environment favorable to the plant growth (Klute, 1982). Among different crop production factors, tillage contributes up to 20% (Khurshid et al., 2006). According to Lal (1979a, 1983), it is defined as physical, chemical or biological soil manipulation to optimize the conditions for germination, seedling establishment and crop growth. According to Antapa and Angen (1990), tillage is any operation or practice carried out to prepare the soil surface for the purpose of crop production. Ahn and Hintze (1990) state that tillage is nothing but physical loosening of the soil by a range of cultivation operations which could be either manually or mechanized. In the past, the soil tillage has been associated with an increased fertility, which originated from the mineralization of soil nutrients as a consequence of tillage operations. In the long term, this process would lead to a reduction of soil organic matter. Therefore, most soils degrade under prolonged intensive arable agriculture. This structural degradation of the soils would result in the formation of crusts and compaction and further would lead to soil erosion. Acharya and Sharma (1994) and Pagliai et al., (1995) reported that the structure of Ap horizon is largely influenced by the soil tillage system and the implements used for tillage operations. Soil tillage has a major influence on the water intake, storage, evaporation and absorption of water from the soil by plant roots, biological activity, and organic matter break down, which influence the soil aeration, soil moisture and soil temperature (Kathirval et al., 1992). Kovac and Zak (1999) found that the changes in soil physical properties were influenced by different tillage treatments but the changes were small and insignificant. Some authors pointed out that the tillage treatments affected the soil physical properties, especially, when the same tillage system has been practiced for a longer time (Jordhal and Karlen, 1993; Mielke Wilhelm, 1998). The proper use of tillage could improve soil related constrains, while an improper tillage would cause destruction of the soil structure, accelerated erosion, depletion of organic matter and fertility, and disruption in cycles of water, organic carbon and plant nutrients (Lal, 1993). Appropriate tillage practices are those that would avoid the degradation of soil properties but would maintain crop productivity as well as ecosystem stability (Lal, 1981b, c, 1982, 1984b, 1985a; Greenland, 1981).

Conventional soil management practices resulted in losses of soil, water and nutrients in the field, and degraded the soil with low organic matter content and a fragile physical structure, which in turn led to low crop yield, low water and fertilizer use efficiency. Conventional tillage overturns the soil layer, which breaks the structure of soil and as a result, decreases the permeability of soil (Kribaa et al., 2001). Annual disturbance and pulverizing caused by the conventional tillage produced a finer and loose soil structure as compared to conservation and no-tillage method which would leave the soil intact (Rashidi and Keshavarzpour, 2007). This difference results in a change of number, shape, continuity and size distribution of the pores network, which would control the ability of a soil to store and transmit air, water and agricultural chemicals. This in turn would control erosion, runoff and crop performance (Khan et al., 2001).

According to the Conservation Technology Information Center in West Lafayette, Indiana, USA, conservation tillage could be defined as "any tillage or planting system in which at least 30% of the soil surface is covered by plant residue after planting to reduce the erosion by water; or where.;p soil erosion by wind is the primary concern, with at least 1120 kg ha⁻¹ flat small grain residue on the surface during the critical wind erosion period." No tillage, minimum tillage, reduced tillage and mulch tillage are terms synonymous with conservation tillage as observed by Willis and Amemiya, (1973); Lal (1973, 1974, 1976b); Phillips et al., (1980); Greenland (1981); Unger et al., (1988); Antapa and Angen (1990); Opara-Nadi (1990); Ahn and Hintze (1990).In recent years, interest in conservation tillage systems has increased in response to the need to limit the erosion and promote water conservation (Hulugalle et al., 1986; Unger et al., 1988). Conservation tillage provides the best opportunity for halting degradation, restoring and improving soil productivity (Lal,

1983; Parr et al., 1990). It has the potential to aggrade the soil quality and reduce the soil loss by providing protective crop residue on soil surface and improving water conservation by decreasing evaporation losses (Carter, 1991).Conservation tillage leads to positive changes in the physical, chemical and biological properties of a soil (Bescanca et al., 2006). The effect of conservation tillage was to reduce the volume fraction of large pores and to increase the volume fraction of small pores relative to the conventional tillage (Bhattacharya et al., 2008). Soil organic matter was increased because of straw recycling, which can increase soil porosity (Lal et al., 1980 and Blanco et al., 2007). Many soil-surface modifications would influence the components in the WUE equation viz. manipulation of the soil surface by tillage and surface residue management or mulching, can increase soil water retention capacity, improve the ability of roots to extract more water from the soil profile, or decrease leaching losses (Hatfield et al., 2001). Soil physical properties that are influenced by conservation tillage include bulk density, infiltration and water retention (Osunbitan et al., 2005). The improved infiltration of rainwater into the soil increases water availability to plants reduces surface runoff and improves the groundwater recharge (Lipiec et al., 2005).Many studies showed that under edapho-climatic conditions, conservation tillage can lead to improvements in the water storage in the soil profile (Pelegrín et al., 1990; Moreno et al., 1997, 2001). Therefore, currently there is a significant interest and emphasis on the shift to the conservation tillage methods for the purpose of controlling erosion process (Iqbal et al., 2005). Under these conditions, improvements were also obtained in the crop development and yield, especially in dry years (Pelegrín, 1990; Murillo, 1998, 2001; Du Preez, 2001). Under arid or semi-arid climatic conditions, high temperatures limit the accumulation of organic carbon at the soil surface (Franzluebbers, 2002a, 2002b; Mrabet, 2002).

World soils, an important pool of active C, play a major role in the global C cycle and contribute to changes in the concentration of GHGs in the atmosphere (Lal et al., 1998). Intensive agriculture is believed to cause some environmental problems, especially related to water use, water contamination, soil erosion and greenhouse effect (Houghton et al., 1999; Schlesinger, 1985; Davidson and Ackerman, 1993). Minimizing the increase in ambient CO2 concentration through soil C management, reduces the production of GHGs and minimizes potential for climate change. In fact, agricultural practices have the potential to store more C in the soil than agriculture releases through land use change and fossil fuel combustion (Lal et al., 1998).

Improved soil and crop management practices, such as reduced tillage and increased cropping intensity, however, would increase SOC as compared to conventional practices (Halvorson et al., 2002a; Sherrod et al., 2003; Sainju et al., 2007). Many studies have reported that implementation of minimum tillage has occasionally caused yield losses, especially in the no tillage method (Rao, 1996; Kirkegaard et al., 1995; Silgram and Shepherd, 1999). As Warkentin (2001) pointed out, the global experience with minimum tillage, or direct drilling, results in equal and even slightly smaller, harvests than traditional tillage (by using mouldboard plough). Since the late sixties, many studies of the effects of conservation tillage systems on soil properties and crop yield have been conducted in many parts of the world.

A complete review is beyond the scope of this presentation. The objective of this study is to give an overview of the early studies on conservation tillage systems, discuss some results from present-day studies and outline research needs and goals for the future aimed at enhancing and sustaining the crop production through conservation tillage systems.

## 2. Agriculture's contribution to greenhouse gas emissions

Agricultural eco-systems represent an estimated 11% of the earth's land surface and include some of the most productive and carbon-rich soils. As a result, they play a significant role in the storage and release of C within the terrestrial carbon cycle (Lal et al., 1995). The primary sources of greenhouse gases in agriculture are the production of nitrogen based fertilizers; the combustion of fossil fuels; and waste management (Fig. 1). Livestock enteric fermentation or the fermentation that takes place in the digestive systems of the ruminant animals, results in methane emissions. The major considerations of the soil C balance and the emission of greenhouse gases from the soil are: potential increase of $CO_2$ emissions from soil contributing to the increase of the greenhouse effect, the potential increase in other gas emissions (e.g., $N_2O$ and $CH_4$) from soil as a consequence of land management practices and fertilizer use, and the potential for increasing C (as $CO_2$) storage into soils, which equals $1.3 – 2.4 \times 10^9$ metric tons of carbon per year, and to help reduce the future increases of $CO_2$ in the atmosphere.

Carbon dioxide is removed from the atmosphere and converted to organic carbon through the process of photosynthesis. As organic carbon decomposes, it is converted back to carbon dioxide through the process of respiration. During 2005, agriculture accounted for 10 to 12 percent of the total global human caused emissions of greenhouse gases, according the Inter-governmental Panel on Climate Change (IPCC, 2007). In the United States, greenhouse gas from agriculture accounts for 8 percent of all emissions and has increased since 1990 (Congressional Research Service, 2008).Conservation tillage, organic fertilizers, cover cropping and crop rotations would drastically increase the amount of carbon stored in the soils.

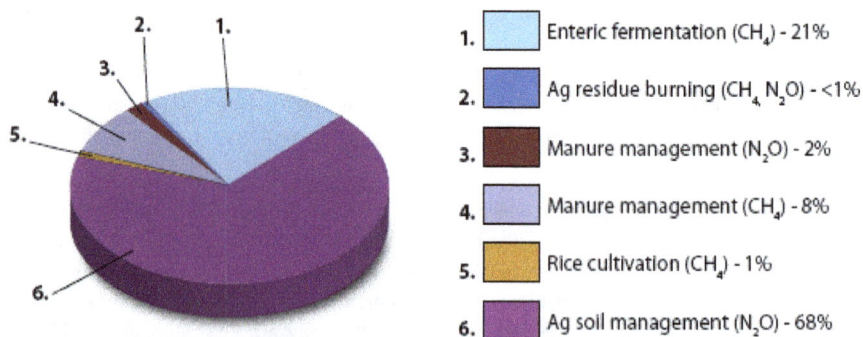

1. Enteric fermentation ($CH_4$) - 21%

2. Ag residue burning ($CH_4$, $N_2O$) - <1%

3. Manure management ($N_2O$) - 2%

4. Manure management ($CH_4$) - 8%

5. Rice cultivation ($CH_4$) - 1%

6. Ag soil management ($N_2O$) - 68%

**Figure 1.** Agricultural green house gas emission (Average 2001-2006), Source: EPA, 2007

Soils store a significant amount of carbon. It has been estimated that global soils contain approximately $1.5 \times 10^{12}$ metric tons of carbon. As a component of the carbon cycle (Fig. 2), soils can be either net sources or net sinks of the atmospheric carbon dioxide. Changes in the land use and agricultural activities during the past 200 years have made the soils act as net sources of atmospheric $CO_2$. Evidence from the long-term experiments suggests that the carbon losses due to oxidation and erosion could be reversed with appropriate soil management practices that would minimize the soil disturbance and optimize plant yield through fertilization. The soil tillage systems would have considerable impact on the environment by influencing the soil structure, which would further substantially affect the water quality, nutrients, sediments, pesticides and air quality and greenhouse effect (Holland, 2004; Hobbs, 2007).

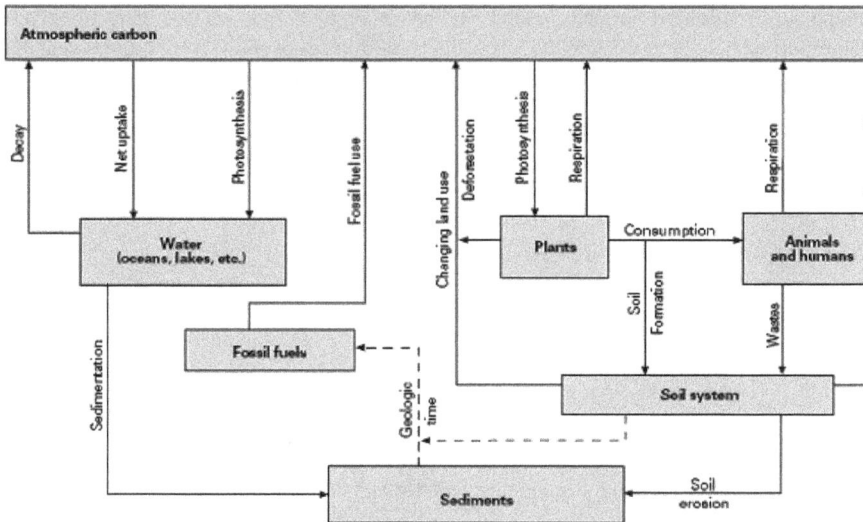

**Figure 2.** Carbon cycle aspects (modified from Paul and Clark)

# 3. Conservation tillage - the effects on soil properties

## 3.1. Organic matter

The amount of organic matter in a soil is often used as an indicator of the potential sustainability of a system. Soil organic matter plays a key role in nutrient cycling and can help improve soil structure. The soil organic matter is the second biggest carbon pool of the planet after the oceans. The soil organic matter is essential to control erosion, water infiltration and conservation of nutrients, and is related with the soil quality. Moist, hot and well-aerated conditions favour rapid decay of organic additions. If the rate of organic matter addition is greater than the rate of decomposition, the organic fraction in a soil will increase (Fig. 3). Reicosky (1997) reported that moldboard plow lost 13.8 times more $CO_2$ as the soil not tilled while conservation tillage systems averaged about 4.3 times more $CO_2$ loss. Reicosky et al. (2002) found that 30 years of fall moldboard plowing reduced the SOC whether the above ground corn biomass was removed for silage or whether the stover was returned and plowed into the soil. Their results suggest that no form of residue management will increase SOC content as long as the soil is moldboard plowed. Hooker et al. (2005) also found that within a tillage treatment, residue management had little effect on SOC in the surface soil layer (0-5 cm). Tillage tended to decrease the SOC content, although only no till combined with stover return to the soil resulted in an increase in SOC in the surface layer compared with moldboard plowed treatments. Walling (1990) reported that over the last 40 years the amount of organic matter being returned to the soil has declined, primarily as a consequence of more intensive soil cultivation, the removal of crop residues, the replacement of organic manures with inorganic fertilizer, and the loss of grass leys from rotations. In addition, organic matter is being eroded from arable land to rivers disproportionately to its availability. Over this period losses of soil C were estimated at 30–50% and a large proportion of arable soils now contain less than 4% C.

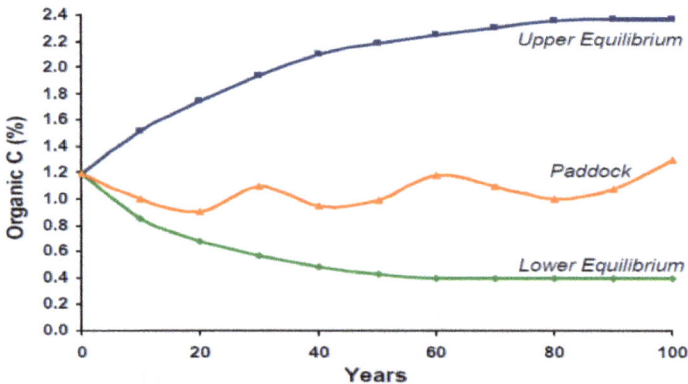

**Figure 3.** Organic carbon levels over time under different management systems

Murillo et al., (2001) compared the traditional tillage (TT) and conservation tillage (CT) under rainfed conditions in south-west Spain. The results indicated that CT improved the soil quality by reaching a greater soil resistance 'SR' (> 2) than that in TT (< 2). This fact could be due to the better water infiltration and storage in the soil profile under CT that would facilitate the uptake of water and nutrients by the plant in the periods of most droughts. In general, in any edapho-climatic context, a high stratification ratio is a good index of soil quality, since ratios above 2 are not common in the degraded systems (Franzluebbers, 2002a, 2002b; Mrabet, 2002). Application of conservation tillage to sandy loam and loamy soil type in the dryland, central rift valley of Ethiopia for five years, markedly improved the organic matter content, N concentration and soil moisture content (Worku, B. et al., 2006).

### 3.2. Bulk Density (BD)

Abu-Hamdeh (2004) studied the effect of tillage treatments (moldboard ploughing MB; chisel ploughing CS; and disk ploughing DP) for comparison of axle load on a clay loam soil. He reported that the dry bulk density from 0 to 20 cm was affected by the tillage treatments and from 20 to 40 cm by axle load. The MB treatment caused the maximum percentage increase of dry bulk density at all depths. These results reflect a more compact soil layer at the 0-10 cm depth than at the 10-20 cm depth. According to Lhotsky (1991), soil BD above 1.50 Mgm$^{-3}$ in the plough horizon on medium heavy soils has a negative effect on the growth and development of agricultural crops and was regarded as the threshold value of adverse soil compaction. Sharma et at (2011) observed that intensive tillage condition increased the bulk density (4.7 %) of a sandy loam soil as compare to the reduced tillage in rainfed condition. Osunbitan et al. ( 2005) observed greater bulk density in no-till system in the 5 to 10 cm soil depth. Logsdon et al. (1999) found no differences in bulk density between the different tillage systems. Jabro et al., (2009) in a 22 years study on a sandy loam soil found that the tillage practices [no-till (NT), spring till (ST), and fall and spring till (FST)] apparently had not significantly influenced the soil BD and only slight differences were observed in BD (Table 1). These findings are in agreement with those of Anken et al., (2004), and Lampurlanes and Cantero-Martinez (2003) but differ from results reported by Hill and Cruse (1985) and McVay et al., (2006). Based on 8 years studies Zhang et al., (2003) reported that the mean soil bulk density was 0.8-1.5% lower in ST (sub-soiling with retention of all surface plant residues) and NT treatments (consisted of zero tillage; planting was through the previous plant residues.) than in CT (consisted of manually removing all plant residues from the soil surface, followed by mouldboard ploughing). The crop residue retention has been reported to increase soil organic carbon and biotic activity (Lal 1989; Karlen et al. 1994, Tiarks et al., 1974; Schjonning et al., 1994)., thereby decreasing bulk density, particularly near the soil surface in the ST and NT plots under investigation.

| Parameter | BD (Mg m$^{-3}$) | GWC (g g$^{-1}$) |
|---|---|---|
| **Tillage** | | |
| NT | 1.59 | 0.141 |
| ST | 1.58 | 0.139 |
| FST | 1.61 | 0.135 |
| **Soil depth (cm)** | | |
| 0 - 5 | 1.49[a] | 0.144 |
| 5 - 10 | 1.68[b] | 0.136 |
| 10 – 15 | 1.60[c] | 0.135 |
| Tillage (T) | 0.439 | 0.515 |
| Soil depth (D) | 0.0001 | 0.253 |
| T × D | 0.990 | 0.161 |

**Table 1.** Effect of tillage and depth on bulk density (BD), and gravimetric water content (GWC)

## 3.3. Soil porosity

Porosity is a measure of the total pore space in the soil. This is measured as a volume or percent. The amount of porosity in a soil depends on the minerals that make up the soil and the amount of sorting that occurs within the soil-structure. For example, a sandy soil will have larger porosity than silty soil because the silt will fill in the gaps between the sand particles. Porosity characteristics differ among tillage systems (Benjamin, 1993). Soil porosity characteristics are closely related to the soil physical behavior, root penetration and water movement (Pagliai and Vignozzi 2002; Sasal et al. 2006). Previous researches showed that the straw returning could increase the total porosity of soil (Lal et al., 1980), while minimal and no tillage would decrease the soil porosity for aeration, but increase the capillary porosity; as a result, it enhances the water holding capacity of soil along with bad aeration of soil (Wang et al., 1994; Glab and Kulig, 2008). However, Børresen (1999) found that the effects of tillage and straw treatments on the total porosity and porosity size distribution were not significant. Allen et al., (1997) indicated that the minimal tillage could increase the quantity of big porosity. Zhang et al., (2003) compared the mean aeration porosity at two locations viz, Dazing and Changping in China in the top 0-0.30 m between conservation tillage treatments and conventionally tilled soil. The results illustrated an improvement in the soil porosity under conservation tillage (ST, sub-soiling with retention of all surface plant residues; and NT consisted of zero tillage and planting was through the previous plant residues) was most probably related to the beneficial effects of soil organic matter caused by minimum tillage and residue cover (Table 2). The increased porosity is especially important for the crop development since it may have a direct effect on the soil aeration and enhances the root growth (Oliveira and Merwin, 2001). The improved root growth would hence increase plant water as well as nutrient uptake. Within the conservation tillage treatments, ST produced more aeration porosity than NT, but the effect on capillary porosity appeared to be reversed in the 0-0.30 m soil layer. Husnjak and Kosutic (2002) reported that higher BD reduced the total porosity and changed the ratio of water holding capacity to air capacity in favour of water holding capacity. Total porosity below 45% on medium heavy soils had a negative effect on the plant growth (Lhotsky, 1991).

| Treatment | Total porosity | Aeration porosity (> 60 [micro]m) | Capillary porosity (> 60 [micro]m) |
|---|---|---|---|
| **Dazing** | | | |
| ST | 52.36 a | 42.64 a | 9.72 a |
| NT | 51.86 a | 41.19 a | 10.67a |
| CT | 45.58 b | 37.24 b | 8.34 a |
| **Changping** | | | |
| ST | 54.25 a | 46.32 a | 7.93 a |
| NT | 53.01 a | 42.99 a | 10.02a |
| CT | 45.74 b | 39.59 b | 6.15 a |

**Table 2.** Soil porosity for ST, NT, and CT treatments at 0-0.30 m depth in Dazing and Changping (Values within a column followed by the same letter are not significantly different at $p < 0.05$)

## 4. Infiltration rate and gravimetric water content

Infiltration is the process by which water on the ground surface enters into the soil. Infiltration is governed by two forces viz; gravity, and capillary action. Tillage disturbs the natural channels that have formed in a soil. The increase in porosity when soil is tilled may not result in an increase in the infiltration rate because of disruption of the vertical continuity of the pores (Kooistra et al., 1984). The plant roots are important in forming new channels (Parker and Jenny, 1945). Tillage plays a vital role in the conservation of soil moisture at different depths in the rainfed cultivation. It would also improve the soil condition by altering the mechanical impedance to root penetration, hydraulic conductivity and water holding capacity (Dexter, 2004). Lal (1978) measured the infiltration rates of 480 mm $h^{-1}$ for no-till and 150 mm $h^{-1}$ for the ploughed treatment after a field had been planted with maize (Zea mays L.) for 5 years. They found that surface residues prevented surface seal in the no-till treatments. Meek et al., (1989) measured a 17% increase in the infiltration rate in the field when soil was packed lightly before the first flood irrigation compared with no packing. Compacting loads of 335 kPa at field capacity on a sandy loam soil reduced infiltration rates to < 1% of the rate obtained when the soil was compacted air dry (Akram and Kemper, 1979). Increases in the bulk density usually result in large decreases in water flow through the soil. Antapa and Angen (1990) reported that retaining crop residues on the soil surface with conservation tillage would reduce evapo-transpiration, increase infiltration rate, and suppress weed growth. Numerous studies shown that the soil moisture and efficiency of moisture use tended to be higher under reduced tillage systems than conventional tillage system. Abu-Hamdeh (2004) observed that mould board plough caused a maximum decrease in the infiltration rate, while with Chiesl plough, CS treatment had the lowest effect. Sharma et al. (2011) observed increase in soil moisture content (12.4%, 16.6%) in minimum tillage (MT) in maize and wheat rotation respectively,in rainfed farming as compared to conventional tillage (Fig 4 & 5).

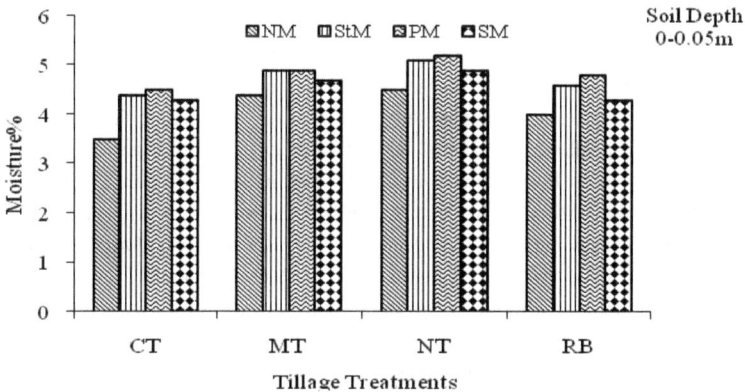

**Figure 4.** Effect of tillage and water management practices on soil water content at maize harvesting (CT= Conventional tillage,MT= Minimum tillage, NT= No till, RB= Raised bed; NM No mulch nad SM=straw mulch)

Jabro et al. (2009), in a long term study evaluated that tillage, soil depth and their interaction had no significant effect on the soil water content (Table 1). Not surprisingly, NT plots resulted in wetter soil to a depth of 10 cm in this study. The NT plots had greater grvimatric water content (GWC, 0.141 g g⁻¹), followed by ST having 0.139 g g⁻¹, and followed by FST with a mean of 0.135 g g⁻¹. Zhang et al., (2009) showed soil mean GWC values averaged across three tillage systems were 0.144, 0.136, and 0.135 g g⁻¹ at 0 to 5 cm, 5 to 10 cm, and 10 to 15 cm depths, respectively. The soil GWC generally was found to decrease with soil depth across the three tillage practices. This could be attributed to greater residues and organic matter in the soil surface than the subsurface proportions of the soil.

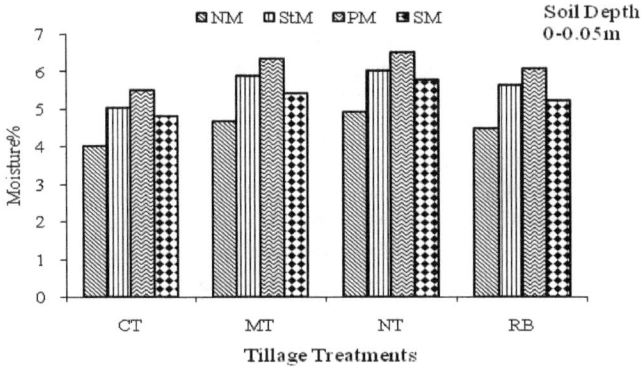

**Figure 5.** Effect of tillage and water management practices on soil water content at wheat harvesting (CT= Conventional tillage,MT= Minimum tillage, NT= No till, RB= Raised bed; NM No mulch nad SM=straw mulch)

## 5. Hydraulic conductivity (Ks)

The hydraulic conductivity of a soil is a measure of the soil's ability to transmit water when submitted to a hydraulic gradient. Iqbal et al., (2005) reported that the mean increase in saturated hydraulic conductivity observed was 4.5, 9.1 and 34.1% in the minimum, conventional and deep tillage treatments, respectively compared to zero tillage indicating that deep tillage increases the saturated hydraulic conductivity compared to other tillage methods. Kribba et al., (2001) reported that hydraulic conductivity values were significantly different between treatments of fallow soil tilled with chisel and disc ploughed fallow, and both treatments yielded higher values than untilled fallow. Mahboubi et al., (1993) found that no-tillage resulted in higher saturated hydraulic conductivity compared with conventional tillage after 28 years of tillage on a silt loam soil in Ohio. Whereas, Chang and Landwell (1989) did not observe any changes in the saturated hydraulic conductivity after 20 years of tillage in a clay loam soil in Alberta. Heard et al., (1988) reported that saturated hydraulic conductivity of silt clay loam soil was higher when subjected to 10 years of tillage than no-tillage in Indiana. They attributed the higher hydraulic conductivity of tilled soil to the greater number of voids and abundant soil macropores caused by the tillage implementation. Jabro et al., (2009) reported that the Soil $K_s$ was slightly influenced by tillage and varied from 3.295 mm h⁻¹ for intensive tillage (FST) to 5.297 mm h⁻¹ for no tillage

(NT), thus, soil $K_s$ decreased with increased intensity of soil manipulation by tillage practices (Fig. 6). Furthermore, previous research demonstrated that continuous tillage of 11 years had developed a compacted layer that impeded water movement at a depth of approximately 10 to 15 cm (Pikul and Aase, 1999; 2003). Soil macropores and aggregations under NT formed by decayed roots can be preserved under NT whereas conventional tillage breaks up the continuity of these macropores. Macropores generally occupy a small fraction of the soil volume but their contribution to water flow in soil is high. Patel and Singh (1981) reported that if the bulk density in a course-textured soil was increased from 1.7 to 1.9 Mg m$^{-3}$, hydraulic conductivity decreased by a factor of 260.

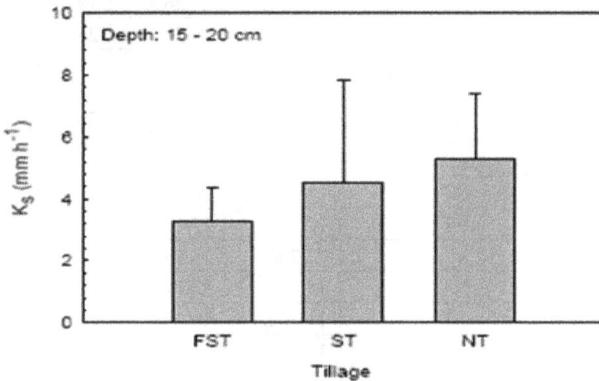

**Figure 6.** Soil saturated hydraulic conductivity, $K_s$, as affected by three tillage practices.

## 6. Soil structure

Good soil structure is important in allowing crop plants to yield well and resist erosion caused by the action of rainfall, melting snow in the early spring and wind. Conservation tillage practices were associated with a greater percentage of macro-aggregates (> 0.25 mm) than conventional tillage (Zhang et al., 2009). Mean macro-aggregates in 0-0.30 m soil depth at Daxing were 22.1% and 12.0% greater under ST (shallow tillage) and NT (no tillage) than CT, and the improvements at Changping were 18.9% under ST and 9.5% under NT (Table 3). These results were consistent with the increase in aggregation occurring as a result of greater biological activity in minimum tilled soils, demonstrated by Tisdall and Oades (1982), and with a reduction in the breakdown of surface soil aggregates as a result of residue cover of soil surface and the absence of tillage (Oyedele et al., 1999).

## 7. C sequestration

Soil carbon or organic matter in general, is important because it affects all soil quality functions (Fenton et al., 1999). The sequestration of atmospheric C in the soil and biomass would not only reduce greenhouse effect, but also helps to maintain or restore the capacity of a soil to perform its production and environmental functions on a sustainable basis. Thus, there is a great interest in the research on sequestration of atmospheric C into the soils for maintaining or restoring soil fertility and mitigating carbon dioxide emissions to the atmosphere.

| Location | Soil depth (m) | Soil Treatment | Aggregate size classes (mm) | | | |
|---|---|---|---|---|---|---|
| | | | > 2 | 2-1 | 1-0.25 | < 0.25 |
| Daxing | 0-0.10 | ST | 13.11 a | 23.14 a | 20.09 a | 43.66 a |
| | | NT | 11.56 a | 18.26 a | 19.37 a | 50.81 a |
| | | CT | 6.42 b | 10.37 b | 26.35 b | 56.86 b |
| | 0.10-0.20 | ST | 20.42 a | 13.74 a | 19.73 a | 46.11 a |
| | | NT | 17.05 a | 13.21 a | 19.35 a | 50.39 a |
| | | CT | 10.03 b | 12.36 a | 21.72 a | 55.89 b |
| | 0.20-0.30 | ST | 19.76 a | 17.34 a | 27.15 a | 35.75 a |
| | | NT | 18.52 a | 16.35 a | 26.35 a | 38.78 a |
| | | CT | 12.35 b | 17.28 a | 26.04 a | 44.33 b |
| Changping | 0-0.10 | ST | 7.23 a | 13.16 a | 40.63 a | 38.98 a |
| | | NT | 13.11 b | 23.14 b | 20.09 b | 43.66 b |
| | | CT | 11.56 b | 18.26 b | 19.37 b | 50.81 b |
| | 0.10-0.20 | ST | 12.53 a | 12.30 a | 38.96 a | 36.21 a |
| | | NT | 8.95 b | 11.36 a | 38.02 a | 41.67 b |
| | | CT | 5.97 b | 10.28 a | 37.96 a | 45.96 b |
| | 0.20-0.30 | ST | 9.59 a | 11.48 a | 38.26 a | 40.67 a |
| | | NT | 7.53 b | 10.23 a | 37.12 a | 45.12 b |
| | | CT | 6.85 b | 9.55 a | 35.23 a | 48.38 b |

**Table 3.** Soil wet stable aggregate size classes for ST, NT, and CT treatments at 0-0.10, 0.10-0.20, and 0.20-0.30 m depths (%) at Daxing and Changping
(Values within a column followed by the same letter are not significantly different at $p < 0.05$)

Maintaining or increasing SOC under dryland cropping systems remains a challenge in the northern Great Plains (Aase and Pikul, 1996). This is because the crop biomass yields and C inputs are often lower in drylands than in the humid regions due to limited precipitation and a shorter growing season. As a result, it often takes more time to enrich SOC (Halvorson et al., 2002a; Sherrod et al., 2003). Many studies have identified the potential of soils cultivated with different conservation practices (e.g., no-till) to sequester large amounts of carbon (C). It is estimated that conservation tillage practices across the United States may drive large-scale sequestration in the order of 24–40 Tg C yr$^{-1}$ (Tg: teragram; 1 Tg = 1012 g), and that additional C sequestration of 25–63 Tg C yr$^{-1}$ can be achieved through other modifications of the traditional agricultural practices. In the northern Great Plains, traditional farming systems, such as conventional tillage with wheat -fallow, have resulted in a decline in soil organic C (SOC) by 30 to 50% of their original levels in the last 50 to 100 years (Haas et al., 1957; Mann, 1985; Peterson et al., 1998). The data in Fig. 7 indicate the yield of wheat decreased with reduction in the SOC pool and increased with increase in the SOC pool.Intensive tillage increases the oxidation of SOC (Bowman et al., 1999; Schomberg and Jones, 1999). Halvorson et al., (2002a) observed that no-till with continuous cropping increased C sequestration in the drylands of the northern Great Plains by 233 kg ha$^{-1}$ yr$^{-1}$ compared to a loss of 141 kg ha$^{-1}$ yr$^{-1}$ in conventional tillage. The use of no-till has allowed producers to increase cropping intensity in the northern Great Plains (Aase and Pikul, 1995; Aase and Schaefer, 1996; Peterson et al., 2001) because no-till conserves surface residues and retains water in the soil profile more than the conventional tillage (Farhani et al., 1998). The reduced tillage and increased cropping intensity could conserve C and N in a dryland soil; and crop residues better than the traditional conventional tillage with wheat- fallow system in northern Great Plains (Sainju et al., 2007). The no tillage practice on Indiana crop land stores five times more carbon than conventional tillage. It is also of particular importance to reduce tillage on organic soils. Because of their high carbon content, these soils were found to emit more C when disturbed as compared to mineral soils (Table 4).

| Management system | Tons of Carbon stored /acre |
|---|---|
| Cropland | 0.107 tons C/acre |
| CRP/Grassland conversion | 0.397 tons C/acre |
| Trees/Wetland conversion | 0.209 tons C/acre |
| Cultivation of organic soils | -3.52 tons C/acre |
| **By tillage systems** | |
| Intensive Tillage | 0.042 tons C/acre |
| Moderate Tillage | 0.169 tons C/acre |
| No- Tillage | 0.223 tons C/acre |

**Table 4.** Carbon stored in Indiana Croplands in 1999 (Smith et al 2002).

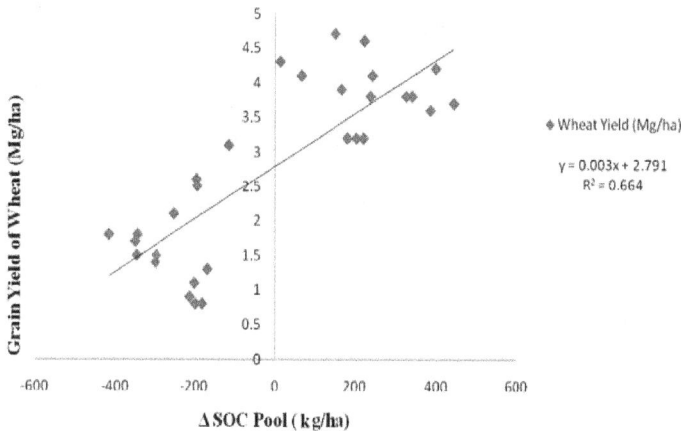

**Figure 7.** Effect of changes in soil organic carbon ($\Delta$SOC) pool in the root zone on grain yield of wheat in Australia (redrawn and recalculated from Farquharson et al., 2003).

There exists a strong relationship between the agronomic production and the SOC pool, especially under low-input agriculture (none or low rate of fertilizer input). An optimal level of the SOC pool is an essential determinant of soil quality because of its positive impact on the soil structure and aggregation, water and nutrient retention, biotic activity including the microbial biomass, erosion control, nonpoint-source pollution abatement, C sequestration, increase in the use efficiency, and increase in biomass production. The increase in aggregation and available water capacity are among important benefits of SOC (Emerson, 1995; Huntington, 2003).

## 8. Conclusion

In future, soil conservation efforts would need greater focus in the Peninsular and central India because of their projected high runoff and soil losses associated with global climate change. A decreasing trend of runoff and soil loss is ordered when we move from tropics to temperate region. A significant tenet of organic agriculture is to build up soil fertility by increasing the levels of organic carbon compounds in a soil. This is primarily achieved by using photosynthesis to convert atmospheric carbon dioxide, and by using management techniques that convert these plant materials into soil organic matter. 'Sufficient organic material should be regenerated and/or returned to the soil to improve, or at least maintain, humus levels. Conservation and recycling of nutrients is a major feature of any organic farming system' (National Standard 2005). Data from the Rodale Institute's long-term comparison of organic and conventional cropping systems (Rodale, 2003) confirms that the organic methods are effective at removing $CO_2$ from the atmosphere, and for fixing it as beneficial organic matter in the soil.The ambiguous nature of research findings document the need for additional studies of the effect of long-term tillage on soil physical properties under various tillage practices in order to optimize the productivity and maintain sustainability of soils. Moreover, there are a few studies that have examined the changes in soil physical properties in response to long term tillage and frequency management (> 20 yr) in the northern Great Plains. Global cereal production must be increased by ~50% by 2050. The crop yields in sub-Saharan Africa and South Asia have either stagnated or declined since the 1990s because of the widespread use of extractive farming practices and problems of soil and environmental degradation. Most degraded and depleted soils of agro-ecosystems contain a lower soil organic carbon (SOC) pool than in those under natural ecosystems. Thus, restoring the SOC pool is essential for improving soil quality, eco-efficiency and numerous ecosystem services. Increasing the SOC pool in the root zone can enhance agronomic production. Thus, the concept of eco-efficiency is important to produce more and more from less and less. Eco-efficiency is related to both "ecology" and "economy," and denotes both efficient and sustainable use of resources in the farm production and land management (Wilkins, 2008). Eco-efficiency is increased by those farming systems that would increase agronomic production by using fewer resources through reduction in losses of input, apart from sustaining and enhancing the production potential of land. Yet it is not enough to develop agricultural practices that would merely minimize the adverse environmental impact. Because of the increasing population and rising standards of living, it is essential to develop those agricultural practices that would maximize agricultural production and also enhance ecosystem services (Firbank, 2009).

## Author details

Vikas Abrol
*Division of Soil Science & Ag-Chemistry, SKUAST-J, India*

Peeyush Sharma
*Dryland Research SubStation, RakhDhiansar, SKUAST-J, India*

## 9. References

Aase, J.K., Schaefer, G.M., 1996. Economics of tillage practices and spring wheat and barley crop sequence in northern Great Plains. J. Soil Water Conserv. 51,167–170.

Abu-Hamdeh, N. H., 2004.The effect of tillage treatments on soil waterholding capacity and on soil physical properties. ISCO 2004 -13th International Soil Conservation Organisation Conference Conserving Soil and Water for Society: Sharing Solutions 532 – Brisbane, July 2004

Acharya,C. L., Sharma, P. D., 1994. Tillage and mulch effects on soil physical environment, root growth, nutrient uptake and yield of maize and wheat on an Alfisol in north-west India. Soil & Till. Res. 32, 291-302

Ahn, P.M. and Hintze, B., 1990. No tillage, minimum tillage, and their influence on soil properties. In: Organic-matter Management and Tillage in Humid and Sub-humid Africa. pp. 341-349. IBSRAM Proceedings No.10. Bangkok: IBSRAM.

Akram, Mohd., and W.D. Kemper. 1979. Infiltration of soils as affected by the pressure and water content at the time of compaction Soil Sci. Soc. Am. J. 43,1080-1086.

Allen, M., Lachnicht, S.L., McCartney, D., Parmelee, R.W.1997. Characteristics of macroporosity in a reduced tillage agroecosystem with manipulated earthworm populations: implications for infiltration and nutrient transport. Soil Biology and Biochemistry, 29,493–498.

Anken, T., Weisskopf, P., Zihlmann, U., Forrer, H., Jansa, J. , Perhacova, H. 2004. Long term tillage system effects under moist cool conditions in Switzerland. Soil Till Res, 78, 171–183.

Antapa, P.L. and Angen, T.V. 1990. Tillage practices and residue management in Tanzania. In: Organic-matter Management and Tillage in Humid and Sub-humid Africa. p. 49-57. IBSRAM Proceedings No.10. Bangkok: IBSRAM.

Benjamin, J.G. 1993. Tillage effects on near-surface soil hydraulic properties. Soil and Tillage Research 26, 277–288.

Bescansa, P., Imaz, M. J., Virto, I., Enrique, A., Hoogmoed, W. B., 2006. Soil water retention as affected by tillage and residue management in semi-arid Spain. Soil and Till. Res. 87, 19-27

Bhattacharyya R., Kundu S., Pandey S.C., Singh K.P., Gupta H.S.,2008. Tillage and irrigation effects on crop yields and soil properties under the rice-wheat system in the Indian Himalayas. Agril. Water Manag. 95, 993–1002.

Blanco-Canqui, H., Lal, R., 2007. Soil structure and organic carbon relationships following 10 years of wheat straw management in no-till. Soil and Till. Res. 95, 567 240–254.

Borresen, T. 1999. The effect of straw management and reduced tillage on soil properties and crop yields of spring-sown cereals on two loam soils in Norway.Soil and Tillage Research 51, 91–102.

Bowman, R.A., Vigil, M.F., Nielsen, D.C., Anderson, R.L.,1999. Soil organic matter changes in intensively cropped dryland systems. Soil Sci. Soc. Am. J. 63,186–191.

Carter, M.R., 1991. Evaluation of shallow tillage for spring cereals on a fine sandy loam. 1. Growth and yield components, N accumulation and tillage economics. Soil and Tillage Res.21 23–35

Chang, C., Lindwall. C.W.,1989. Effect of long term minimum tillage practices on some physical properties of Chernozemic clay loam, Can. J. Soil Sci. 69, 433–449.

Congressional Research Service., 2008. Climate Change: The Role of the U.S. Agriculture Sector. Renee Johnson. ttp://fpc.state.gov/documents/organization/81931.pdf

Davidson, E.A., Ackerman, I.L., 1993. Changes in soil carbon inventories following cultivation of previously untilled soils.Biogeochemistry 20, 161–193.

Dexter, A. R., and M. Birkas. 2004. Prediction of the soil structures produced by tillage. Soil Till. Res. 79, 233–238.

Dupreez,C.C.,Steyn, J.T., Kotze, E., 2001. Long-term effects of wheat residue management on some fertility indicators of a semi-arid plinthosol. Soil Till Res. 63, 25-33.

Emerson, W.W. 1995. Water retention, organic carbon and soil texture. Aust. J. Soil Res. 33,241–251.

Farhani, H.J., Peterson, G.A., Westfall, D.G. 1998. Dryland cropping intensifi cation: A fundamental solution to effi cient use of precipitation. Adv. Agron. 64, 197–223.

Farquharson, R.J., G.D. Schwenke, and J.D. Mullen. 2003. Should we manage soil organic carbon in Vertisols in the northern grains region of Australia. Aust. J. Exp. Agric. 43,261–270.

Fenton, T. E., Brown, J. R. and Maubach. M. J. , 1999. Effects of long-term cropping on organic matter content of soils: Implication for soil quality. Soil and Water Con. J.95–124.

Firbank, L., 2009. Commentary: It's not enough to develop agriculture that minimises environmental impact. International Journal of Agricultural Sustainability 7, 151-152

Franzluebbers, A. J., 2002a. Soil organic matter stratification ratio as an indicator of soil quality. Soil Till Res. 66, 95-106.

Franzluebbers, A. J., 2002b. Water infiltration and soil structure related to organic matter and its stratification with depth. Soil Till Res. 66, 197-205.

Glab, T., Kulig, B. 2008, Effect of mulch and tillage system on soil porosity under wheat (Triticum aestivum).Soil and Tillage Research, 99, 169–178.

Greenland, D.J., 1981. Soil management and soil degradation. J. Soil Science 32, 301-322.

Haas, H.J., Evans, C.E., Miles, E.F., 1957. Nitrogen and carbon changes in Great Plains soils as infl uenced by cropping and soil treatments. USDA Tech. Bull. 1164. U.S. Gov. Print. Offi ce, Washington, DC.

Halvorson, A.D., Peterson, G.A.,, Reule, C.A., 2002a. Tillage system and crop rotation effects on dryland crop yields and soil carbon in the central Great Plains. Agron. J. 94,1429–1436.

Hatfield, J.L., T.J. Sauer, and J.H. Prueger., 2001. Managing soils to achieve greater water use efficiency: A review. Agron. J. 93, 271-280.

Heard. JR., E.J. Kladivko. and J.V. Mantiering. 1988.Soil niacroporosin h ydraulic conductivity, and air permeability of silty soils under ls:tig-teeits conservation.

Hill, R .L., Cruse, R. M. 1985. Tillage effects on bulk density and soil strength of two Mollisols. Soil Sci Soc Am J. 49, 1270–1273.

Hobbs, P. R. 2007. Conservation agriculture: what is it and why is it important for future sustainable food production? J. Agric. Sci. 145, 127-137.

Holland, J.M., 2004. The environmental consequences of adopting conservation tillage in Hollinger, S.E., Bernacchi, C.J., Meyers, T.P. 2005. Carbon budget of mature no-till ecosystem in North Central Region of the United States. Agricultural and Forest Meteorology. 130,59–69.

Hooker, B.A., T.F. Morris, R. Peters and Z.G. Cardon. 2005. Long-term effects of tillage and cornstalk returnon soil carbon dynamics. Soil Sci. Soc. Am. J. 69:188-196.

Houghton, R.A., J.L. Hackler and K.T. Lawrence. 1999. The U.S. carbon budget: Contributions from land-usechange. Science 285:574-577.

Hulugalle, N.R., Lal, R. and Opara-Nadi, O.A., 1986. Effect of spatial orientation of mulch on soil properties and growth of yam (Dioscorea rotundata) and cocoyam (Xanthosoma sagittifolium) on an Ultisol. J. Root Crops 12, 37-45.

Huntington, T.G. 2003. Available water capacity and soil organic matter. In R. Lal (ed.) Encyclopedia of soil sciences. Vol. 1. CRC Press, Boca Raton, FL.

Husnjak, S., filipovic, D., Kosutic, S. 2002.Influence of different tillage systems on soil physical properties and crop yield. Rostlinna Vyroba, 48, 6, 249–254.

IPCC. ,2007. Climate Change 2007: The Physical Science Basis. Contribution of Working Group I to the Fourth Assessment Report of the Intergovernmental Panel on Climate Change [Solomon, S., D. Qin, M. Manning (eds)].

Iqbal, M., Ul-Hassan, A., Ali, A., Rizwanullah, M. 2005. Residual effect of tillage and frm manure on some soil physical properties and growth of wheat. Int. J. Ag. & Bio.7(1), 54–57.

Jabro, J.D., Stevens, W.B., Evans, R.G., Iversen, W.M., 2009. Tillage effects on physical properties in two soils of the Northern Great Plains. Applied Engineering in Agriculture 25, 377-382

Jordhal, J. L., Karlen, D.L., 1993. Comparison of alternative farming systems. III. Soil aggregate stability. Am. J. Altern. Agric. 8, 27-33.

Karlen, D.L., Wollenhaupt, N.C., Erbach, D.C., Berry, E.C., Swan, J.B., Eash, N.S., Jordahl, J.L. 1994. Long-term tillage effects on soil quality. Soil & Tillage Research 32, 313-327.

Kathirvel, M., G., Balasubramanian, M. Gopolan, and C. V. Sivakmar., 1992. Effect of seed treatment with botanical and chemicals for the control of root-knot nematodes, M. incognita infesting okra. Indian Journal of Plant Protection 20, 191-194.

Keshavarzpour F. and Rashidi M., 2008. Effect of different tillage methods on soil physical properties and crop yield of watermelon (Citrullus vulgaris). World Appl. Sci. J. 3, 359-364.

Khan, F.U.H., Tahir, A.R. and Yule, I.J., 2001. Intrinsic implication of different tillage practices on soil penetration resistance and crop growth. Int. J. Agri. Biol. 3, 23-26.

Khurshidi, K., Iqbal, M., Arif, M. S., Nawaz, A., 2006 Effect of tillage and mulch on soil physical properties and growth of maize . International. J. Ag. & Biology 8 (5),593–596

Kirkegaard, J.A., Munn, S. R., James, R.A., Gardner, P.A., Angus, J.F., 1995. Reduced growth and yield of wheat with conservation cropping. II. Soil biological factors limit growth under direct drilling. Aust J Agric. Res. 46, 75-88.

Klute, A., 1982. Tillage effects on hydraulic properties of soil. A review. In: Predicting Tillage Effects on Soil Physical Properties and Processes. P.W. Unger and Van Doren, D.M. (eds.) ASA Special Publication No.44:29-43.

Kooistra, M.J., J. Bouma, O.H. Boersma, and A. Jager. 1984. Physical and morphological characterization of undisturbed and disturbed ploughpans in a sandy loam soil. Soil Tillage Res. 4, 405-417

Kovác, K., Žák, Š., 1999. Vplyv rôznych spôsobov obrábania pôdy na jej fyzikálne a hydrofyzikálne vlastnosti. Rostlinná Výroba, 45, 359–364.

Kribaa, M., Hallaire, V., Curmi, P., 2001. Effect of various cultivation methods on the structure and hydraulic properties of a soil in a semi-arid climate. Soil and Tillage Res. 60, 43–53.

Lal, R. ,1976b. Soil erosion problems on an Alfisol in Western Nigeria and their control. IITA Monograph No. 1. Ibadan, Nigeria.

Lal, R. 1979a. Importance of tillage systems in soil and water management in the tropics. In: Soil Tillage and Crop Production. R. Lal (ed.). pp. 25-32. IITA Proc. Ser. 2.

Lal, R. and Hahn, S.K., 1973. Effects of methods of seedbed preparation, mulching and time of planting on yam in Western Nigeria. Symp. Proc. Int. Soc. Trop. Root Crops. Inst. Trop. Agric. Ibadan, Nigeria. 2-12 December 1973.

Lal, R., 1974. No-tillage effects on soil properties and maize (Zea mays L.) production in Western Nigeria. Plant and Soil 40, 321-331.

Lal, R., 1978. Influence of within and between row mulching on soil temperature, soil moisture, root development and yield of maize (Zea mays) in a tropical soils. Field Crops Res. 1, 127–39

Lal, R., 1980. Crop residue management in relation to tillage techniques for soil and water conservation. In: Organic recycling in Africa 74-79. Soils Bulletin 43. FAO, Rome.

Lal, R., 1981b. Soil conditions and tillage methods in the tropics. Proc. WARSS/WSS Symposium on No-tillage and Crop Production in the Tropics (Liberia 1981).

Lal, R., 1981c. Soil management in the tropics. In: Characterisation of Soils of the Tropics: Classification and Management. D.J. Greenland (ed.). Oxford University Press, UK.

Lal, R., 1982. Tillage research in the tropics. Soil and Tillage Research 2,305-309.

Lal, R., 1983. No-till farming: Soil and water conservation and management in the humid sand sub-humid tropics. IITA Monograph No. 2, Ibadan, Nigeria.

Lal, R., 1985a. No-till in the lowland humid tropics. In: The Rising Hope of our Land Conference (Georgia, 1985). pp. 235-241.

Lal, R., 1993. Tillage effects on soil degradation, soil resilience, soil quality and sustainability. Soil and Tillage Res., 27: 1–8

Lal, R., 1995. Tillage and mulching effects on maize yield for seventeen consecutive seasons on a tropical alfisol. J. Sustain. Agric. 5, 79–93

Lal, R., de Vleeschawer, D., Nganje, R.M., 1980. Changes in properties of a newly cleared tropical alfisol as affected by mulching. Soil Science Society of America Journal, 44, 827–833.

Lal, R., J.M. Kimble, R.F. Follett and V. Cole. 1998. Potential of U.S. cropland for carbon sequestration andgreenhouse effect mitigation. USDA-NRCS, Washington, D.C. Ann Arbor Press, Chelsea, Ml.

Lal, R., 1989. Conservation tillage for sustainable agriculture. Advances in Agronomy 42, 85-197.

Lampurlanes, J. , Cantero-Martinez, C., 2003. Soil bulk density and penetration resistance under different tillage and crop management systems and their relationship with barley root growth. Agron J. 95, 526–536.

Lhotsky, J., 1991.Komplexniagromelioreni soustavy prozhutnele pudy. Met. Zavad. Vysl. Vyzk. Praxe.UVTIZ, Praha, 20.

Lipic, J., Kus, J., Slowinska-Jurkiewicz, A., Nosalewicz, A., 2005. Soil porosity and water infiltration as influenced by tillage methods. Soil and Till. Res., 89, 210-220

Logsdon, S.D., Kasper, T.C., Camberdella. C.A., 1999. Depth incremental soil properties under no-till or chisel management. Soil Sci. Soc. Am. J. 63,197-200.

Mahboubi, A.A., Lal, R., Fausey. N.R., 1993.Twenty-eight years of tillage effects on two soils in Ohio. Soil Sci. Soc. Am. J. 57, 506-512.

Mann, L.K. , 1985. Changes in soil carbon storage after cultivation. Soil Sci. 142, 279–288.

Marbet, R., 2002. Stratification of soil aggregation and organic matter under conservation tillage systems in Africa. Soil Till Res. 66, 119-128.

McVay, K. A., Budde, J. A., Fabrizzi, K., Mikha, M. M., Rice, C. W., Schlegel, A., Peterson, J. D.E., Sweeney, D. W., Thompson, C. 2006. Management effects on soil physical properties in long-term tillage studies in Kansas. Soil Sci Soc Am J. 70, 434–438.

Meek, B.D., E.A. Rechel, L.M. Carter, and W.R. DeTar. 1989. Changes in infiltration under alfalfa as influenced by time and wheel traffic. Soil Sci. Am. J. 53, 238-241.

Mielke, L.N., Wilhelm, W.W., 1998. Comparisons of soil physical characteristics in long-term tillage winter wheat fallow tillage experiments. Soil Tillage Res. 49(1-2), 29-35.

Murillo, J. M., Moreno, F., Pelegrin, F., 2001. Respuesta del trigo y girasol al laboreo tradicional y de conservación bajo condiciones de secano (Andalucía Occidental). Invest Agr: Prod Prot Veg. 16, 395-406.

Murillo, J. M., Moreno, F., Pelegrin, F., Fernandez, J.E., 1998. Responses of sunflower to traditional and conservation tillage under rainfed conditions in southern Spain. Soil Till Res. 49, 233-241.

National Standard (2005) National Standard for Organic and Bio-Dynamic Produce, Edition3.1, As Amended January 2005, Organic Industry Export Consultative Committee, c/oAustralian Quarantine and Inspection Service, GPO Box 858, Canberra, ACT, 2601

Oliveira, M.T., Merwin, I.A.2001. Soil physical conditions in a New York orchard after eight years under different groundcover management systems. Plant and Soil 234, 233-237

Osunbitan, J.A., Oyedele, D.J., Adekalu. K.O., 2005. Tillage effects on bulk density, hydraulic conductivity and strength of a loamy sand soil in southwestern Nigeria. Soil Till. Res. 82,57-64.

Oyedele, D. J., Schjonning, P., Sibbesen, E., Debosz, K. 1999. Aggregation and organic matter fractions of three Nigerian soils as affected by soil disturbance and incorporation of plant material. Soil & Tillage Research 50, 105-114.

Pagliai, M., Paglione, M. Panini, T., Meletta, M., Lamarca, M., 1995. The structure of two alluvialsoils in Italy after 10 years of conventional and minimum tillage. Soil & Till. Res. 34, 209-223.

Pagliai, M., Vignozzi, N. 2002. Soil pore system as an indicator of soil quality. Advances in Geoecology,35, 69–80.

Parker, E.R., and H. Jenny. 1945. Water infiltration and related soil properties as affected by cultivation and organic fertilization. Soil Sci. 60,353-376.

Parr, J.F., Papendick, R.I., Hornick, S.B. and Meyer, R.E., 1990. The use of cover crops, mulches and tillage for soil water conservation and weed control. In: Organic matter Management and Tillage in Humid and Sub-humid Africa. pp. 246-261.

Patel, M.S., Singh, N.T.. 1981. Changes in bulk density and water intake rate of a coarse textured soil in relation to different levels of compaction. J. Indian Soc. Soil Sci. 29,110-112.

Pelegrin, F., Moreno, F., Martin-Aranda, J., Camps, M., 1990. The influence of tillage methods on soil physical properties and water balance for a typical crop rotation in SW Spain. Soil Till Res. 16,345-358.

Peterson, G.A., Halvorson, A.D., Havlin, Jones, O.R., Lyon, D.G., Tanaka, D.L., 1998. Reduced tillage and increasing cropping intensity in the Great Plains conserve soil carbon. Soil Tillage Res. 47, 207–218.

Peterson, G.A., Westfall, D.G., Peairs, F.B., Sherrod, L., Poss, D., Gangloff, W., Larson, K., Thompson, D.L., Ahuja, L.R., Koch, M.D., Walker. C.B., 2001. Sustainable dryland agroceosystem management. Tech. Bull. TB01–2. Agric. Exp. Stn., Colorado State Univ., Fort Collins.

Phillips, R.E., Blevins, R.L., Thomas, G.W., Frye, W.W. and Phillips, S.H., 1980. No798 tillage agriculture. Science 208,1108-1113.

Pikul, J.L., Jr., Aase. J.K., 1999. Wheat response and residual soil properties following subsoiling of a sandy loam in eastern Montana. Soil Till. Res. 51,61-70.

Pikul, J.L., Jr., Aase. J.K., 2003. Water infiltration and storage affected by subsoiling and subsequent tillage. Soil Sci. Soc. Am. J. 67,859-866.

Raos., 1996. Evaluation of nitrification inhibitors and urea placement in no-tillage winter wheat. Agron. J. 88, 212-216

Rashidi, M. and Keshavarzpour, F., 2007. Effect of different tillage methods on grain yield and yield components of maize (Zea mays L.). Int. J. Agri. Biol. 9, 274-277.

Reicosky, D.C. 1997. Tillage-induced CO2 emission from soil. Nutrient Cycling Agroecosystems. 49:273-285.

Reicosky, D.C., S.D. Evans, C.A. Cambardella, R.R. Allmaras, A.R. Wilts and D.R. Huggins. 2002. Con-tinuous corn with moldboard tillage: Residue and fertility effects on soil carbon. J. Soil Water Conserv.57(5):277-284.

Rodale, 2003. Eco-efficient approaches to land management: a case for increased integration of crop and animal production systems by R J Wilkins Volume: 363, Issue: 1491, Publisher: The Royal Society, Pages: 517-525

Sainju, U.M., Lenssen, A., Caesar-Tonthat, T., Waddell, J. , 2007. Dryland plant biomass and soil carbon and nitrogen fractions on transient land as infl uenced by tillage and crop rotation. Soil Tillage Res. 93,452–461.

Sasal,M.C., Andriulo,A.E., Taboada , M.A. 2006. Soil porosity characteristics and water movement under zero tillage in silty soils in Argentinian Pampas. Soil and Tillage Research 87, 9–18.

Schjonning P., Christensen B.T., Carstensen B.1994. Physical and chemical properties of a sandy loam receiving animal manure, mineral fertilizer or no fertilizer for 90 years. European Journal of Soil Science 45, 257-268.

Schlesinger, W.H. 1985. Changes in soil carbon storage and associated properties with disturbance and recov-ery. In I.R. Trabalha and D.E. Reichle (ed), "The changing carbon cycle: A global analysis," SpringerVerlag, New York: 194-220.

Schomberg, H.H., Jones, O.R., 1999. Carbon and nitrogen conservation in dryland tillage and cropping systems. Soil Sci. Soc. Am. J. 63,1359–1366.

Sharma, Peeyush, Abrol, Vikas, Sharma, R. K., 2011. Impact of tillage and mulch management on economics, energy requirement and crop performance in maize– wheat rotation in rainfed subhumid inceptisols, India Europ. J. Agronomy 34, 46–51

Sherrod, L.A., Peterson, G.A., Westfall, D.G., Ahuja, L.R. 2003. Cropping intensity enhances soil organic carbon and nitrogen in a no-till agroecosystem. Soil Sci. Soc. Am. J. 67,1533–1543.

Silgram, M., Shepherd, M.A., 1999. The effects of cultivation on soil nitrogen mineralization. Adv. Agron. 65, 267-311.

Tans, P. P., Fung, Y., Takashi, T., 1990. Observational constraints on the global atmosphericcarbon dioxide budget. Science 247,1431-1438.

Tiarks A.E., Mazurak A.P., Chesnin L., 1974. Physical and chemical properties of soil associated with heavy applications of manure from cattle feedlots. Soil Science Society of America Proceedings 38, 826-830.

Tisdall, J.M., Oades, J.M. 1982. Organic matter and water stable aggregates in soil. European Journal of Soil Science 33, 141-163.

Unger, P.W., Langdale, G.W. and Papendick, R.I., 1988. Role of crop residues improving water conservation and use. In: Cropping Strategies for Efficient Use of Water and Nitrogen. W.L. Hargrove (ed.) pp. 69-100. American Society of Agronomy Special Publication No.51.

Walling, D.E., 1990. Linking the field to the river: sediment delivery from agricultural land. In: Boardman, J., Foster, I.D.L., Dearing, J.A. (Eds.), Soil Erosion on Agricultural Land. Wiley, Chichester, pp. 129–152.

Wang, D.W., Wen, H.D. 1994. Effect of protective tillage on soil pore space status and character of micro morphological structure. Journal of Agricultural University of Hebei 17, 1–6.

Warkentin B.P., 2001. The tillage effect in sustaining soil functions. J Plant Nutr. Soil Sci. 164, 345-350.

Wilkins RJ., 2008. Eco-efficient approaches to land management: a case for increased integration of crop and animal production systems Philos Trans R Soc Lond B Biol Sci. 12, 517-25.

Willis, W.O. and Amemiya, M., 1973. Tillage management principles. In: Conservation Tillage (Iowa, 1973) pp. 22-42. Washington, DC: SCSA.

Worku, B., Sombat, C., Rungsit, S., Thongchai, M., Sunanta, J., 2006. Conservation Tillage and Crop Rotation: Win-Win Option for Sustainable Maize Production in the Dryland, Central Rift Valley of Ethiopia. Kamphaengsaen Acad. J. 4(1), 48 – 60

Zang Y., Gao H.W., Zhou X.X. 2003. Experimental study on soil erosion by wind under conservation tillage. Transactions of the American Society of Agricultural Engineers 19, 56-60.

Zhang, X., Li., H., Jin He, Wang, Q., Golabi, M. H., 2009. Influence of conservation tillage practices on soil properties and crop yields for maize and wheat cultivation in Beijing, China. Australian Journal of Soil Research

# Permissions

The contributors of this book come from diverse backgrounds, making this book a truly international effort. This book will bring forth new frontiers with its revolutionizing research information and detailed analysis of the nascent developments around the world.

We would like to thank Dr. Vikas Abrol and Dr. Peeyush Sharma, for lending their expertise to make the book truly unique. They have played a crucial role in the development of this book. Without their invaluable contribution this book wouldn't have been possible. They have made vital efforts to compile up to date information on the varied aspects of this subject to make this book a valuable addition to the collection of many professionals and students.

This book was conceptualized with the vision of imparting up-to-date information and advanced data in this field. To ensure the same, a matchless editorial board was set up. Every individual on the board went through rigorous rounds of assessment to prove their worth. After which they invested a large part of their time researching and compiling the most relevant data for our readers. Conferences and sessions were held from time to time between the editorial board and the contributing authors to present the data in the most comprehensible form. The editorial team has worked tirelessly to provide valuable and valid information to help people across the globe.

Every chapter published in this book has been scrutinized by our experts. Their significance has been extensively debated. The topics covered herein carry significant findings which will fuel the growth of the discipline. They may even be implemented as practical applications or may be referred to as a beginning point for another development. Chapters in this book were first published by InTech; hereby published with permission under the Creative Commons Attribution License or equivalent.

The editorial board has been involved in producing this book since its inception. They have spent rigorous hours researching and exploring the diverse topics which have resulted in the successful publishing of this book. They have passed on their knowledge of decades through this book. To expedite this challenging task, the publisher supported the team at every step. A small team of assistant editors was also appointed to further simplify the editing procedure and attain best results for the readers.

Our editorial team has been hand-picked from every corner of the world. Their multi-ethnicity adds dynamic inputs to the discussions which result in innovative outcomes. These outcomes are then further discussed with the researchers and contributors who give their valuable feedback and opinion regarding the same. The feedback is then collaborated with the researches and they are edited in a comprehensive manner to aid the understanding of the subject.

Apart from the editorial board, the designing team has also invested a significant amount of their time in understanding the subject and creating the most relevant covers. They scrutinized every image to scout for the most suitable representation of the subject and create an appropriate cover for the book.

The publishing team has been involved in this book since its early stages. They were actively engaged in every process, be it collecting the data, connecting with the contributors or procuring relevant information. The team has been an ardent support to the editorial, designing and production team. Their endless efforts to recruit the best for this project, has resulted in the accomplishment of this book. They are a veteran in the field of academics and their pool of knowledge is as vast as their experience in printing. Their expertise and guidance has proved useful at every step. Their uncompromising quality standards have made this book an exceptional effort. Their encouragement from time to time has been an inspiration for everyone.

The publisher and the editorial board hope that this book will prove to be a valuable piece of knowledge for researchers, students, practitioners and scholars across the globe.

# List of Contributors

M.A. Martin-Luengo, E. Sáez Rojo, A.M. Martínez Serrano, M. Dia, L. Medina Trujillo and S. Nogales
Department of New Architectures, Institute of Materials Science of Madrid, CSIC, Spain

M. Yates, F. Plou, E. Sáez Rojo, M. Diaz, S. Nogales and R. Lozano Pirrongelli
Institute of Catalysis and Petroleochemistry, CSIC, Spain

L. Vega Argomaniz
Institute of Catalysis and Petroleochemistry, CSIC, Spain
Department of New Architectures, Institute of Materials Science of Madrid, CSIC, Spain

M. Ramos and A.M. Martínez Serrano
Centre for Biomedical Technology, Polytechnical University of Madrid, Spain

J.L. Salgado
AIZCE Technical Committee, Interprofesional Asociation of Juices and Citric Concentrates, Spain

J.L. Lacomba and and A. Civantos
Institute of Biofunctional Studies, Complutense University of Madrid, Madrid, Spain

G. Reilly
Department of Materials Science and Engineering, Kroto Research Institute, Broad Lane, University of Sheffield, Sheffield, United Kingdom

C. Vervaet
Laboratory of Pharmaceutical Technology, Gent, Belgium

Ahmad T. Yuliansyah
Dept. of Chemical Engineering, Faculty of Engineering, Gadjah Mada University, Yogyakarta,Indonesia

Tsuyoshi Hirajima
Dept. of Earth Resources Engineering, Faculty of Engineering, Kyushu University, Nishi-ku, Fukuoka, Japan

Nicolas Greggio, Pauline Mollema, Marco Antonellini and Giovanni Gabbianelli
Interdepartmental Centre for Environmental Science Research (C.I.R.S.A.), University of Bologna, Ravenna, Italy

Daniele Zaccaria
Division of Land and Water Resources Management, Mediterranean Agronomic Institute of Bari (CIHEAM-IAMB), Valenzano, Bari Italy

**Giuseppe Passarella**
Water Research Institute (IRSA), National Research Council (CNR), Bari, Italy

**Hugo Ernesto Flores López, Celia De La Mora Orozco, Álvaro Agustín Chávez Durán and José Ariel Ruiz Corral**
Instituto Nacional de Investigaciones Forestales Agrícolas y Pecuarias, México

**Humberto Ramírez Vega and Víctor Octavio FuentesHernández**
Universidad de Guadalajara, México

**Peeyush Sharma**
Division of Soil Science & Ag-Chemistry, SKUAST-J, India

**Vikas Abrol**
Dryland Research Substation, Rakh Dhiansar, SKUAST-J, India

**Shrdha Abrol**
GGM Science College, Jammu, J&K, India

**Ravinder Kumar**
KVK, Rampur, Sardar Vallabhbhai Patel University of Agriculture and Technology, Meerut, India

**K.L. Sharma and B. Venkateswarlu**
Central Research Institute for Dryland Agriculture, Hyderabad (Andhra Pradesh), India

**Biswapati Mandal**
Bidhan Chandra Krishi Vishwavidalaya, Mohanpur, Nadia (West Bengal), India

**G.R. Maruthi Sankar, K.L. Sharma, N. Ashok Kumar, B. Venkateswarlu A. Girija and P. Ravi**
All India Coordinated Research Project for Dryland Agriculture (AICRPDA), CRIDA, Santoshnagar, Hyderabad, Andhra Pradesh, India

**Y. Padmalatha, K. Bhargavi, M.V.S. Babu and P. Naga Sravani**
AICRPDA, Acharya NG Ranga Agricultural University, Anantapur, Andhra Pradesh, India

**B.K. Ramachandrappa and G. Dhanapal**
AICRPDA, University of Agricultural Sciences, Bangalore, Karnataka, India

**Sanjay Sharma and H.S. Thakur**
AICRPDA, College of Agriculture, Indore, Madhya Pradesh, India

**A. Renuka Devi and D. Jawahar**
AICRPDA, Tamil Nadu Agricultural University, Kovilpatti, Tamil Nadu, India

**V.V. Ghabane**
AICRPDA, Punjabrao Deshmukh Krishi Vidyapeeth, Akola, Maharastra, India

**Vikas Abrol, Brinder Singh and Peeyush Sharma**
AICRPDA, Sher-e-Kashmir University of Agriculture and Technology, Rakh Dhiansar, Jammu & Kashmir, India

**A.K. Singh**
Krishi Anusandhan Bhavan, ICAR, Pusa, New Delhi, India

**Ioan Tenu, Petru Carlescu, Petru Cojocariu and Radu Rosca**
University of Agricultural Sciences and Veterinary Medicine, Iasi, Romania

**Vikas Abrol**
Division of Soil Science & Ag-Chemistry, SKUAST-J, India

**Peeyush Sharma**
Dryland Research SubStation, Rakh Dhiansar, SKUAST-J, India